MODERN MAGNETIC MATERIALS

Properties and Applications

MODERN MAGNETIC MATERIALS

Properties and Applications

Edited by

Iuliana Stoica, PhD

Ann Rose Abraham, PhD

A. K. Haghi, PhD

APPLE ACADEMIC PRESS

First edition published 2024

Apple Academic Press Inc.
1265 Goldenrod Circle, NE,
Palm Bay, FL 32905 USA

760 Laurentian Drive, Unit 19,
Burlington, ON L7N 0A4, CANADA

CRC Press
2385 NW Executive Center Drive,
Suite 320, Boca Raton FL 33431

4 Park Square, Milton Park,
Abingdon, Oxon, OX14 4RN UK

Apple Academic Press exclusively co-publishes with CRC Press, an imprint of Taylor & Francis Group, LLC

Library and Archives Canada Cataloguing in Publication

Title: Modern magnetic materials : properties and applications / edited by Iuliana Stoica, PhD, Ann Rose Abraham, PhD, A.K. Haghi, PhD.
Names: Stoica, Iuliana, editor. | Abraham, Ann Rose, editor. | Haghi, A. K., editor.
Description: First edition. | Includes bibliographical references and index.
Identifiers: Canadiana (print) 2023046985X | Canadiana (ebook) 20230469876 | ISBN 9781774912997 (hardcover) | ISBN 9781774913000 (softcover) | ISBN 9781003372066 (ebook)
Subjects: LCSH: Magnetic materials. | LCSH: Magnetism.
Classification: LCC TK454.4.M3 M63 2024 | DDC 621.34—dc23

Library of Congress Cataloging-in-Publication Data

CIP data on file with US Library of Congress

ISBN: 978-1-77491-299-7 (hbk)
ISBN: 978-1-77491-300-0 (pbk)
ISBN: 978-1-00337-206-6 (ebk)

About the Editors

Iuliana Stoica, PhD
Department of Physical Chemistry of Polymers, "Petru Poni" Institute of Macromolecular Chemistry, Romania

Iuliana Stoica, PhD, is a Scientific Researcher in Physics at the Romanian Academy, "Petru Poni" Institute of Macromolecular Chemistry, Department of Physical Chemistry of Polymers. She received her PhD from Romanian Academy, in the Department of Polymer Physics and Structure at the same institute. She joined a postdoctoral fellowship program at Politehnica University of Bucharest, Faculty of Applied Chemistry and Materials Science, Department of Bioresources and Polymer Science (2014–2015) as well as a postdoctoral fellowship awarded by L'Oreal - UNESCO National Program "For Women in Science" (2020–2021). Her area of scientific activity is focused on characterization of a wide range of polymers, copolymers, polymeric composites, and polymeric mixtures. She has been an author or coauthor for over 115 papers in peer-reviewed ISI journals. She has edited three books and has contributed several book chapters on polymer and materials science. She has also been a member of the organizing and program committees of several scientific conferences. She has been a reviewer for a number of prestigious journals in the field of polymer science. She has an 18 Hirsch index, according to the Web of Science Core Collection.

Ann Rose Abraham, PhD
Sacred Heart College (Autonomous), Thevara, Kochi, Kerala, India

Ann Rose Abraham, PhD, is currently an Assistant Professor at the Department of Physics, Sacred Heart College (Autonomous), Thevara, Kochi, Kerala, India. Dr. Ann received her MSc, MPhil, and PhD degrees in Physics from School of Pure and Applied Physics, Mahatma Gandhi University, Kerala, India. Her PhD thesis was titled "Development of Hybrid Mutliferroic Materials for Tailored Applications." She has expertise in the field of condensed matter physics, nanomagnetism, multiferroics, and polymeric nanocomposites, etc. She has research experience at various reputed national institutes like Bose Institute, Kolkata, India, SAHA Institute of Nuclear Physics, Kolkata, India, UGC-DAE CSR Centre, Kolkata, India and collaborations with various international laboratories. She is a recipient of a Young Researcher Award in the area of physics and Best Paper Awards–2020, 2021, a prestigious forum for showcasing intellectual capability. She served as assistant professor and examiner, at the Department of Basic Sciences, Amal Jyothi College of Engineering, under APJ Abdul Kalam Technological University, Kerala, India. Dr. Ann is a frequent speaker at national and international conferences. She has a good number of publications to her credit in many peer-reviewed high impact journals of international repute. She has authored many book chapters and edited more than 10 books with Taylor and Francis, Elsevier, etc.

A. K. Haghi, PhD
Coimbra University, Portugal

A. K. Haghi, PhD, has published over 250 academic research-oriented books as well as over 1000 research papers published in various journals and conference proceedings. He has received several grants, consulted for several major corporations, and is a frequent speaker to national and international audiences. He is founder and former editor-in-chief of the *International Journal of Chemoinformatics and Chemical Engineering,* published by IGI Global (USA) as well as the *Polymers Research Journal,* published by Nova Science Publishers (USA). Professor Haghi has acted as an editorial board member of many international journals. He has served as a member of the Canadian Research and Development Center of Sciences and Cultures (CRDCSC) and the Research Chemistry Centre, Coimbra, Portugal. Dr. Haghi holds a BSc in urban and environmental engineering from the University of North Carolina (USA), an MSc in mechanical engineering from North Carolina A&T State University (USA) and an MSc in applied mechanics, acoustics, and materials from the Université de Technologie de Compiègne (France), and a PhD in engineering sciences from Université de Franche-Comté (France).

Contents

Contributors .... *ix*

Abbreviations .... *xiii*

Preface .... *xvii*

1. **Explanations of Mesoparticles' Magnetic Properties Changes during Red-Ox Processes and the Polarization Influence at the Polymeric Compositions Modification by Mesoparticles 1**
   V. I. Kodolov and V. V. Kodolova-Chukhontzeva

2. **X-Ray Photoelectron Study of the Formation of the Chemical Bond and the Atomic Magnetic Moment in Nickel–Carbon Nanocomposites Modified by *d*-Metal Oxides 13**
   N. S. Terebova, V. I. Kodolov, and I. N. Shabanova

3. **Red-Ox Synthesis of Metal/Carbon Mesocomposites in Nanosized Reactors of Polymeric Matrices 23**
   V. V. Kodolova-Chukhontzeva and V. I. Kodolov

4. **Mechanochemical Modification of Metal-Carbon Mesocomposites 51**
   V. I. Kodolov, V. V. Trineeva, R. V. Mustakimov, N. S. Terebova, T. M. Makhneva, and I. N. Shabanova

5. **Investigation of Copper/Carbon Nanocomposites Modified with Phosphorus-Containing Groups as Inhibiting Additives in Mineral Oil 63**
   I. N. Shabanova, S. M. Reshetnikov, V. I. Kodolov, N. S. Terebova, R. V. Mustakimov, F. F. Chausov, and S. G. Bystrov

6. **Investigation of the Formation of the Electron Structure of Metallocarbonic Nanoforms 73**
   I. N. Shabanova, V. I. Kodolov, and N. S. Terebova

7. **Dependence of the Value of the Atomic Magnetic Moment of *d*-Metals on the Chemical Structure of Nanoforms 85**
   I. N. Shabanova, N. S. Terebova, and V. I. Kodolov

8. **Some Aspects of Magnetic Metal Carbon Mesoscopic Composites With Regulated Magnetic Characteristics 95**
V. I. Kodolov, V. V. Kodolova–Chukhontseva, Yu. V. Pershin, R. V. Mustakimov, I. N. Shabanova, and N. S. Terebova

9. **Metal-Carbon Mesocomposites Application Possibilities as Magnetic Mesoscopic Materials... 111**
V. I. Kodolov, V. V. Kodolova-Chukhontzeva, Yu. V. Pershin, and R. V. Mustakimov

10. **Epoxy Composites with 5 wt.% Nanodispersed Magnetites and Ferroxides: Strength, Heat Resistance, and Morphology................151**
D. Starokadomsky, M. Reshetnyk, N. Bodul, and L. Kokhtych

11. **Magnetic Particles and Their Role in Polymer Composites: From Molecular Modeling to Applications...167**
Raluca Marinica Albu

12. **Magnetism Towards Smart Materials...191**
V. N. Archana, N. G. Divya, and Reyha Benedict

13. **Magnetic Nanoparticles in Biomedical Applications..............................205**
Namitha Binu, Ruby Varghese, and Yogesh B. Dalvi

14. **A Technical Note on Metal-Carbon Mesocomposites Magnetic Characteristics Growth 223**
V. I. Kodolov and V. V. Kodolova-Chukhontzeva

15. **Synthesis and Applications of Magnetic Nanoparticles: A Review 229**
Vandana Chauhan, Rajpreet Kaur, Anita Gupta, and Poonam Khullar

16. **Electrospinning Technique and Magnetic Nanofibers.............................247**
Roop Varghese Rubert, Maria Mathew, and Rony Rajan Paul

17. **Magnetic Polymer Composites and Their Role in Engineering Applications 291**
Simona Luminita Nica

Index...309

Contributors

Raluca Marinica Albu
"Petru Poni" Institute of Macromolecular Chemistry, Laboratory of Physical Chemistry of Polymers, Iasi, Romania

V. N. Archana
Mar Athanasius College, Kothamangalam, Ernakulam, Kerala, India

Reyha Benedict
St. Teresa's College, Ernakulam, Kerala, India

Namitha Binu
Pushpagiri Institute of Medical Sciences and Research Center, Tiruvalla, Pathanamthitta, Kerala, India

N. Bodul
Ukrayinsky Phyzico-Mathematic Litseum, T. Schevchenko Kyiv National University, Ukraine

S. G. Bystrov
Udmurt Federal Research Center, UB RAS, Russian Federation

Vandana Chauhan
Department of Chemistry, Amity University, Noida, Uttar Pradesh, India

F. F. Chausov
Udmurt Federal Research Center, UB RAS, Russian Federation

Yogesh B. Dalvi
Pushpagiri Institute of Medical Sciences and Research Center, Tiruvalla, Pathanamthitta, Kerala, India

N. G. Divya
Bharat Mata College, Thrikkakkara, Ernakulam, Kerala, India

Anita Gupta
Department of Chemistry, Amity University, Noida, Uttar Pradesh, India

Rajpreet Kaur
Department of Chemistry, Amity University, Noida, Uttar Pradesh, India; Department of Chemistry, BBK DAV College for Women, Amritsar, Punjab, India

Poonam Khullar
Department of Chemistry, BBK DAV College for Women, Amritsar, Punjab, India

V. I. Kodolov
Basic Research–Research Center of Chemical Physics and Mesoscopic, Udmurt Federal Research Center, Russian Academy of Sciences, Izhevsk, Russia; M.T. Kalashnikov Izhevsk State Technical University, Izhevsk, Russia

V. V. Kodolova-Chukhontzeva
Basic Research–High Educational Center of Chemical Physics and Mesoscopic, UFRC RAS, Izhevsk, Russia; St. Petersburg Institute of Macromolecular Compounds, Russian Academy of Sciences, St. Petersburg, Russia

L. Kokhtych
School of Engineering and Architecture, Lucerne University of Applied Sciences and Arts, Horw, Switzerland; Institute of Physics, NAS, National Academy of Sciences (NAS) of Ukraine, Ukraine

T. M. Makhneva
Basic Research–High Educational Center of Chemical Physics and Mesoscopic, Udmurt Federal Research Center, Russian Academy of Sciences, Izhevsk, Russia; Institute of Mechanics, Udmurt Federal Research Center, Russian Academy of Sciences, Izhevsk, Russia

Maria Mathew
Department of Chemistry, CMS College, Kottayam, Kerala, India

R. V. Mustakimov
Basic Research–High Educational Center of Chemical Physics and Mesoscopic, Udmurt Federal Research Center, Russian Academy of Sciences, Izhevsk, Russia; Izhevsk Electromechanical Plant "KUPOL," M.T. Kalashnikov Izhevsk State Technical University, Izhevsk, Russia

Simona Luminita Nica
"Petru Poni" Institute of Macromolecular Chemistry, Department of Physical Chemistry of Polymers, Iasi, Romania

Rony Rajan Paul
Department of Chemistry, CMS College, Kottayam, Kerala, India

Yu. V. Pershin
Basic Research – High Educational Center of Chemical Physics and Mesoscopic, UD, RAS, Izhevsk, Russia; M.T. Kalashnikov Izhevsk State University, Izhevsk, Russia

S. M. Reshetnikov
Udmurt Federal Research Center, UB RAS, Izhevsk, Russian Federation; Udmurt State University, Izhevsk, Russian Federation

M. Reshetnyk
National Natural History Museum NAS, Ukraine; M. P. Semenenko Institute of Geochemistry, Mineralogy and Ore Formations, National Academy of Sciences (NAS) of Ukraine, Ukraine

Roop Varghese Rubert
Department of Chemistry, CMS College, Kottayam, Kerala, India

I. N. Shabanova
Basic Research–High Educational Center of Chemical Physics and Mesoscopic, Udmurt Federal Research Center, Russian Academy of Sciences, Izhevsk, Russia; Physicotechnical Institute, Ural Branch, Udmurt Federal Research Center, Russian Academy of Sciences, Izhevsk, Russia

D. Starokadomsky
M. P. Semenenko Institute of Geochemistry, Mineralogy and Ore Formations, National Academy of Sciences (NAS) of Ukraine, Ukraine; Chuyko Institute of Surface Chemistry, NAS, Ukraine

N. S. Terebova
Basic Research–High Educational Center of Chemical Physics and Mesoscopic, Udmurt Federal Research Center, Russian Academy of Sciences, Izhevsk, Russia; Physicotechnical Institute, Udmurt Federal Research Center, Russian Academy of Sciences, Izhevsk, Russia

V. V. Trineeva
Basic Research–High Educational Center of Chemical Physics and Mesoscopic, Udmurt Federal Research Center, Russian Academy of Sciences, Izhevsk, Russia; Institute of Mechanics, Udmurt Federal Research Center, Russian Academy of Sciences, Izhevsk, Russia

Ruby Varghese
Department of Chemistry, School of Sciences, Jain Deemed to be University, Bangalore, Karnataka, India

Abbreviations

AA	acetylacetone
AC	alternating current
AFM	atomic force microscopy
AG	agarose
APPh	ammonium polyphosphate
B_2O_3	boron oxide
BC	bacterial cellulose
CBT	cyclic butylene terephthalate
CLL-37	cathelicidin LL-37
CNFs	carbon nanofibers
CNTs	carbon nanotubes
CS	chitosan
DC	direct current
DIW	direct ink writing
DMF	dimethylformamide
DMS	dilute magnetic semiconductors
EG	expanded graphite
EM	electromagnetic
EMI	electromagnetic interference
EPD	electrophoretic deposition
EPR	electron paramagnetic resonance
ER	electrorheological
ERs	epoxy resins
FAS	fluoroalkyl silane
GO	graphene oxide
HMCs	hard magnetic composites
HM-PLLA	high molecular weight poly-L-lactic acid
IGF-1R	insulin-like growth factor receptor
LBL	layer by layer
MCs	mesocomposites
ME	magnetoelectric
Me–C MC	metal-carbon mesocomposite
MGCE	magnetic glass carbon electrode
MHT	magnetic hyperthermia treatment

MIONs	magnetic iron oxide nanoparticles
MNFs	magnetic nanofibers
MNPs	magnetic nanoparticles
MR	magnetorheological
MRI	magnetic resonance imaging
MSP	merozoite surface protein
NFe	nickel spinel ferrite
NMR	nuclear magnetic resonance
NPs	nanoparticles
ODA	4,4′-oxydianiline
P(VDF–TrFE)	poly(vinylidene fluoride-co-trifluoroethylene)
PAA	polyacrylic acid
PCHMA	poly(cyclohexyl methacrylate)
PCL	poly(ε-caprolactone)
PCU	polycarbonate urethane
PDA	polydopamine
PDMS	polydimethylsiloxane
PEI	polyethyleneimine
PEN	polyethylene naphthalene
PEO	poly(ethylene oxide)
PEPA	polyethylene polyamine
PET	polyethylene terephthalate
PLGA	poly(lactic-co-glycolic acid)
PLLA	poly(L-lactic acid)
PMDA	pyromellitic dianhydride
PMMA	poly(methyl methacrylate)
PmT	poly(m-toluidine)
PS	polystyrene
PVA	polyvinyl alcohol
PVDF	poly(vinylidene-fluoride)
PVP	poly(vinyl pyrrolidone)
QSAR	quantitative structure-activity relationships
RF	radiofrequency
RH	relative humidity
RIU	refractive-index units
SAED	selected area electron diffraction
SE	shielding effectiveness
SEM	scanning electron microscope
SMCs	soft magnetic composites

SPMNPs	superparamagnetic nanoparticles
SPNs	super-paramagnetic
SPP	surface plasmon polaritons
SPR	surface plasmon resonance
SPS	spark plasma sintering
SUs	structural units
TBA	tetrabutylammonium
TEM	transition electron microscopy
TENGs	triboelectric nanogenerators
VSM	vibrating sample magnetometer
XPS	X-ray photoelectron spectroscopy

Preface

This new book will introduce postgraduate students to a range of advanced topics in physics and magnetic materials science covering basic physical concepts, experimental methods, and applications.

It covers basic concepts and applications of magnetism science and focuses on state-of-the-art magnetic materials. It also presents the advanced treatment of materials that can hold a magnetic field and describes recent research developments in magnetic materials, including fabrication, characterization, applications, and many more related subjects which are covered in detail.

This is a valuable and comprehensive research-oriented book for post-graduate students seeking to understand this rather complicated and difficult subject.

Meanwhile, at a practical level, this new research-oriented book covers several novel case studies. In this new title, selected fields for modern technical applications and a large diversity of magnetic materials are discussed in detail.

Chapter 1 of this book is devoted to the changes in mesoparticles' magnetic properties during red-ox processes and the polarization influence at the modification of the polymeric composition by mesoparticles.

Chapter 2 is focused on the X-ray photoelectron study of the formation of the chemical bond and the atomic magnetic moment in nickel-carbon nanocomposites modified by *d*-metal oxides.

Chapter 3 is dedicated to the explanation of the base of mesoscopic physics principles for metal-carbon mesocomposites obtained in the nano-sized reactors (mesoreactors) of polymeric matrices.

Mechanochemical modification of metal-carbon mesocomposites is presented in Chapter 4. The chapter presents materials at the practical rather than theoretical level allowing for a physical, quantitative, measurement-based understanding of magnetism among readers.

An investigation of copper/carbon nanocomposites modified with phosphorus-containing groups as inhibiting additives in mineral oil is reviewed in Chapter 5.

The formation of the electron structure of metallocarbonic nanoforms is studied in Chapter 6. This chapter contains useful research ideas for readers approaching magnetism for the first time, as well as a quick reference for anybody seeking information about notions and facts pertaining to magnetism and magnetic materials.

The dependence of the value of the atomic magnetic moment of d-metals on the chemical structure of nanoforms is reviewed in Chapter 7.

In Chapter 8, we have shown how mesoscopic physics explains the red-ox synthesis of metal-carbon mesocomposites within polymeric matrices mesoreactors.

In Chapter 9, metal-carbon mesocomposites application possibilities as magnetic mesoscopic materials are studied in detail.

Epoxy composites with 5 wt.% nanodispersed magnetites and ferroxides with details of their strength, heat resistance, and morphology are presented in Chapter 10.

It conveys the main physical ideas behind the macroscopic phenomenology of magnetism in the materials and their ultimate realization in actual technological applications.

Chapter 11 is focused on important aspects related to magnetic particles and their contribution to the performance of magnetic composites with polymer matrix.

In Chapter 12, magnetism towards smart materials is presented in detail. Biomedical applications of magnetic nanoparticles are discussed in Chapter 13.

Metal-carbon mesocomposites' magnetic characteristics growth are studied in Chapter 14. Synthesis and applications of magnetic nanoparticles are reviewed in Chapter 15. Chapter 16 is devoted to magnetic nanofibers in the electrospinning process.

Magnetic polymer composites and their role in engineering applications are discussed in Chapter 17 in detail. By reading these interesting chapters, postgraduate students and materials scientists will enjoy the presentation of the different processing methods and their impact on magnetic properties. Meanwhile, readers will profit from the survey, starting from the basics of magnetism down to the applications.

CHAPTER 1

Explanations of Mesoparticles' Magnetic Properties Changes during Red-Ox Processes and the Polarization Influence at the Polymeric Compositions Modification by Mesoparticles

V. I. KODOLOV[1,2] and V. V. KODOLOVA-CHUKHONTZEVA[1,3]

[1]*BRHEC of Chemical Physics and Mesoscopics, Udmurt Federal Research Center, RAS, Izhevsk, Russia*

[2]*M.T. Kalashnikov Izhevsk State Technical University, Izhevsk, Russia*

[3]*Peter Great St. Petersburg Polytechnic University, St. Petersburg, Russia*

ABSTRACT

This chapter presents a new scientific trend (chemical mesoscopic) with examples of the reduction-oxidation reactions peculiarities with nanostructures and mesoparticles participation and modifying the polymeric composition by mesoparticles. In these cases, such notions as interference (the chemical bond formation at the phase coherence), and annihilation, which stimulates the electron shift and leads to the chemical bonds' formation accompanied by atom magnetic moment growth, are discussed. The explanation of the magnetic characteristics with simultaneous growth unpaired electrons on the carbon fibers of mesoparticles is possible only owing to the annihilation of negative and positive charges quants in the red-ox processes with the participation of metal-carbon mesocomposites (MCs) and the oxidizers containing phosphates, silicates, and metal oxides.

Modern Magnetic Materials: Properties and Applications. Iuliana Stoica, Ann Rose Abraham, & A. K. Haghi (Eds.)

The hypothesis of Me–C MC interaction with modified agents (oxidizers) on the solid phase's boundary is proposed. The obtained effect is concluded in the growth of metal cluster atomic magnetic moments because of unpaired *d* electron formation. The polarization influence is discussed in the observation of polymeric compounds and materials modification by Me–C MC.

1.1 INTRODUCTION

Mesoscopic Physics and later appeared mesoscopic chemistry can be presented as the basis of the new scientific trend – chemical mesoscopics. The above scientific trends are near to such scientific trends and theories as synergetics (self-organization), fractal theory (self-similarity), and Theories of Chemical Kinetics and Catalysis [1–3]. All these theories are used for the description of mesoscopic particles (or nanostructures) in the different media and at the various conditions changes. Therefore, the above new trend is very near to Chemical Physics on the considered objects and also on the phenomena and particularities of the various reactions and processes at the changes of conditions of their realization. However, the basic aim of this advanced trend is appeared in the investigation of nanostructures (or mesoparticles) reactivity in various media and at different, changed conditions.

Especial attention is devoted to the interactions of metal (Copper, Nickel) carbon mesoparticles with such oxidizers as ammonium polyphosphate (APPh) and aluminum oxide. In these reactions, it is observed the reduction of Phosphorus and Aluminum, and also the growth of mesocomposite (MC) metal atomic magnetic moment, and the simultaneous increasing of spin quantity on the MC carbon shell. The introduction of "annihilation" notion explains the increasing of magnetic characteristics.

The polymeric compositions polarization influence appeared under the action of Me–C MC is considered at the polymeric compounds' modification.

1.2 ABOUT THE ANNIHILATION PHENOMENON AT RED-OX REACTIONS OF MESOPARTICLES AND OXIDIZERS (MODIFIED AGENTS)

The metal atomic magnetic moments values of modified copper– and nickel–carbon MCs may be caused by the polarization of Met–O bonds participated in Red-Ox processes. In this case, the influence of correspondent modified agents (oxidizers) on the changes of metal atomic magnetic moments remains constant.

In Table 1.1, the examples of metal atomic magnetic moments changes (in Born magnetons) and quantities of unpaired electrons (in spin/g) for mesoparticles modified by APPh or silica (SiO_2) after the mechanochemical modification processes proceeding are given.

TABLE 1.1 The Values of Copper (Nickel) Atomic Magnetic Moments and the Quantity of Unpaired Electrons at the Interaction of Mesocomposite Cu–C with Silica and Ammonium Polyphosphate

Systems Cu–C MC – Substances	μ_{cu}	Quantity of Unpaired Electrons (spin/g)
Cu–C mesocomposite	1.3	1.2×10^{17}
Cu–C MC – Silica	3.0	3.4×10^{19}
Cu–C MC – APPh	2.0	2.8×10^{18}
Cu–C MC – APPh, relation 1:0.5	4.2	4.3×10^{20}

The presence of reduced forms of Carbon and Phosphorus is determined on the results of X-ray photoelectron spectroscopy (XPS) and electron sound micro-roentgen spectroscopic analysis of element composition [4–6]. It's noted that the phosphorus is reduced because of the interaction with unpaired electrons which are found on the MC carbon shell. It's possible that the Phosphorus atom is disposed between the carbon fibers of Cu–C mesoscopic carbon shell.

The reduction process and the mesoparticle structure change may be presented by the following multi-stage scheme: the first step – the negative charges quants are directed to the positive charged nucleus of Phosphorus atom, near which there is the cloud of positive charge quants; the second step – the interaction of negative charged quants with positive charged quants, which leads to the annihilation phenomena creation; the third step – at the annihilation the inner electromagnetic (EM) field is formed which stimulates the Phosphorus reduction process; the fourth step – at the reduction process the water vapors are formed and are evaporated, that leads to C=P bond formation (132.6 eV on X-ray photoelectron spectra [6]); the fifth step – the EM radiation formed stimulates across the coordination bonds d orbital of Copper atom that leads to the formation of unpaired electrons and the growth of Copper atomic magnetic moment; the sixth step – the fifth step explains the increasing of the spins number on the carbon cover of Copper cluster. The last two steps are accompanied by changes of MC carbon cover structure. According to X-ray photoelectron spectra data it's observed the decrease of C–H bonds with simultaneous growth of the carbine fragments. These structure changes lead to the MCs reactivity increasing.

A similar explanation of the mechanism of the reduction-oxidation reaction is possible for the interaction of Aluminum oxide with nickel–carbon mesoscopic composite. In this case, Aluminum is reduced completely with simultaneous growth of Nickel atomic magnetic moment from 1.3 to 4.8 Bohr magnetons. These changes are corresponded with the increasing of spin number on the Carbon shell from 10^{17} spin/g to 10^{20} spin/g [4]. The comparison of atomic magnetic moments in the series of copper and nickel–carbon MCs shows that the difference between series is equaled 1 Bohr magnetons. It is interesting to note that at the using of such oxidizers as metal oxides, the increasing of atomic magnetic moments of the MC metal and the oxide metal is observed (Table 1.2).

TABLE 1.2 The Metal Atomic Magnetic Moments Changes at the Ni–C Mesocomposites Modification

Mesocomposite – Modified Agent (Oxidizer)	μ_{Ni}, μ_B	μ_{Cu}, μ_B	μ_{Fe}, μ_B
Ni/C – NiO (0.5)	3.0	–	–
Ni/C – NiO	4.5	–	–
Ni/C – CuO	2.0	2.0	–
Ni/C – Fe_2O_3	2.5	–	3.0
Ni/C – Al_2O_3 (0.2)	4.8	–	–
Ni–C MC – SiO_2	4.0	–	–
Ni–C MC – APPh	3.0	–	–

The significant growth ofNickel atomic magnetic moment at the decreasing of modified agent layer when the modified agent is Aluminum oxide takes place. In these conditions the Aluminum oxide is reduced completely.

Thus, proposed hypothesis about the processes of unpaired electrons forming and shift of metal electrons on the carbon cover of MCs is confirmed by the experimental data. In this case, the electrons quantity, which participates in the process, determines the value of metal atomic magnetic moment growth.

The hypothesis is that the negative charge quants move to reducing element nucleus to provoke the positive charge quants and the creation of annihilation phenomenon. Therefore, the direct EM field formed promotes to d electrons shift on higher energetic levels including the carbon shell of nanocomposite nanogranyl. The picture of this phenomenon may be presented:

i. At the joint grinding of modified agent with metal/carbon nanocomposites nanogranule the nanoreactor (multiplet) is formed on the boundaries of which the chemical potentials difference creates.
ii. Because of potentials difference, the delocalized electrons, which are disposed on the nanogranule shell, move to positive charged atoms of the modified agent and stimulate the atoms positive charges quantization.
iii. The annihilation phenomenon arises at the superposition of negative and positive charges on the boundary of reagents interaction. In this case the electron magnetic field activates the delocalization of d electrons of nanogranule metals and the electron shift to nanogranule carbon shell.
iv. The electron shift process within nanogranule is facilitated by modified agent medium polarization growth.

The hypothesis explains the growth of metal atomic magnetic moments which may be bigger than 4.5 μ_B because of the process of delocalization d electrons. The shift of electrons on nanogranule carbon shell restored the electron balance after Red-Ox synthesis. There is a correlation between electron number participated in Red-Ox synthesis and the value of atomic magnetic moment as well as between the polarization degree of reactive systems and the increasing of atomic magnetic moments of nanogranule metals.

Thus, it's possible to change the magnetic properties of mesoparticles and the number of unpaired electrons using the red-ox reactions of Me–C MC with different oxidizers. In other words, there is the possibility of the direct change of mesoparticles magnetic-electric properties.

1.3 THE NANOSTRUCTURED COMPOSITION POLARIZATION IS CONDITION FOR THE MODIFICATION POLYMERIC SYSTEMS WITH POSITIVE RESULTS

The MCs activity in the different media (materials) is changed in the dependence on polarity or polarization of their media. The reactivity of considered MCs can be explained by the radiation of negatively charged quants from mesoparticles. Therefore, the modification conditions for the different materials can be differed from each other. It's noted [7–25], that the mechanism of polymer modification by the nanostructures or mesoparticles differs from the correspondent mechanism at the micro-particles using. The nanostructures radiate the quants of negative charges which increase polarization of the medium and lead to the self-organization of medium molecules accompanied the density growth.

The possible polarization leads to the increasing of medium (material) density owing to the regular orientation of material fragments or the macromolecules self-organization with the creation of super molecular and crystalline structures. At the different cases there is the change of character of quants radiation wave propagation from 2D (in the surface plane) to 3D (in the space field at surface).

The composition polarization is possible because of the presence of the charge quantization with the wave expansion on polar functional groups of media. These phenomena are determined by the IR spectra lines intensities changes. In Table 1.3, the instance of fine dispersed suspension Cu–C MC (hardener for epoxy resins (ERs)) is shown. According to data of Table 1.3, the decreasing of MC quantity to 0.001% leads to the intensity increasing of some fields in IR spectra. This effect is explained by the negative charged quants flow propagation freedom increasing in the considered medium. However, in these cases, the great influence on the suspensions IR spectra intensities can show the conditions of suspension preparation, for example, the ultrasound treatment (Table 1.4).

TABLE 1.3 The Change of Peaks Intensity Depending on Cu–C Mesocomposite Concentration

SL. No.	ν (cm^{-1})	I_1/I_Θ	$I_{0.01}/I_\Theta$	$I_{0.001}/I_\Theta$	Atomic Groups
1.	1,050	1,235	1,411	1,686	C–O–C st
2.	1,450	1,179	1,590	1,744	C–H
3.	1,776	1,458	1,347	1,691	C=O st as
4.	1,844	1,463	1,412	1,678	C=O st sy
5.	2,860–3,090	1,182	1,545	1,750	C–H

At the second day of that suspension existence the floccules are formed, and peaks intensity sharply drops. However, the suspension activity can be increased with the use of ultrasound treatment (Table 1.4). For the instance, it's established on the IR spectra data that the Cu–C MC suspension ultrasound treatment optimal duration determined as 7 minutes (the peak intensity in IR spectra is increased in 2–4 times). At the MCs suspension ultrasound treatment phase, coherence disturbance is possible. In this case, the phase coherence disturbance is observed at the duration equaled to 10 minutes.

The charge (electron) quantization should lead to the macromolecule electron structure change and, as a corollary, to change sub-molecular structures of polymeric substances. Therefore, the special raw of nanostructured materials such as polyvinyl alcohol (PVA), polymethyl metacrylate, polycarbonate, which contain Me–C MC in the minute quantities (10^{-1}–10^{-5}%) are prepared.

TABLE 1.4 The Changes of Peaks Intensity in IR Spectrum of Cu–C Mesocomposite Depending on the Duration of Ultrasound Treatment

ν (cm^{-1})	I_7/I_θ	I_{10}/I_θ	Atomic Groups
1776.6	3.7932	0.7574	C=O st as
1844.1	2.5065	0.9115	C=O st sy
3039.1	2.3849	0.9589	C–H st

The samples obtained are studied by the XPS and by the atomic force microscopy (AFM). The samples investigations by XPS show that the samples based on polycarbonate have more changes of electron structure at the minute quantities introduction of copper–carbon MC in comparison with other polymeric samples because they are more polarized.

According to the results of C1s spectra for polycarbonate, contained the different minute quantities of Cu–C MC, can note that after concentration equaled to 10^{-2}% of Cu/C MC the peaks correspondent to sp^2 and sp^3 peaks are appeared in spectra. In other words, the "stamp" of MC which is used during modification is appeared.

For the decision of question about the nanostructure influence on sub molecular composition structures the AFM method is applied. Polycarbonate nanostructured samples surface are presented in Figure 1.1. Polycarbonate is modified by Cu–C MC minute quantities (from 10^{-1} to 10^{-4}%). As follow from AFM images (Figure 1.1), the surface layers structure strongly changes at the concentration of Cu–C MC equaled to 10^{-4}% the transition from two-dimensional level to three-dimensional level of sub-molecular structures orientation. This fact is confirmed by the growth of sp^3 in comparison with sp^2 peak from X-ray photoelectron C1s spectra.

It's necessary to note that the results obtained by AFM are confirmed by the x-RPES data (on C1s spectra). The peculiarities of structure surface on Figure 1.1(d) are, possibility, explained by the Blodgett effect [19].

The changes of samples materials structure are caused by the actions of Cu–C MC radiation flow of the negative charged quants on the polymeric compositions that in turn influence on the modified materials properties.

Thus, the self-organization mechanism for polymeric compositions modified by the Me–C MC minute quantities is concluded in the composition polarization, which leads to the great change of electron and sub-molecular structures of materials.

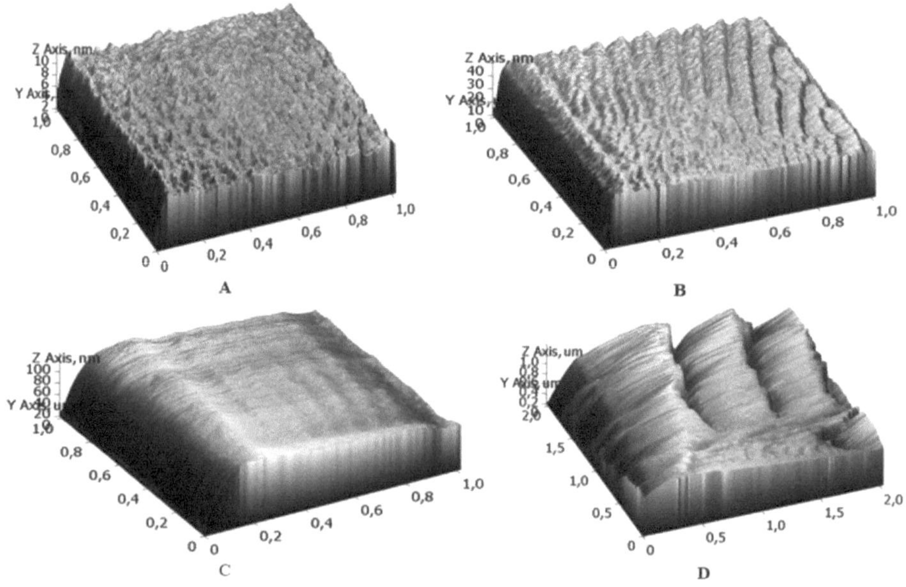

FIGURE 1.1 AFM images of polycarbonate nanostructured films surface: (a) 0.1% Cu–C MC; (b) 0.01% Cu–C MC; (c) 0.001% Cu–C MC; and (d) 0.0001% Cu–C MC.

1.4 CONCLUSION

The hypothesis of Red-Ox synthesis with the using of chemical mesoscopics idea about the charge quantization, annihilation, and interference phenomena is proposed. The hypothesis concludes in that the negative charge quants move to positive charged element nucleus to provoke the positive charge quants and the creation of annihilation phenomenon. Therefore, the direct EM field formed promotes to d electrons shift on higher energetic levels including the carbon shell of MC.

These processes take place when such MCs are included in the polymeric compositions and the modification of polymeric materials with the change of the structure and properties of materials happens.

KEYWORDS

- **atomic magnetic moments**
- **charges quantization**
- **interference and annihilation**
- **mesocomposites reactivity**
- **polarization**
- **polymeric composition modification**
- **red-ox reactions**
- **unpaired electrons quantity**

REFERENCES

1. Ivashkin, Yu. A., (2008). *The Crystal Structure Defects of Deformed Cubic Crystals* (p. 137). Bryansk: Publisher BSPTL.
2. Kodolov, V. I., & Trineeva, V. V., (2017). New scientific trend – chemical mesoscopics. *Chemical Physics & Mesoscopy, 19*(3), 454–465.
3. Shabanova, I. N., Kodolov, V. I., Terebova, N. S., & Trineeva, V. V., (2012). *X Ray Photoelectron Spectroscopy at the Investigation of Metal Carbon Nanosystems and Nanostructed Materials* (p. 252). M. – Izhevsk: Publ. Udmurt University.
4. Lipanov, A. M., Kodolov, V. I., Mel'nikov, M. Ya., Trineeva, V. V., & Pergushov, V. I., (2016). *The Influence of Small Quantities of Metal Carbon Nanosystems on Polymeric Materials Properties* (Vol. 466. No. 1. pp. 15–17). Doklagy RAS.
5. Kodolov, V. I., Trineeva, V. V., & Vasil'chenko, Yu. M., (2014). The calculating experiment for metal/carbon nanocomposites synthesis with the application Avrami equation. *Nanostructure, Nanomaterials and Nanotechnologies to Nanoindustry* (pp. 105–118, 436). Toronto-New Jersey: Apple Academic Press.
6. Wang, J. Q., Wu, W. H., & Feng, D. M., (1992). *The Introduction to Electron Spectroscopy (XPS/XAES/UPS)* (p. 640). Beijing: National Defense Industry Press.
7. Kodolov, V. I., Lipanov, A. M., Trineeva, V. V., et al., (2013). *The Changes of Properties of Materials Modified by Metal/Carbon Nanocomposites* (pp. 327–373). Ibid.
8. Kodolov, V. I., & Trineeva, V. V., (2013). Theory of modification of polymeric materials by super small quantities of metal/carbon nanocomposites. *Chemical Physics & Mesoscopy, 15*(3), 351–363.

9. Kodolov, V. I., & Trineeva, V. V., (2012). Perspectives of idea development about nanosystems self-organization in polymeric matrixes. In: *Problems of Nanochemistry for the Creation of New Materials* (pp. 75–100). Torun, Poland: IEPMD.
10. Akhmetshina, L. F., Lebedeva, G. A., & Kodolov, V. I., (2012). Phosphorus containing metal/carbon nanocomposites and their application for the modification of intumescent fireproof coatings. *Journal of Characterization and Development of Novel Materials, 4*(4), 451–468.
11. Chashkin, M. A., (2012). *Peculiarities of Modification by Metal/Carbon Nanocomposites for Cold Hardened Epoxy Compositions and the Investigation of Properties of Polymeric Compositions Obtained* (p. 17). Thesis of cand. Diss. – Perm: PNSPU.
12. Kodolov, V. I., & Kodolova, V. V., (Trineeva), Semakina, N. V., Yakovlev, G. I., Volkova, E. G., et al., (2008). *Patent 2337062 Russia Technique of Obtaining Carbon Nanostructures from Organic Compounds and Metal Containing Substances.*
13. Kodolov, V. I., Trineeva, V. V., Kovyazina, O. A., & Vasil'chenko, Yu. M., (2012). Production and application of metal/carbon nanocomposites. In: *Problems of Nanochemistry for the Creation of New Materials* (pp. 23–36). Torun, Poland: IEPMD.
14. Kodolov, V. I., Khokhriakov, N. V., Trineeva, V. V., & Blagodatskikh, I. I., (2008). Activity of nanostructures and its display in nanoreactors of polymeric matrixes and in active media. *Chem. Phys. & Mesoscopy, 10*(4), 448–460.
15. Kodolov, V. I., Khokhriakov, N. V., Trineeva, V. V., & Blagodatskikh, I. I., (2010). Problems of nanostructures activity estimation, nanostructures directed production and application. In: *Nanomaterials Yearbook – 2009: From Nanostructures, Nanomaterials and Nanotechnologies to Nanoindustry* (pp. 1–18). N.Y.: Nova Science Publ., Inc.
16. Nikolaeva, O. A., Kodolov, V. I., Zakharova, G. S., et al., (2004). *Method of Obtaining Carbon-Metal-Containing Nanostructures.* Patent of the RF 2225835.
17. Kodolov, V. I., & Trineeva, V. V., (2015). The metal/carbon nanocomposites influence mechanisms on media and compositions. In: *Nanostructures, Nanomaterials and Nanotechnologies to Nanoindustry* (pp. 171–185). Toronto, New Jersey: Apple Academic Press.
18. Shabanova, I. N., Kodolov, V. I., Terebova, N. S., et al., (2016). X-ray photoelectron study of the influence of the amount of carbon nickel containing nanostructures on the degree of the poly methyl methacrylate modification. In: *Multifunctional Materials and Modeling* (pp. 211–218). Canada-USA: AAP.
19. Kodolov, V. I., & Trineeva, V. V., (2016). Self-organization in processes under action of super small quantities of metal/carbon nanocomposites: Review on investigation results. In: *Multifunctional Materials and Modeling* (pp. 263–331, 343). Canada-USA: AAP.
20. Kodolov, V. I., Trineeva, V. V., Pershin Yu. V., et al., (2020). *Method of Metal Carbon Nanocomposites Obtaining from Metal Oxides and Polyvinyl Alcohol.* Pat. RU 2018122 001.
21. Mustakimov, R. V., Kodolov, V. I., Shabanova, I. N., & Terebova, N. S., (2017). Modification of copper carbon nanocomposites with the use of ammonium polyphosphate for the application as modifiers of epoxy resins. *Chemical Physics & Mesoscopics, 19*(1), 50–57.
22. Kodolov, V. I., Trineeva, V. V., Kopylova, A. A., et al., (2017). Mechanochemical modification of metal carbon nanocomposites. *Chemical Physics & Mesoscopics, 19*(4), 569–580.

23. Kodolov, V. I., Trineeva, V. V., Terebova, N. S., et al., (2018). The change of electron structure and magnetic characteristics of modified copper carbon nanocomposites. *Chemical Physics & Mesoscopics, 20*(1), 72–79.
24. Karavaeva, N. M., Pershin, Yu. V., Kodolov, V. I., et al., (2019). Change of morphology and swelling of cured epoxy compositions upon their modification with minute quantities. *Pol. Sci., Ser. D., 12*(2), 179–181.
25. Kodolov, V. I., Semakina, N. V., & Trineeva, V. V., (2018). Introduction in science about nanomaterials. *Monograph* (p. 476). – Izhevsk: Publisher – M.T. Kalashnikov Izhevsk State Technical University.

CHAPTER 2

X-Ray Photoelectron Study of the Formation of the Chemical Bond and the Atomic Magnetic Moment in Nickel–Carbon Nanocomposites Modified by *d*-Metal Oxides

N. S. TEREBOVA[1], V. I. KODOLOV[2], and I. N. SHABANOVA[1]

[1]*Udmurt Federal Research Center, Ural Branch, Russian Academy of Sciences, Izhevsk, Russia*

[2]*Kalashnikov Izhevsk State Technical University, Izhevsk, Russia*

ABSTRACT

X-ray photoelectron spectroscopy (XPS) is used to study the modification of Ni/C nickel-carbon nanocomposites with oxides (NiO, CuO, and Fe_2O_3) in the ratio of 0.5–1.0. Depending on the type of modifier used and its content, the magnetic moment of Ni atoms changes in the nanocomposite, i.e., during the Red-Ox process, the nanocomposite structure changes, and the interaction between the carbon skeleton and the modifier metal is enhanced. In the presence of electron acceptors, the paired *d* electrons of the nanocomposite metal become unpaired and move to higher orbitals. This is indicative of self-organization processes in nanostructures, which leads to an increase in the atomic magnetic moment of the metal.

Modern Magnetic Materials: Properties and Applications. Iuliana Stoica, Ann Rose Abraham, & A. K. Haghi (Eds.)

2.1 INTRODUCTION

The mechanical, thermal, electrical, magnetic, and other properties of *Me*/C (*Me* is metal) nanocomposites provide their diverse applications [1–3]. In order to improve the dispersion and solubility of nanocomposites along with the magnetic, electrical, and strength properties and to prevent beam coagulation, *Me*/C nanocomposites are modified by *sp* [4] and *d* elements. To examine the formation of a chemical bond between atoms, X-ray photoelectron study of the core levels (C1*s*, O1*s*, *Me*2*p*, *Me*3*s*) of modified nanocomposites is performed. X-ray photoelectron *Me*3*s* spectra are used to estimate the spin state of 3*d* metal atoms in the nanocomposites under study because they have a multiplet structure due to spin-spin interactions [5–10].

The parameters of multiplet splitting in the *Me*3*s* spectra correlate with the number of uncompensated *d* electrons of metal atoms and its spin magnetic moment. It is also proved that the relative intensity of peaks of multiplets in the 3*s* spectra correlates with the number of unpaired *d* electrons of atoms in 3*d* metal systems, and the distance between the multiplet peaks gives information about the exchange interaction of 3*s*–3*d* shells [11, 12]. From changes occurring in the 3*d* shell (an increase in the degree of localization or hybridization of *d* electrons of metal atoms), it is possible to gain information about a change in the distance between neighboring atoms and the nearest environment of 3*d* metal atoms. An increase in the number of uncompensated *d* electrons is explained by the participation of *d* electrons of modifier metal atoms in the hybridized chemical bond with *p* electrons of carbon atoms in the nanocomposites.

The model developed is applied to determine a change in the atomic magnetic moment of the metal in metal-carbon nanocomposites compared to bulk metal samples [13]. A change in the relative intensity of peaks of the multiplet splitting and the distance between them in the nanocomposites is shown in comparison with the bulk samples. The results obtained indicate an increase in the number of uncompensated *d* electrons in *d* metal atoms and the appearance of copper atoms on their shells in copper–carbon mesocomposites (MCs) [13].

The aim of the work is the X-ray photoelectron spectroscopy (XPS) study of the effect of modifying Ni–C MCs by metal oxides on a change in the atomic magnetic moment of the metal.

2.2 EXPERIMENTAL METHODS

Nickel-carbon MCs with a modifier were mechanically and chemically treated. The red-ox saturated shell of the nickel-carbon nanocomposites.

The ratio of the modifiers (NiO, CuO, Fe_2O_3) relative to the Ni/C nanocomposites varied from 0.5 to 1.0. To activate the process, the reaction mixture was moistened. After the Ni–C MC and oxide were ground together in a mechanical mortar at an energy consumption of about 260–270 kJ/mol, the mesoscopic product formed was dried and kept in a closed crucible at about 150°C, and then in vacuo at 100–150°C for no more than 3 min. X-ray photoelectron spectroscopic analysis was conducted on a magnetic spectrometer with a resolution of 10–4, instrument luminosity of 0.085%, Al*K*α excitation energy of 1486.5 eV in a vacuum chamber with a residual pressure of 10^{-8}–10^{-10} Torr [14].

At present, commercial electrostatic electron spectrometers are used internationally. X-ray photoelectron magnetic spectrometers were earlier designed in Sweden under the guidance of Nobel laureate K. Siegbahn [15]; however, after his death this work ceased. In Russia, it has continued under the leadership of K. Siegbahn's disciple Prof. V. A. Trapeznikov [14]. The advantages of X-ray photoelectron magnetic spectrometers are due to the fact that the magnetic energy analyzer is meaningfully separated from the vacuum chamber of the spectrometer. This enables the application of technological impacts on the sample, which are accompanied by aggressive gas release. Moreover, magnetic spectrometers are characterized by more contrast spectra than electrostatic spectrometers.

2.3 RESULTS AND DISCUSSION

2.3.1 Ni–C MESOCOMPOSITE (MC) MODIFIED BY THE OXIDE NiO

Ni–C MC modified by NiO oxide in the ratios Ni/C:NiO = 1:0.5 and 1:1 was studied. Figure 2.1 presents the Ni2*p* spectrum of the MC sample modified by nickel oxide NiO in the 1:1 ratio.

The sample surface contains oxygen contaminants, and the peak of the Ni2*p* spectrum corresponds to an energy of 854.2 eV (curve *1*). After cleaning the sample in the spectrometer chamber, the spectral peak shifts to a lower binding energy of 852 eV (curve *2*), which is characteristic of unoxidized nickel and indicates reduction of the modifier metal during mechanical and chemical treatment.

Table 2.1 summarizes the data obtained from the Ni3*s* spectra on the atomic magnetic moments of the metal for unmodified Ni/C nanocomposites and two samples of nanocomposites modified by NiO. It is seen that the Ni atomic magnetic moment varies from 1.8 μB for the unmodified MC [13] to

3.0 μB in the modified sample of the composition Ni/C: NiO = 1: 0.5 and to 4.0 μB in the sample with the 1: 1 composition.

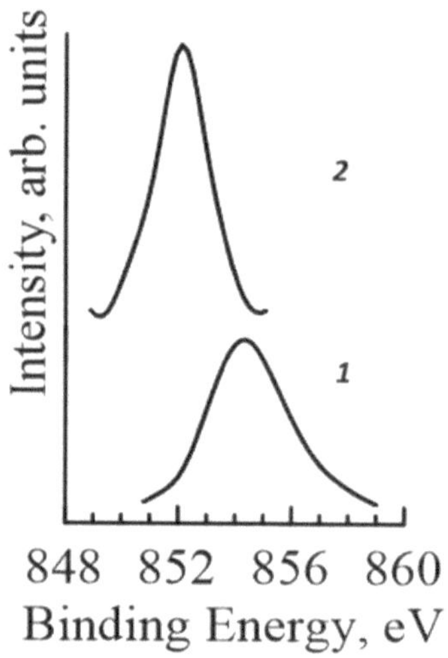

FIGURE 2.1 X-ray photoelectron Ni2*p* spectra of NiO (*1*); and Ni (*2*).

Figure 2.2 depicts the X-ray photoelectron Ni3*s* spectra of two samples.

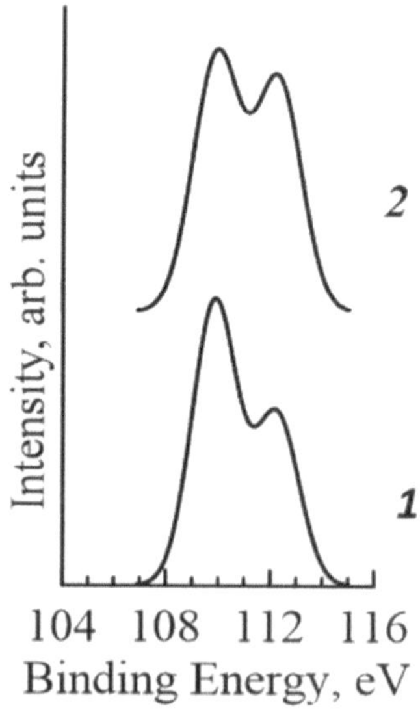

FIGURE 2.2 X-ray photoelectron Ni3*s* spectra of Ni/C: NiO =1: 0.5 (*1*); and 1: 1 (*2*).

TABLE 2.1 Parameters of the *Me*3*s* Spectra and Magnetic Moments of Metal *d* Atoms in the Reference Samples and Samples of Nanostructures Ni/C:NiO = 1:0.5; Ni/C:NiO = 1:1; Ni/C:CuO = 1:1; and Ni/C:Fe_2O_3 = 1:1

Sample	I_2/I_{1Ni}	Δ_{Ni} (eV)	μ_{Ni}, $\mu_Б$	I_2/I_{1Cu}	Δ_{Cu} (eV)	μ_{Cu}, $\mu_Б$	I_2/I_{1Fe}	Δ_{Fe} (eV)	μ_{Fe}, $\mu_Б$
Ni/C	0.32	3.0	1.8	–	–	–	–	–	–
Cu/C	–	–	–	0.2	3.5	1.3	–	–	–
Fe/C	–	–	–	–	–	–	0.5	4.0	2.5
Ni/C + NiO = 1:0.5	0.6	2.6	3.0	–	–	–	–	–	–
Ni/C + NiO = 1:1	0.8	2.2	4.0	–	–	–	–	–	–
Ni/C + CuO = 1:1	0.5	3.0	2.5	0.4	3.0	2.0	–	–	–
Ni/C + Fe_2O_3 = 1:1	0.5	2.4	2.5	–	–	–	0.6	2.0	3.0

Note: I_2/I_1 is the intensity ratio of the multiplet splitting peaks; Δ is the energy distance between the multiplet-splitting peaks in the 3*s* spectra, which determines the exchange integral between 3*s* and 3*d* electrons of the metal atom and depends on the overlap of its 3*s* and 3*d* shells.

The C1*s* spectrum of the reference sample of the unmodified Ni/C MC (Figure 2.3, curve *1*) exhibits three components: 285 (C–H), 284 (C–C, sp^2 hybridization), and 286 eV (C–C, sp^3 hybridization of carbon atom electrons).

In the C1*s* spectrum of the sample with the composition Ni/C MC: NiO=1:0.5, the 284-eV component (C–C, sp^2) increases. In the C1*s* spectrum of the sample Ni/C MC: NiO = 1:1 (Figure 2.3, curve *3*), the 286-eV component of the C–C bond with the sp^3 hybridization of carbon-atom valence electrons increases. Enhancement of the atomic magnetic moment of nickel in the nanocomplexes is caused by a change in their structures, which is evidenced by alterations in the C1*s* spectra. An increase in the C–C-bond component with the sp^3 hybridization of valence electrons in addition to C–C bonds with the sp^2 hybridization leads to appreciable enhancement of the metal atomic magnetic moment in the modified MC.

Thus, when the nickel-carbon MC is modified by nickel oxide, the atomic magnetic moment of nickel considerably increases. The largest nickel atomic magnetic moment is observed when the MC is modified by nickel oxide (NiO) in the 1:1 ratio. Nickel reduction from the oxide is seen in the Ni2*p* spectra. Released oxygen is bonded to hydrogen atoms from C–H fragments of the MC. The component characteristic of the C–C bond with the sp^3 hybridization of valence electrons simultaneously increases in the C1*s* spectrum. Electron consumption from the surface of MC C–C fragments increases due to the bond with atoms of the reduced modifier metal.

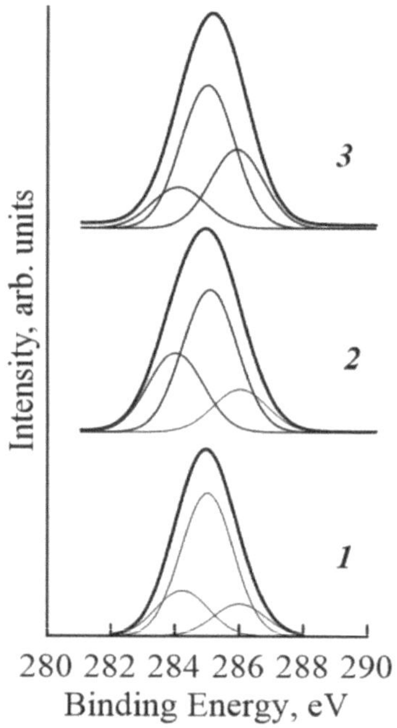

FIGURE 2.3 X-ray photoelectron C1*s* spectra of the nanostructures: Ni/C (*1*); Ni/C MC:NiO = 1:0.5 (*2*); and Ni/C MC:NiO = 1:1 (*3*).

2.3.2 Ni/C MESOCOMPOSITE (MC) AND THE CuO MODIFIER

Figure 2.4 shows the X-ray photoelectron Ni3*s* and Cu3*s* spectra of the sample.

Table 2.1 lists data on the metal atomic magnetic moments for the Ni/C MC modified by CuO. It is seen that the Ni atomic magnetic moment varies from 1.8 μB for the unmodified MC to 2.5 μB in the modified sample; also a copper atomic magnetic moment of 2.0 μB appears in the modified sample of the composition Ni/C MC: CuO = 1:1, which is noticeably larger than that (1.3 μB) in the Cu/C nanostructures studied previously [13].

The C1*s* spectrum of the Ni/C: CuO sample consists of three components: 284 (C–C, sp^2), 286 (C–C, sp^3), and 285 eV, which corresponds to C–H. Thus, modification of the Ni/C MC by copper oxide, in comparison with modification of this sample by nickel oxide, results in lower enhancement of the nickel atomic magnetic moment, but at the same time, the magnetic moment of reduced copper atoms increases.

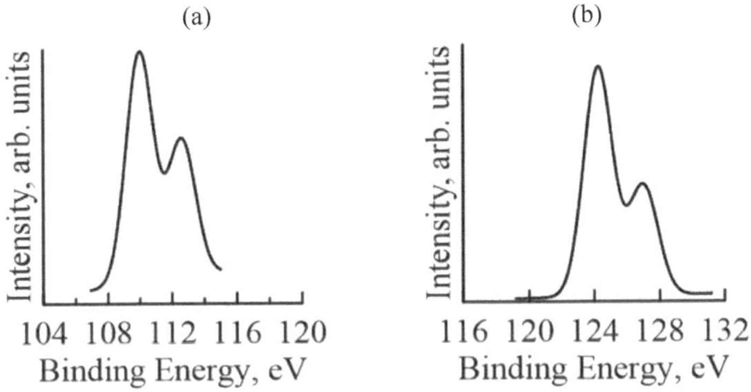

FIGURE 2.4 X-ray photoelectron Ni3*s* (a); and Cu3*s* (b) spectra of Ni/C MC: CuO = 1:1.

2.3.3 Ni/C MESOCOMPOSITE (MC) AND THE Fe_2O_3 MODIFIER (1:1)

When this sample is heated to 150°C, iron oxide is retained on the surface and decomposes in the spectrometer chamber upon heating to 500°C, and iron atoms incorporate into the Ni/C MC, interacting with carbon atoms having free bonds. The results of the multiplet splitting of the Ni3*s* and Fe3*s* spectra are summarized in Table 2.1 and Figure 2.5. A change in the Ni atomic magnetic moment is observed from 1.8 μB in the unmodified Ni/C MC to 2.5 μB in the modified one along with a change in the iron atomic magnetic moment from 2.5 μB in the Fe/C MC [13] to 3.0 μB in the sample studied.

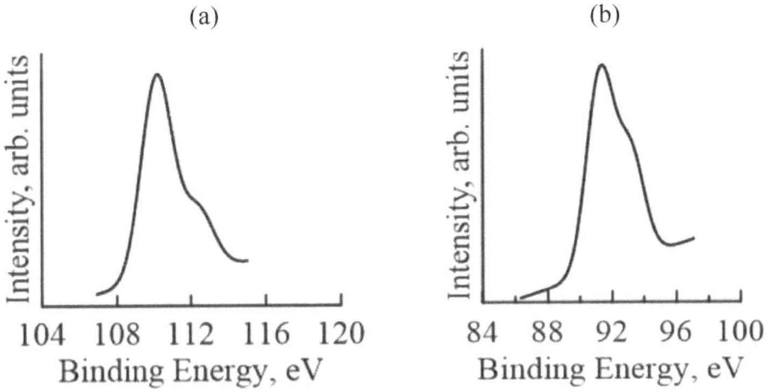

FIGURE 2.5 X-ray photoelectron Ni3*s* (a); and Fe3*s* (b) spectra of Ni/C MC:Fe_2O_3 = 1:1.

Thus, when nickel-carbon nanocomposites are modified by copper or iron oxide, the nickel atomic magnetic moment increases insignificantly.

This can be explained by the fact that copper or iron oxides are reduced in steps. In the case of copper oxide, copper(I) oxide must form first, with the possible equilibrium transfer of this oxide to copper and initial oxide arising. A sufficiently active oxygen atom of copper(I) oxide interacts with C–H bonds of the polyacetylene moiety, pushing it to the surface, whereas a part of the C–C bonds "sinks." If this occurs, then the displacement of electrons from the metal of the nanostructure to the carbon shell is insignificant, and enhancement of the nickel atomic magnetic moment should not be expected.

2.4 CONCLUSIONS

Hence, during the red-ox process, the fraction of nickel-carbon nanostructures increases and its form changes, i.e., positively charged metal atoms are reduced, hydrocarbon groups are dehydrogenated, and the interaction between the carbon framework and metal is enhanced. The modifier interacts with carbon atoms of the MC where free bonds are present. In the presence of electron acceptors, the paired *d* electrons of the MC metal become unpaired and move to higher orbital [16]. This indicates self-assembly processes in the nanostructures, which results in an increase in the metal atomic magnetic moment.

The red-ox process on the surface of the metal-carbon MC facilitates a significant change in the chemical bond of metal atoms. By varying the composition and structure of nanosized forms, it is possible to control the physical characteristics of mesoscopic materials. Consequently, the possibility arises to manage the magnetic properties and obtain mesomaterials with a super high atomic magnetic moment upon mechanical and chemical modification by metal oxides of metal-carbon MCs.

KEYWORDS

- **atomic magnetic moment**
- **carbon atoms**
- **nanocomposites**
- **nickel-carbon nanocomposites**
- ***sp*, sp^2, and sp^3 hybridization**
- **valence electrons**
- **X-ray photoelectron spectroscopy**

REFERENCES

1. Gubin, S. P., Koksharov, Yu. A., Khomutov, G. B., et al., (2005). *Usp. Khim., 74*(6), 489.
2. Eletskii, A. V., (2007). *Phys.–Usp., 50*(3), 225.
3. Bocharov, G. S., Eletskii, A. V., Zakharenkov, A. V., et al., (2018). *J. Surf. Invest.: X-ray, Synchrotron Neutron Tech., 12*, 27.
4. Shabanova, I. N., Terebova, N. S., Sapozhnikov, G. V., & Kodolov, V. I., (2017). *Phys. Solid State, 59*, 174. https: doi.org/10.1134/S1063783417010279.
5. Van, V. J. H., (1934). *Phys. Rev., 45*, 405.
6. Vasudevan, S., Vasan, H. N., & Rao, C. N. R., (1979). *Chem. Phys. Lett., 65*(3), 444.
7. Fadley, C. S., Shirley, D. A., Freeman, A. J., et al., (1969). *Phys. Rev. Lett., 23*, 1397.
8. Briggs, D., & Seah, M. P., (1983/1987). *Practical Surface Analysis by Auger and X-Ray Photoelectron Spectroscopy.* New York, Wiley; Mir, Moscow.
9. Van, A. F., Stadnik, Z. M., Fuggle, J. C., et al., (1988). *Phys. Rev. B, 37*, 6827.
10. Selwood, P. W., (1943/1958). *Magnetochemistry.* Interscience, New York; Inostr. Lit., Moscow.
11. Fadley, C. S., & Shirley, D. A., (1970). *Phys. Rev. A, 2*, 1109.
12. Lomova, N. V., & Shabanova, I. N., (2004). *J. Electron Spectrosc. Relat. Phenom., 137–140*, 511.
13. Shabanova, I. N., & Terebova, N. S., (2010). *Surf. Interface Anal., 42*(6, 7), 846.
14. Trapeznikov, V. A., Shabanova, I. N., Dobysheva, L. V., et al., (1986). *Izv. Akad. Nauk SSSR, Ser. Fiz., 50*(9), 1677.
15. Siegbahn, K., Nordling, C., Fahlman, A., & Nordberg, R., (1967/1971). *Atomic, Molecular and Solid-State Structure Studied by Means of Electron Spectroscopy.* Royal Society of Sciences of Uppsala, Uppsala, Sweden; Mir, Moscow.
16. Al Ma'Mari, F., Moorsom, T., Teobaldi, G., et al., (2015). *Nature, 524*, 69. https: doi.org/10.1038/nature14621.

CHAPTER 3

Red-Ox Synthesis of Metal/Carbon Mesocomposites in Nanosized Reactors of Polymeric Matrices

V. V. KODOLOVA-CHUKHONTZEVA[1,2] and V. I. KODOLOV[1,3]

[1]*BRHEC of Chemical Physics and Mesoscopics, Udmurt Federal Research Center, RAS, Izhevsk, Russia*

[2]*Peter Great St. Petersburg Polytechnic University, St. Petersburg, Russia*

[3]*M.T. Kalashnikov Izhevsk State Technical University, Izhevsk, Russia*

ABSTRACT

This chapter explains the base of mesoscopic physics principles for metal-carbon mesocomposites (MCs) obtained in polymeric matrices' nanosized reactors (mesoreactors). The conditions of mesoscopic physics principles applications are proved. These conditions include such phenomena as interference, spectrum, and charge quantization, which take place at the action of mesoscopic (nano) particles having sizes from 1,000 to 0.1 nm on different media. In addition to that, the correspondent particles can only vibrate and conduct motions. It's shown that these particularities of clusters or mesoparticles are possible during the red-ox synthesis of Me–C MC in mesoreactors of polymeric matrices.

3.1 INTRODUCTION

At present, some hypothesis and the investigations results are explained by means of mesoscopic physics principles [1, 2], and also by the fractal theory equations [3–5]. Self-organization in system and self-similarity (image) are

Modern Magnetic Materials: Properties and Applications. Iuliana Stoica, Ann Rose Abraham, & A. K. Haghi (Eds.)

the ground notions in mesoscopic physics as well as in the fractal theory and technology for nanosized particles or nanostructures.

Therefore, it is interesting in this chapter to consider the application of these disciplines for the synthesis of nanostructures, including metal-carbon mesocomposites (Me–C MCs). Mesoscopic Physics deals with mesoscopic particles which have linear size from 1,000 to 0.1 nm, because the length of coherence wave equals to 1,000 nm, and the electron wavelength in different materials changes from 100 to 0.1 nm. For these particles such particularities are known as the interference, spectrum quantization and charge quantization [1].

The application Mesoscopic Physics principles is possible, if the mesoscopic particle will be established within small volume space, for instance, mesoscopic reactor (mesoreactor), in which this mesoparticle is coordinated on active groups of mesoreactor walls in correspondence with realized conditions. In these cases, the correspondent particle can only vibrate and conduct the electrons across oneself. Therefore, the mesoreactor can be presented as the nanosized hollow or the limited free volume in polymeric matrix. Then chemical particles are directly self-organized in this hollow for the creation of transitional (mesoscopic) state before the mesoscopic product formation. Also, the mesoreactor definition can be proposed as the specific porous nanostructure, in which the distance between walls changes from 1 nm to 100 nm.

3.2 THEORETICAL AND EXPERIMENTAL INVESTIGATIONS OF SYNTHESIS WITHIN NANOREACTORS OF POLYMERIC MATRICES

To our mind, the formation of Me–C MC within mesoreactors of polymeric matrices may be similar to the photography processes. At the beginning, when metal containing phase is mixed up with the polymeric phase, the mesoreactor within polymeric matrix is filled by metal clusters of metal containing phase. The process is accompanied with cluster coordination on active centers of mesoreactor walls. It's similar to photography process. Then during process of xerogel formation the development flows. And then at the heating of obtained product the fixation occurs.

Thus, at the Me–C MC obtaining there are following processes: the coordination reactions, the red-ox reactions and certainly the self-organization processes.

Let us discuss this picture of the nanostructure's formation. Now the active chemical metal containing particle is taken root in the free nanosized hollow of polymeric matrices. It coordinates with functional groups on mesoreactor walls and loses the ability to diffusion and rotation. This muddled particle can be characterized by vibration motion, and also electron motion including the transport electrons across it. However, if the electrons transport across our particle (cluster) will be carried out, the red-ox reactions will be stimulated, and the metal reduction will be occurred. This process is direct and lead to self-organization which may be observed with using AFM investigations. For the further stimulation of Me–C MC formation process it is necessary the regular regime of unfinished product heating.

Consequently, the red-ox synthesis of Me–C MC can be explained by basic principles of Mesoscopic Physics.

The investigation of Red-Ox synthesis of Me–C MC in mesoreactors of polymeric matrices is realized [9, 10] in three stages (Scheme 3.1).

Polymeric Matrices, → for example, Polyvinyl Alcohol	*Nanoreactors of Polymeric Matrices for the filling by Metal Compounds Phase*	*← Metal Compounds, for instance, Clusters of Copper or Nickel Oxides*
Coordination Reactions and, Partially, Red-Ox Processes ↓		
The obtaining of xerogels containing the different mesoreactors in dependence on metal nature		
The estimation of energetic and geometric characteristics of Nanoreactors ↓		
Thermochemical stage (Red-Ox Synthesis of Metal-carbon Mesocomposites)		
DTA-TG investigations for the determination of temperature conditions and the heating duration (the middle temperature for the Copper–Carbon Synthesis films equals to 200°C, at the middle duration equals to 3 hours).		

SCHEME 3.1 The scheme of metal-carbon mesocomposites synthesis within mesoreactors of polymeric matrices.

Before the experimental investigations, the computational designing of mesoreactors filled by metal containing phase as well as quantum chemical modeling of processes within mesoreactors are carried out.

Then the experimental designing and mesoreactors filling by metal containing phase are realized with using two methods:

i. The mixing of salt solution with the solution of functional polymer, for example, polyvinyl alcohol (PVA) solution; and
ii. The common degeneration of polymeric phase with metal containing phase (metal oxides) in active medium (water).

The following stage includes the obtaining of xerogels filled by metal containing phase. The third stage is the properly red-ox synthesis of Me–C

MC in mesoreactors of polymeric matrices (obtained xerogels) at narrow temperature intervals.

The first and second stages concern to preparatory stages. On the second stage the functional groups in nanoreactor walls participate in coordination reactions with metal ions [(i) method] or with the clusters of metal containing phase [(ii) method].

The computational experiment was carried out with Gamess and HyperChem software products, with visualization. The definite result of the computational experiment is obtained stage by stage. Any transformations at the initial stages are taken into account at further stages. The prognostic possibilities of the computational experiment consist in defining the probability of the formation of nanostructures of definite shapes. The optimal dimensions and shape of internal cavity of nanoreactors, optimal correlation between metal containing and polymeric components for obtaining the necessary mesoproducts are found with the help of quantum-chemical modeling [9].

If the method (i) is used, for instance, the interaction of PVA and metal chloride solutions where metals are iron, cobalt, nickel, and copper, than the colored xerogels is formed as follows: iron – brown-red, nickel – pale-green, cobalt – blue and copper – yellow-green. Consequently, PVA interacts with metal ions with the formation of complex compounds.

For the corresponding correlations "polymer-metal containing phase" the dimensions, shape, and energetic characteristics of nanoreactors are found with the help of coordination, the structure and relief of xerogel surface change. The comparison of phase contrast pictures on the corresponding films indicates a greater concentration of the extended polar structures in the films containing copper, in comparison with the films containing nickel and cobalt (Figure 3.1).

The processing of the pictures of phase contrast to reveal the regions of energy interaction of cantilever with the surface in comparison with the background produces practically similar result with optical transmission microscopy [11, 12].

Corresponding to data of AFM [11], the sizes of nanoreactors obtained from solutions of metal chlorides and the mixture of PVA with polyethylene polyamine (PEPA) are determined (Table 3.1).

The results of AFM investigations of xerogels films obtained from metal oxides and PVA [13, 14] is distinguished (Figure 3.2) in comparison with previous data, which testify to the difference in reactivity of metal chlorides and metal oxides.

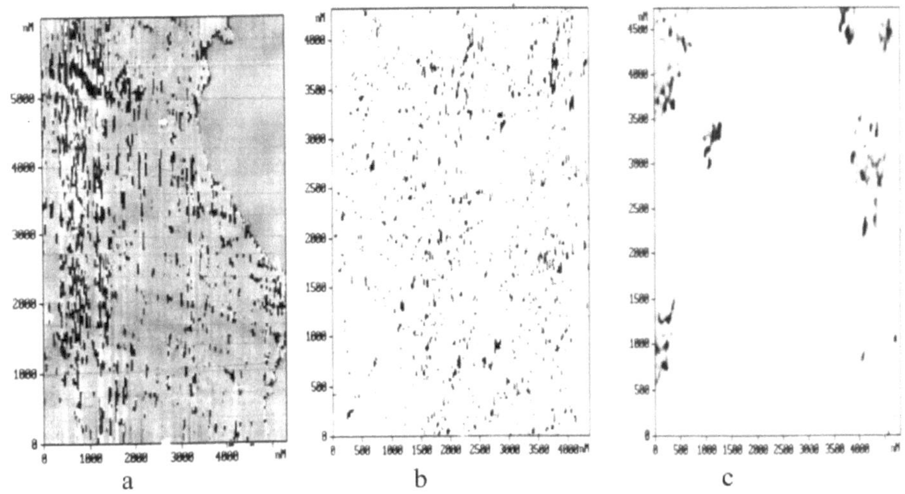

FIGURE 3.1 Pictures of phase contrast of PVA surfaces containing copper (a); nickel (b), and cobalt (c).

TABLE 3.1 Sizes of Nanoreactors Found with the Help of Atomic Force Microscopy

Composition	Sizes of AFM Formations				
	Length	Width	Height	Area	Density
PVA:PEPA:CoCI2 = 2:1:1	400–800	150–400	30–40	60–350	5.5
PVA:PEPA:NiCI2 = 2:1:1	80–100	80–100	25–35	6–12	120
PVA:PEPA:CuCI2 = 2:1:1	80–100	80–100	20–30	6–20	20
PVA:PEPA:CoCI2 = 2:2:1	600–900	300–600	100–120	180–500	3.0
PVA:PEPA:NiCI2 = 2:2:1	40–80	40–60	10–30	2–4	350

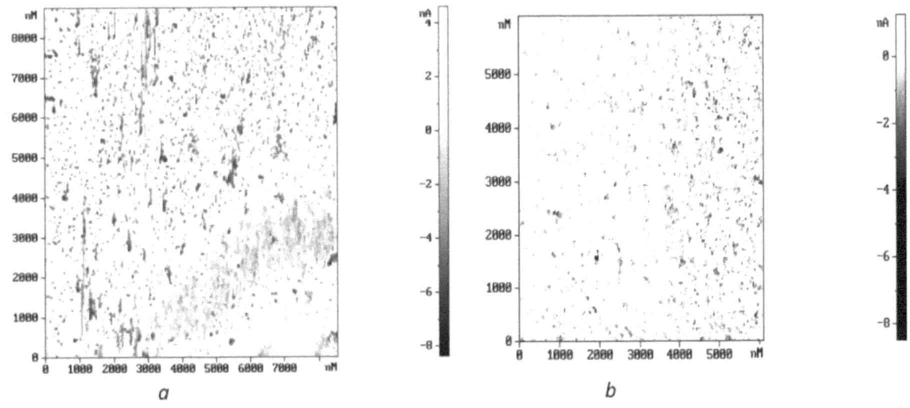

FIGURE 3.2 Phase contrast pictures of xerogels films PVA–Ni (a); and PVA–Cu (b).

According to AFM results investigation the addition of Ni/C MC in PVA leads to more strong coordination in comparison with analogous addition Cu/C MC.

The mechanism of formation of nanoreactors filled with metals was found with the help of IR spectroscopy. Depending on metal coordinating ability and conditions for nanostructure obtaining (in liquid or solid medium with minimal content of liquid), we obtain "embryos" of future nanostructures of different shapes, dimensions, and composition. It is advisable to model coordination processes and further red-ox processes with the help of quantum chemistry apparatus.

The availability of d metal in polymeric matrix results, in accordance with modeling results, in its regular distribution in the matrix and self-organization of the matrix.

At the same time, the metal orientation proceeds in interface regions and nanosized pores of polymeric phase which stipulates further direction of the process to the formation of metal/carbon nanocomposites. In other words, the birth and growth of nanosized structures occur during the process in the same way as known from the macromolecule physics and fractal theory [16].

The mechanism of formation of nanoreactors filled with metals was found with the help of IR spectroscopy. Thus, at the second stage the coordination of metal containing phase and corresponding orientation in mesoreactor take place.

To investigate the processes at the second stage of obtaining metal/carbon nanocomposites X-ray photoelectron spectroscopy (XPS), transmission electron microscopy and IR spectroscopy are applied. The sample for IR spectroscopy was prepared when mixing Me–C MC powder with one drop of Vaseline oil in agate mortar to obtain a homogeneous paste with further investigation of the paste obtained on the appropriate instrument. As the Vaseline oil was applied when the spectra were taken, we can expect strong bands in the range 2,750–2,950 cm^{-1}.

At the third stage, it is required to give the corresponding energy impulse to transfer the "transition state" formed into carbon metal MC of definite size and shape. To define the temperature ranges in which the structuring takes place, DTA-TG investigation is applied.

It is known that small changes of weight loss (TG curve) at invariable exothermal effect (DTA curve) testify to the self-organization (structural formation) in the system. Below typical curves of DTA-TG investigation are shown.

According to data of DTA-TG investigation optimal temperature field for film nanostructure obtaining is 230–270°C, and for spatial nanostructure obtaining – 325–410°C.

It is found that in the temperature range at 200°C nanosized films, from carbon fibers associated with metal phase as well, are formed on metal or metal oxide clusters. When the temperature elevates up to 400°C, 3D nanostructures are formed with different shapes depending on the coordinating ability of the metal.

Assuming that the nanocomposites obtained can be considered as oscillators transferring their oscillations onto the medium molecules, we can determine to what extent the IR spectrum of liquid medium will change.

It is proposed to consider the obtaining of Me–C MC in mesoreactors of polymeric matrixes as self-organization process similar to the formation of ordered phases.

The perspectives of this investigation are looked through in an opportunity of thin regulation of processes and the entering of corrective amendments during processes.

This process can be considered as the crystallization and can be described by the fractal theory with the application of Avrami-Kolmogorov equation. However, this equation was adapted to the conditions of Me–C MC red-ox synthesis. At the beginning let us discuss the possibilities of Avrami equations for the determination of the heating regime conditions at the red-ox synthesis of Me–C MC.

The difference of potentials between the interacting particles and object walls stimulating these interactions is the driving force of self-organization processes (formation of mesoparticles with definite shapes). The potential jump at the boundary "mesoreactor wall-reacting particles" is defined by the wall surface charge and reacting layer size. If we consider the red-ox process as the main process preceding the nanostructure formation, the work for charge transport corresponds to the energy of mesoparticle formation process in the reacting layer. Then the equation of energy conservation for mesoreactor during the formation of mesoparticle will be as follows:

$$nF\Delta\varphi = RT\ln\left(\frac{N_p}{N_r}\right), \tag{1}$$

where; n is the number reflecting the charge of chemical particles moving inside "the mesoreactor;" F is the Faraday number; $\Delta\phi$ is the difference of potentials between the mesoreactor walls and flow of chemical particles; R is the gas constant; T is the process temperature; N_p is the mol share of mesoparticles obtained; and N_r is the mol share of initial reagents from which the mesoparticles are obtained.

Using the above equation, we can determine the values of equilibrium constants when reaching the certain output of mesoparticles, sizes of mesoparticles and shapes of mesoparticles formed with the appropriate equation modification. The internal cavity sizes or mesoreactor reaction zone and its geometry significantly influence the sizes and shapes of nanostructures.

The sequence of the processes is conditioned by the composition and parameters (energy and geometry) of mesoreactors. To accomplish such processes, it is advisable to preliminarily select the polymeric matrix containing the nanoreactors in the form of nanosized pores or crazes as process appropriate. Such selection can be realized with the help of computer chemistry. Further, the computational experiment is carried out with the reagents placed in the mesoreactor with the corresponding geometry and energy parameters.

Examples of such computations were given in Ref. [3]. The experimental confirmation of polymer matrix and mesoreactor selection for the obtaining of the carbon metal containing nanostructures was given in Refs. [4, 5]. Kolmogorov-Avrami equations are widely used for such processes which usually reflect the shares of a new phase produced.

When the metal ion moves inside the mesoreactor with red-ox interaction of ion (mol) with mesoreactor walls, the balance setting in the pair "metal containing-polymeric phase" can apparently be described with the following equation:

$$zF\Delta\varphi = RT\ln K = RT\ln\left(\frac{N_p}{N_r}\right) = RT\ln(1-W) \tag{2}$$

where; z is the number of electrons participating in the process; $\Delta\varphi$ is the difference of potentials at the boundary "mesoreactor wall-reactive mixture;" F is the Faraday number; R is the universal gas constant; T is the process temperature; K is the process balance constant; N_p is the number of moles of the product produced in mesoreactor; N_r is the number of moles of reagents or atoms (ions) participating in the process which filled the mesoreactor; and W is the share of nanosized product obtained in mesoreactor.

In turn, the share of the transformed components participating in phase interaction can be expressed with the equation which can be considered as a modified Avrami equation:

$$W = 1-\exp\left[-\tau^{n}\exp\left(\frac{zF\Delta\varphi}{RT}\right)\right] \tag{3}$$

where; τ is the duration of the process in mesoreactor; and n is the number of freedom degrees changing from 1 to 6.

When "n" equals 1, one-dimensional nanostructures are obtained (linear nanostructures, nanosized fibers). If "n" equals 2 or is changed from 1 to 2, flat nanostructures are formed (nanosized films, circles, petals, wide nanosized bands). If "n" changes from 2 to 3 and more, spatial nanostructures are formed as "n" also indicates the number of freedom degrees. The selection of the corresponding equation recording form depends on the mesoreactor (nanostructure) shape and sizes and defines the nanostructure growth in the mesoreactor.

During the red-ox process connected with the coordination process, the chemical bonds character changes. Therefore, correlations of wave numbers of the changing chemical bonds can be applied as the characteristic of the nanostructure formation process in mesoreactor:

$$W = 1 - \exp\left\{-\tau^n \cdot \frac{\nu_{is}}{\nu_{fs}}\right\} \quad (4)$$

where; ν_{in} corresponds to wave numbers of the initial state of chemical bonds, and ν_f is the wave numbers of chemical bonds changing during the process.

By the analogy with the above calculations, the parameters a in the Eqn. (5) should be considered as a value that reflects the transition from the initial to the final state of the system and represents the ratios of activities of system states.

The experimental modeling of obtaining nanosized films after the alignment of copper compounds with PVA at 200°C revealed that optimal duration, when the share of mesoscopic films approaches 100%, equals 2.5 hours. This corresponds to the calculated value based on the aforesaid Avrami equation.

The calculations are made supposing the formation of copper nanocrystals on the nano-films. It is pointed out that copper ions are predominantly reduced to metal. Therefore, it was accepted for the calculations that n equals 2 (two-dimensional growth), potential of red-ox process during the ion reduction to metal ($\Delta\varphi$) equals 0.34 V, temperature (T) equals 473 K, Faraday number (F) corresponds to 26.81 (A·hour/mol), gas constant R equals 2.31 (W·hour/mol·degree). The analysis of the dimensionality shows the zero dimension of the ratio: $\frac{zF\Delta\varphi}{RT}$.

The calculations are made when changing the process duration with a half-hour increment. Results of calculations on the Eqn. (3) are brought in table.

If nano-films are scrolled together with copper nano-wires β is taken as equaled to 3, the temperature increases up to 400°C, the optimal time when the transformation degree reaches 99.97%, corresponds to the duration of 2 hours, thus also coinciding with the experiment. According to the calculation results, if following the definite conditions of the system exposure, the duration of the exposure has the greatest influence on the value of nanostructure share.

The selection of the corresponding equation form depends upon the shape and sizes of mesoreactor (nanostructure) and defines the nanostructure growth in mesoreactor or the influence distribution of the nanostructure on the structurally changing medium. At the same time ν_{is} corresponds to the frequency of skeleton oscillations of C–C bond at 1,100 cm^{-1}, ν_{fs} is the symmetrical skeleton oscillations of C=C bond at 1,050 cm^{-1}. In this case the equation looks like Eqn. (4).

For the example discussed the content of Nano-films in % will be changing together with the changes in the duration as follows in table.

Under the aforesaid conditions, the linear sizes of copper (from ion radius to atom radius) and carbon-carbon bond (from C–C to C=C) are changing during the process. Apparently, the structure of copper ion and electron interacts with electrons of the corresponding bonds forming the layer with linear sizes $r_i + l_{C-C}$ in the initial condition and the layer with the size $r_a + l_{C=C}$ in the final condition. Then the equation for the content of Nano-films can be written down as follows:

$$W = 1 - \exp\left\{-\tau^n \cdot \frac{r_a + l_{C=C}}{r_i + l_{C-C}}\right\} \tag{5}$$

At the same time r_i for Cu^{2+} equals 0.082 nm, r_a for four-coordinated copper atom corresponds to 0.113 nm, bond energy C–C equals 0.154 nm, and C=C bond – 0.142 nm. Representing the ratio of activities as the ratio of corresponding linear sizes and taking the value n as equaled to 2, at the same time changing τ in the same intervals as before, we get the following change in the transformation degree based on the process duration (Table 3.2).

Modified Kolmogorov-Avrami equations were tested to prognosticate the duration of the processes of obtaining metal-carbon Nano-films in the system "Cu–PVA (PVA)" at 200°C [2]. The calculated time (2.5 hours) correspond to the experimental duration of obtaining carbon Nano-films on copper clusters.

Thus, with the help of Kolmogorov-Avrami equations or their modified analogs we can determine the optimal duration of the process to obtain the required result. It opens up the possibility of defining other parameters of the process and characteristics of nanostructures obtained (by shape and sizes).

TABLE 3.2 The Changes of Nano-films Content Calculated Dependent on the Duration of Synthesis

Duration (hours)		0.5	1.0	1.5	2.0	2.5
Content of nanofilms (%)	Eqn. (3)	22.5	63.8	89.4	98.3	99.8
	Eqn. (5)	23.0	64.9	90.5	98.5	99.9
	Eqn. (6)	23.7	66.0	91.2	98.7	99.9

The methods of optical spectroscopy and XPS give the possibility to determine the energy of the interaction of the chemical particles in the mesoreactors with the mesoreactor walls active centers, which stimulate reduction-oxidation processes.

Depending on the nature of the metal salt and the electrochemical potential of the metal, different metal reduction mesoscopic products in the carbon shells differing in shape are formed. Based on this result we may speak about a new scientific branch – mesoscopic metallurgy. The stages of nanostructures synthesis may be represented by the scheme in Figure 3.3.

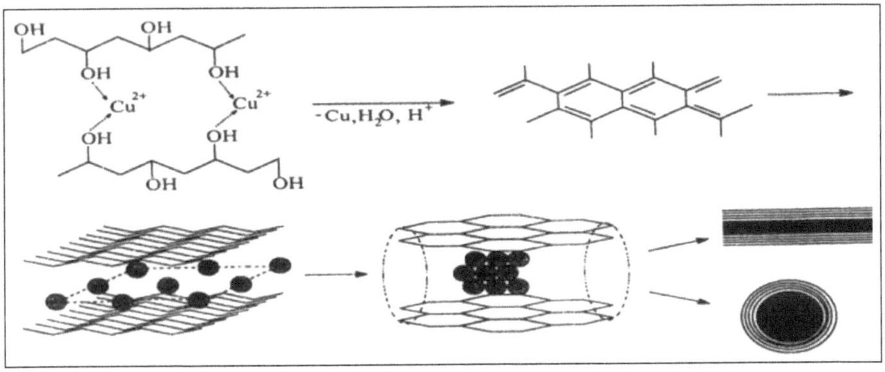

FIGURE 3.3 Scheme of copper/carbon nanostructures obtained from copper ions and polyvinyl alcohol.

Otherwise, as follows from this scheme, the first step (stage) of Me–C MC formation is the growth carbon fibers associated with metal clusters. The second stage is the formation of carbon film which covers metal clusters. And then, this film transforms in the 3D nanostructures with metal containing clusters inside.

The possible ways for obtaining metallic nanostructures in carbon shells have been determined. The investigation results allow considering the possibility of the isolation of metallic and metal-containing nanoparticles (NPs)

in the carbon shells differing in shape and structure. However, there are still problems related to the calculation and experiment because using the existing investigation methods it is difficult unambiguously to estimate the geometry and energy parameters of nanoreactors under the condition of 'erosion' of their walls during the formation of metallic nanostructures in them.

It is interesting that the scheme (Figure 3.4) and the succession of stages on fractal dimension correspond to experimental morphology of nanocomposites obtained after corresponding heating stages (Figure 3.5).

It is found that in the temperature range at 200°C mesoscopic films, from carbon fibers associated with metal phase as well (Figure 3.4), are formed on metal or metal oxide clusters. When the temperature elevates up to 400°C, 3D nanostructures are formed with different shapes depending on the coordinating ability of the metal.

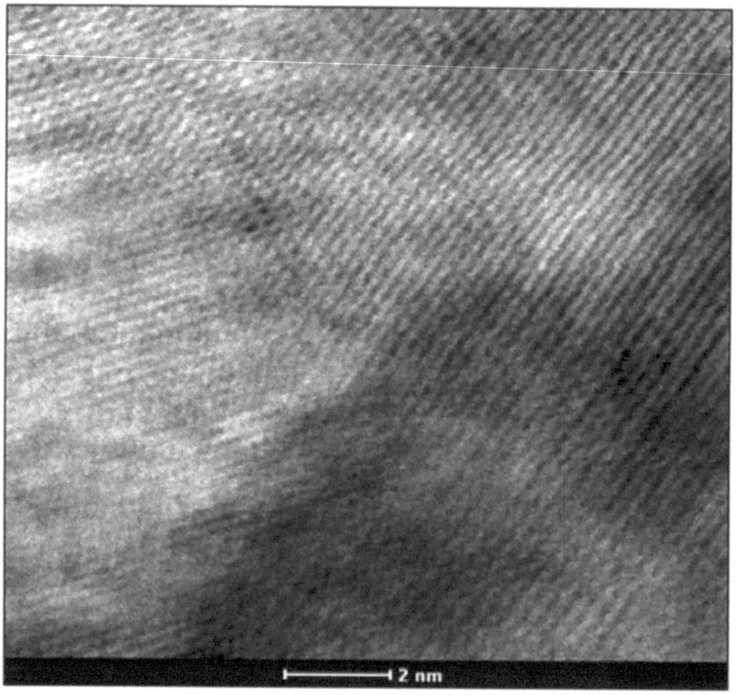

FIGURE 3.4 Microphotographs obtained with the help of transmission electron microscopy – Cu/C mesocomposite.

Two types of nanocomposites rather widely applied during the modification of various polymeric materials were investigated. These were:

copper–carbon MC and nickel–carbon MC specified below. In turn, the mesoscopic powders obtained were tested with the help of high-resolution transmission electron microscopy, electron micro-diffraction, laser analyzer, XPS and IR spectroscopy.

The method of Me–C MC synthesis applied has the following advantages:

- Originality of stage-by-stage obtaining of metal/carbon nanocomposites with intermediary evaluation of the influence of initial mixture composition on their properties.
- Wide application of independent modern experimental and theoretical analysis methods to control the technological process.
- Technology developed allows synthesizing a wide range of metal/carbon nanocomposites depending on the process conditions.
- Process does not require the use of inert or reduction atmospheres and specially prepared catalysts.
- Method of obtaining metal/carbon nanocomposites allows applying secondary raw materials.

So, the mesoreactor walls, containing the functional groups, have different potentials. When the mesoscopic particles (in our case, ions or metal oxide clusters) are coordinated with functional groups of mesoreactor walls, their potentials change too.

Now, according to mesoscopic physics principles, the conditions are created for electrons transport across mesoscopic particles (clusters) in which there are reducers (positively charged metal or metal ions). The red-ox processes are begun and the d metal electron structure changes with the growth of unpaired electron number. *d*-metal becomes paramagnetic. At the same time, the hydrocarbon shells, corresponding to mesoreactor walls, are oxidized with the aromatic cycles or hexagon formation in which π electrons are interacted with metal unpaired electrons. Otherwise, there is the possibility of unpaired electrons formation on the carbon shell of Me–C MC obtained. Let us discuss the properties and especially energetic characteristics of metal/carbon nanocomposites.

3.3 METAL-CARBON MESOCOMPOSITES (ME–C MCS) PROPERTIES, INCLUDING ENERGETIC AND MAGNETIC CHARACTERISTICS

Metal-carbon mesocomposite (Me–C MC) will be more active than carbon or silicon nanostructures because their masses are bigger at identical sizes and shapes. Therefore, the vibration energy transmitted to the medium is also high.

Me–C MC represents metallic clusters stabilized in carbon cover which contains structures [12–14]. In turn, nanosized film structures are formed from carbon amorphous nanosized fibers associated with metal containing phase. As a result of stabilization and association of metal mesoparticles with carbon phase, the metal chemically active particles are stable in the air and during heating as the strong complex of metal mesoparticles with carbon material matrix is formed.

For the corresponding correlations "polymer-metal containing phase" the dimensions, shape, and energy characteristics of nanoreactors are found with the help of AFM [9, 15]. Depending on a metal participating in coordination, the structure and relief of xerogel surface change.

To investigate the processes at the second stage of obtaining metal/carbon nanocomposites XPS, transmission electron microscopy and IR spectroscopy are applied.

In turn, the mesoscopic powders obtained were tested with the help of such methods as high-resolution transmission electron microscopy, electron micro-diffraction, IR spectroscopy, XPS and also with using of laser analysis. The test results of MCs obtained are given in Table 3.3.

The distinctive feature of the considered technique for producing metal/carbon nanocomposites is a wide application of independent, modern, experimental, and theoretical analysis methods to substantiate the proposed technique and investigation of the composites obtained (quantum-chemical calculations, methods of transmission electron microscopy and electron diffraction, method of XPS, X-ray phase analysis, etc.). The technique developed allows synthesizing a wide range of metal/carbon nanocomposites by composition, size, and morphology depending on the process conditions. In its application it is possible to use secondary metallurgical and polymer raw materials. Thus, we can adjust the MC structure to extend the function of its application without pre-functionalization. Controlling the sizes and shapes of nanostructures by changing the metal-containing phase, we can, to some extent, apply completely new, practicable properties to the materials which sufficiently differ from conventional materials.

The essence of the method [11] consists in coordination interaction of functional groups of polymer and compounds of 3D-metals as a result of grinding of metal-containing and polymer phases. Further, the composition obtained undergoes thermolysis following the temperature mode set with the help of thermogravimetric and differential thermal analyzes. At the same time, we observe the polymer carbonization, partial or complete reduction of metal compounds and structuring of carbon material in the form of nanostructures with different shapes and sizes.

TABLE 3.3 Characteristics of Metal-Carbon Mesocomposites (Met/C NC)

Type of Met/C NC	**Cu/C**	**Ni/C**	**Co/C**	**Fe/C**
Composition, metal/carbon (%)	50/50	60/40	65/35	70/30
Density (g/cm^3)	1.71	2.17	1.61	2.1
Average dimension (nm)	20(25)	11	15	17
Specific surface (m^2/g)	160 (average)	251	209	168
Metal mesoparticle shape	Close to spherical, there are dodecahedrons	There are spheres and rods	Nano-crystals	Close to spherical
Caron phase shape (shell)	Nano-fibers associated with metal phase forming Nan coatings	Nano-films scrolled in nanotubes	Nano-films associated with nano-crystals of metal-containing phase	Nano-films forming nano-beads with metal-containing phase
Atomic magnetic moment (metal) [8] (μB)	0.0	0.6	~1.0	2.1
Atomic magnetic moment (nanocomposite) (μB)	0.6	1.6	1.7	2.3

The conditions of metal/carbon nanocomposites red-ox synthesis such as nature of metal containing phase and also polymeric phase (for example, molecular mass and acetate groups number for PVA) influence on the composition and structure of Me–C MC.

Examples for the phase composition changes and the morphology of copper–carbon MCs obtained in different conditions are presented in Figures 3.5(a) and (b); Table 3.4.

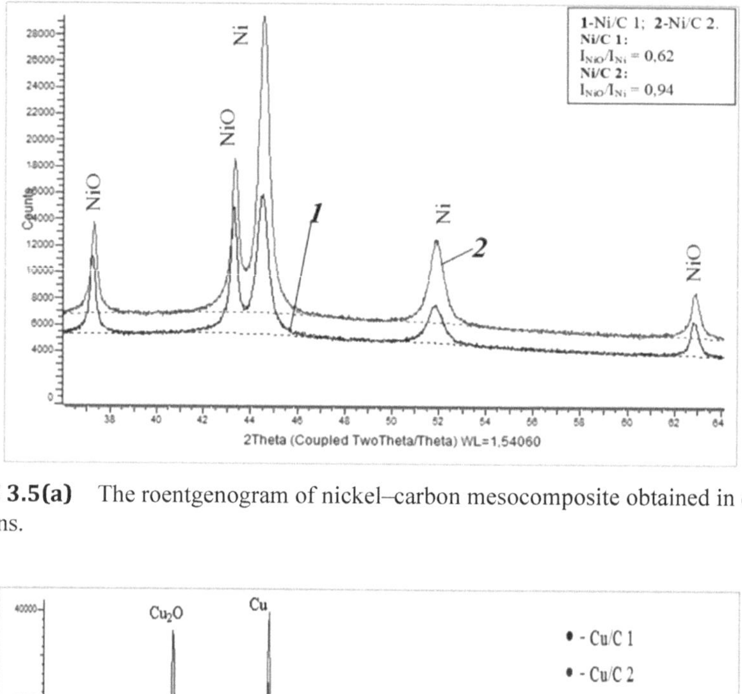

FIGURE 3.5(a) The roentgenogram of nickel–carbon mesocomposite obtained in different conditions.

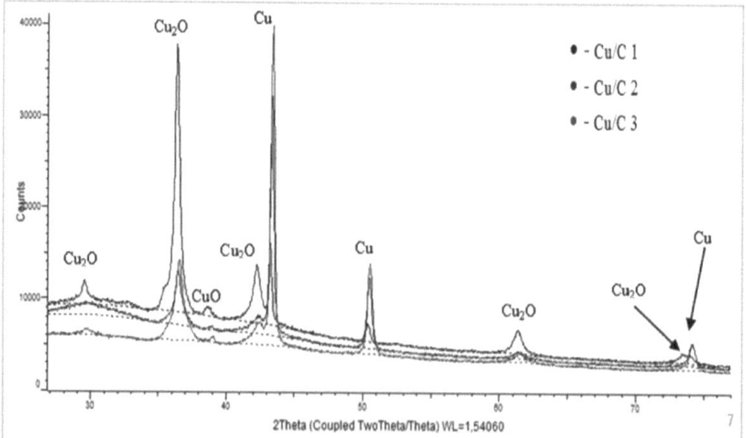

FIGURE 3.5(b) The roentgenograms of copper–carbon mesocomposite obtained in different conditions.

TABLE 3.4 The Example of Copper–Carbon Mesocomposites Morphology Changes Stipulated by Different Conditions

Nanocomposite Code	Reagents of Synthesis	Temperature Regime of Synthesis (°C)	Morphology of Nanocomposites	Particles Sizes of Nanocomposites
Cu/C 1	Copper oxide + PVA	420	Metal crystal phase within carbon nanosized films	Min. – 5 nm Max. – 35 nm Individual mesoparticles near to 100 nm
Cu/c 2	Copper oxide + PVA	390	Metal spherical mesoparticles within carbon film nanostructure	Min. – 2 nm Max. – 40 nm Individual mesoparticles near to 70–100 nm

Here, Me–C MC represents metal mesoparticles stabilized in carbon film nanostructures [12, 13].

In turn, Nan film structures are formed with carbon amorphous Nanofibers associated with metal containing phase. As a result of stabilization and association of metal mesoparticles with carbon phase, the metal chemically active particles are stable in the air and during heating as the strong complex of metal mesoparticles with carbon material matrix is formed.

Figure 3.6 demonstrates the microphotographs of transmission electron microscopy specific for different types of metal/carbon nanocomposites.

If the metal/carbon nanocomposites are considered as supermolecules, their surface energies analogously energy of usual molecules consists of portions of energy which correspond to progressive, rotation, and vibration motions and also electronic motion:

$$\varepsilon_s^{NC} = \varepsilon_{prog} + \varepsilon_{rot} + \varepsilon_{vib} + \varepsilon_{elm} \tag{6}$$

where; ε_s^{NC} is the surface energy of MC; ε_{prog} is the MC surface energy portion which corresponds to the progressive motion; ε_{rot} is the MC surface energy portion which corresponds to the rotation motion; ε_{vib} is the MC surface energy portion which corresponds to the vibration motion; and ε_{elm} is the nanocomposite surface energy portion which corresponds to electron motion for the interactions of nanocomposites with surroundings molecules.

When the MC progressive motion portion increases the mesoparticles diffusion processes importance grows in media that leads to their coagulation with the decreasing of the surface energy aggregate obtained.

The coagulation of mesoparticles increases at the increasing of their quantities in a medium which has little viscosity.

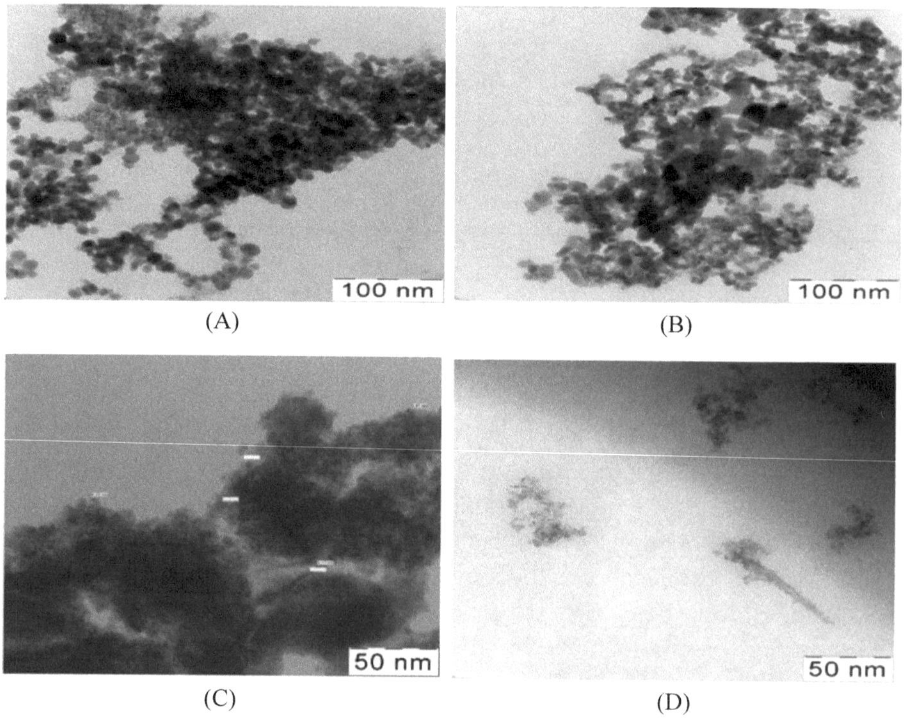

FIGURE 3.6 Microphotographs of metal/carbon nanocomposites: (a) Cu/C; (b) Ni/C; (c) Co/C; and (d) Fe/C.

From the metal/carbon nanocomposites Raman and IR spectra analysis it follows that the skeleton vibration of them on the vibrations frequencies corresponds to ultrasonic vibrations. The MC vibrations energy values are determined by the corresponding mesoparticles sizes and masses. The mass of mesoparticle (Me–C MC) depends on their content and types of metal containing phase clusters.

Usually Me–C MC have the great dipole moment. Therefore, it is possible the proposition that MC is a vibrator which radiates electromagnetic (EM) waves.

The MC vibration emission in the medium is determined by their dielectric characteristics and the corresponding functional groups presence in the medium.

At the metal/carbon nanocomposites obtaining the interaction of polymeric matrix with metal containing phase leads to the formation of metal clusters covered by carbon shells accompanied by metal electron structure changes. In some cases, the medium characteristics influence on nanocomposites throw into the increasing of MC surface energy portion, which concerns with changes of their electron structure and equally with electron structure of medium.

In these cases, the growth of metal atomic magnetic moments is observed (Table 3.5) that corresponds to the unpaired electron number increasing. The considerable changes of metal atomic magnetic moments in MCs proceed when the phosphorus atoms include to carbon shells of MCs.

TABLE 3.5 The Comparison of Atomic Magnetic Moments of Some 3D Metals in Metal-Carbon Mesocomposites and Their Functionalized Phosphorus Containing Analogs [6, 13]

Metal in Mesocomposite	Cu	Ni	Co	Fe
Metal atomic magnetic moment for mesocomposite.	0.6	1.6	1.7	2.3
Metal atomic magnetic moment for Phosphorus containing mesocomposite.	2.0	3.0	2.5	2.5

The appearance or increasing of paramagnetic properties is linked with the possibility of chemical bonds formation, and also the growth of Me–C MC reactivity into polar media.

Vibration portion of surface energy of metal/carbon nanocomposites can be presented as:

$$\varepsilon_{vib} = \frac{mu_{vib}^2}{2} \tag{7}$$

where; m is the mass of Me–C MC, which includes summary mass of carbon or carbon, polymeric shell and cluster of metal containing phase; u_{vib} is the vibration velocity correspondent to product of vibration amplitude on vibration frequency.

Usually, the fine dispersed suspensions or sols of nanostructures are used for the modification of polymeric compositions (matrixes) to distribute NPs into polymeric matrix. The liquids, which are used for polymeric materials production, are represented as dispersion medium. For the obtaining of fine dispersed suspensions small and super small quantities of metal/carbon nanocomposites are introduced into dispersed media at the mechanical and further the ultrasonic mixing.

The MC vibration amplitude is the product of wave number on velocity of light. Therefore, the Eqn. (7) can be written:

$$\varepsilon_{vib} = \frac{(m_{cl} + m_{sh})\,dvc}{2} \tag{8}$$

where; m_{cl} is the mass of metal containing phase; m_{sh} is the mass of carbon or carbon polymeric shell; d is the average linear size of Me–C MC particles; v is the wave number of MC skeleton vibration; and c is the velocity of light.

Under the influence by EM radiation arising at the ultrasonic vibration of MC the self-organization of polymeric compositions macromolecules is possible. According to the Eqn. (8), the MC vibration energy is determined by the average mass and size of MC (Table 3.6).

TABLE 3.6 Energetic Characteristics of Metal-Carbon Mesocomposites [11–13]

Metal in Metal/Carbon Nanocomposites	Cu	Ni	Co	Fe
Summary mass of metal/carbon nanocomposite (au)	36.75	40.01	42.50	42.69
Specific surface of metal/carbon nanocomposite, S (m^2/g)	160	251	209	163
Skeleton vibration frequency of metal/carbon nanocomposite (ν_{vib}, s^{-1})	4×10^{11}	4.2×10^{11}	4.1×10^{11}	4×10^{11}
Average vibration velocity of metal/carbon nanocomposites (u_{vib}, cm/s)	8.5×10^5	4.6×10^5	6.1×10^5	6.9×10^5
Average vibration energy of metal/carbon nanocomposites (ε_{vib}, erg)	1.6×10^{13}	4.3×10^{12}	7.6×10^{12}	9.9×10^{12}

The vibration motion portion energy essentially determines by the vibration amplitude because the MC skeleton vibration frequencies practically equal (the proximity of metal nature).

If the vibration amplitudes and the MC sizes are correlated, the nanocomposites sizes show the great influence on the vibration energy and also on the vibration velocity. Therefore, according to the increasing of vibration energy as well as the increasing of metal/carbon MCs linear sizes, the correspondent MCs (M–C MC) are displaced in the following order:

$$\varepsilon_{vib}\ (\text{Cu/C MC}) > \varepsilon_{vib}\ (\text{Fe/C MC}) > \varepsilon_{vib}\ (\text{Co/C MC}) > \varepsilon_{vib}\ (\text{Ni/C MC})$$

This disposition of nanocomposites on the vibration energy values is possible, when the vibration motion portion is greatly bigger on the comparison with other portions of surface energy. However, the common surface energy of nanocomposites obtained differs from previous row and corresponds to values of specific surface:

$$S_{sp}\ (Ni/C\ MC) > S_{sp}\ (Co/C\ MC) > S_{sp}\ (Fe/C\ MC) > S_{sp}\ (Cu/C\ MC)$$

Certainly, in dependence on methods and conditions of the metal/carbon nanocomposites synthesis, the changes of forms and contents for nanocomposites are possible. At the same time, the fundamentals of the Me–C MC activity open the new perspectives for their interaction prognosis in the different media.

The nanocomposites described above were investigated with the help of IR spectroscopy by the technique indicated above. In this chapter, the IR spectra of Cu/C and Ni/C nanocomposites are discussed (Figures 3.7 and 3.8), which find a wider application as the material modifiers.

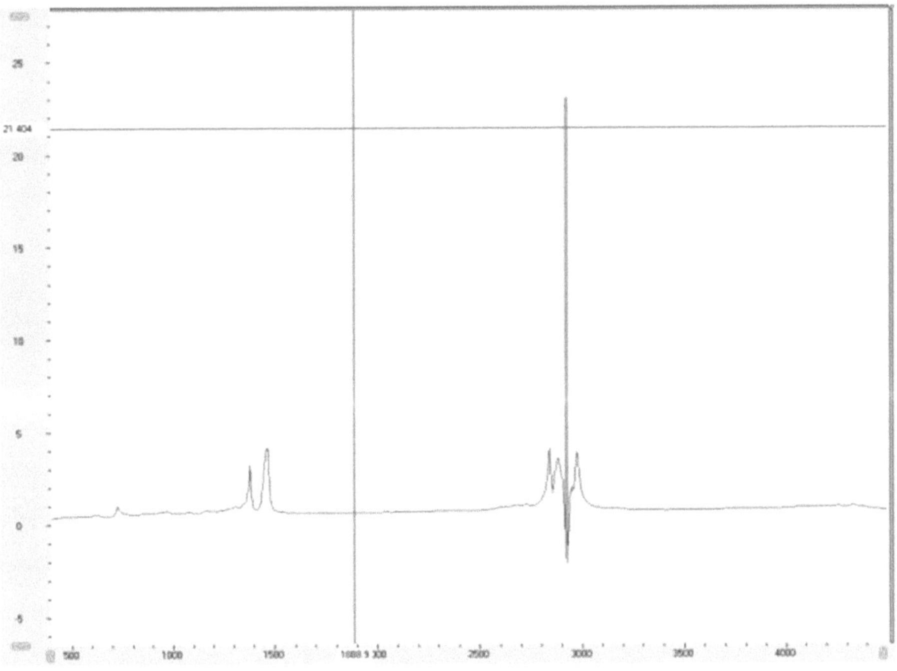

FIGURE 3.7 IR spectra of copper–carbon mesocomposite powder.

On IR spectra of two nanocomposites, the common regions of IR radiation absorption are registered. Further, the bands appearing in the spectra and having the largest relative area were evaluated. We can see the difference in the intensity and number of absorption bands in the range 1,300–1,460 cm^{-1}, which confirms the different structures of composites.

In the range 600–800 cm^{-1} the bands with a very weak intensity are seen, which can be referred to the oscillations of double bonds (π-electrons) coordinated with metals.

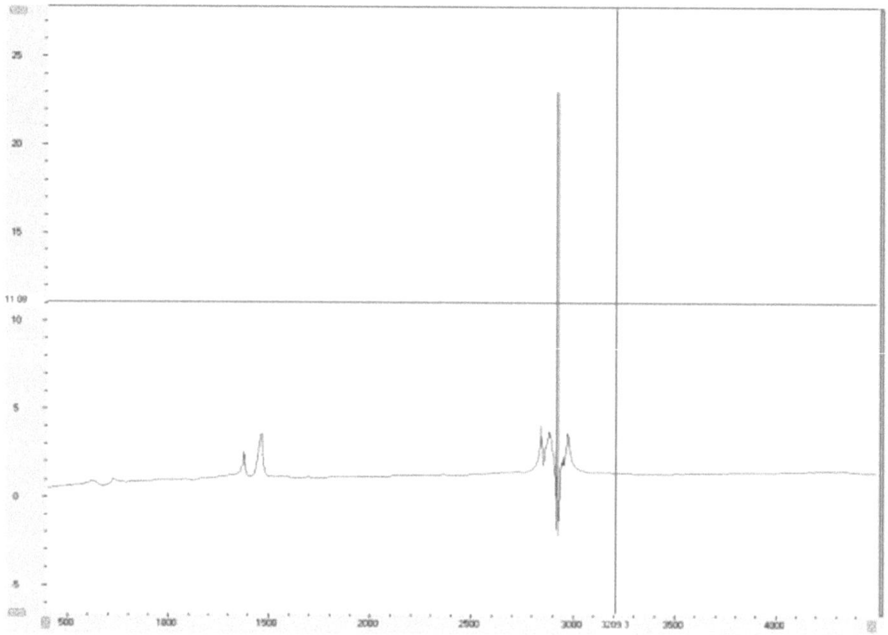

FIGURE 3.8 IR spectra of nickel–carbon mesocomposite powder.

In the spectrum of Cu/C MC a weak absorption is found at 720 cm^{-1}, and in case of Ni/C MC the absorption at 620 cm^{-1} is also observed.

In IR spectrum of copper–carbon MC two bands with a high relative area are found:

i) at 1,323 cm^{-1} (relative area – 9.28)
ii) at 1,406 cm^{-1} (relative area – 25.18).

These bands can be referred to as skeleton oscillations of polyphenylene rings.

In IR spectrum of Nickel/Carbon MC the band mostly appears at 1,406 cm^{-1} (relative area – 14.47). The investigations of Carbon films in Metal/Carbon Nanocomposites and their peculiarities are carried out by Raman spectroscopy with the use of Laser Spectrometer Horiba Lab Ram HR 800.

The Raman spectra and PEM microphotograph of Copper/Carbon (Cu/C) MC are represented in Figure 3.9.

Wave numbers and intensity relation testify to the presence of NPs containing the Copper atoms coordinated. At the same time, the comparison of IR and Raman spectra shows their closeness on wave numbers and intensities relation.

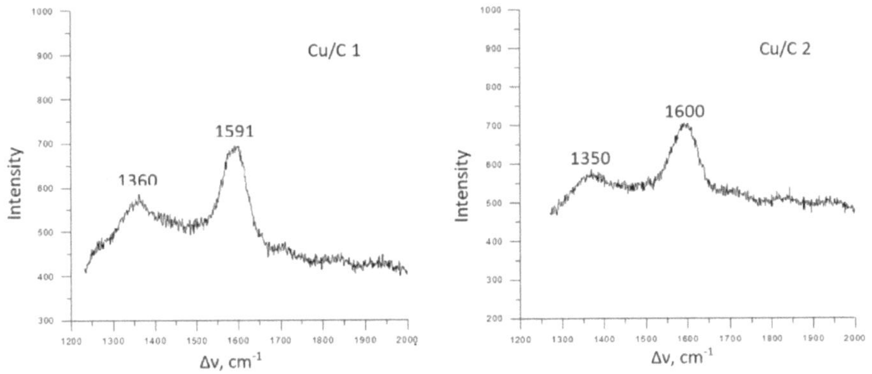

FIGURE 3.9 Raman spectra of Cu–C mesocomposites: Cu/C 1 (28 nm); and Cu/C 2 (25 nm).

According to the investigations with transmission electron microscopy, the formation of carbon Nan film structures consisting of carbon threads is characteristic for copper–carbon MC.

In contrast, carbon fiber structures, including nanotubes, are formed in nickel–carbon MC. There are several absorption bands in the range 2,800–3,050 cm^{-1}, which are attributed to valence oscillations of C–H bonds in aromatic and aliphatic compounds. These absorption bonds are connected with the presence of Vaseline oil in the sample. It is difficult to find the presence of metal in the composite as the metal is stabilized in carbon nanostructure. At the same time, it should be pointed out that apparently nanocomposites influence the structure of Vaseline oil in different ways. The intensities and number of bands for Cu/C and Ni/C nanocomposites are different:

- for copper–carbon mesocomposite in the indicated range – 5 bands, and total intensity corresponds by the relative area – 64.63.
- for nickel–carbon mesocomposite in the same range – 4 bands with total intensity (relative area) – 85.6.

The distribution of mesoparticles in water, alcohol, and water-alcohol suspensions prepared based on the above technique are determined with the help of laser analyzer.

In Figures 3.10 and 3.11, distributions of copper–carbon MC in the media different polarity and dielectric penetration are shown.

At the comparison of Figures 3.10 and 3.11, the difference in distribution of Cu–C mesoparticles is shown which is explained by the polarity values of alcohol and water.

In water solution, the average size of Cu/C nanocomposite equals 20 nm, and in alcohol medium – greater by 5 nm.

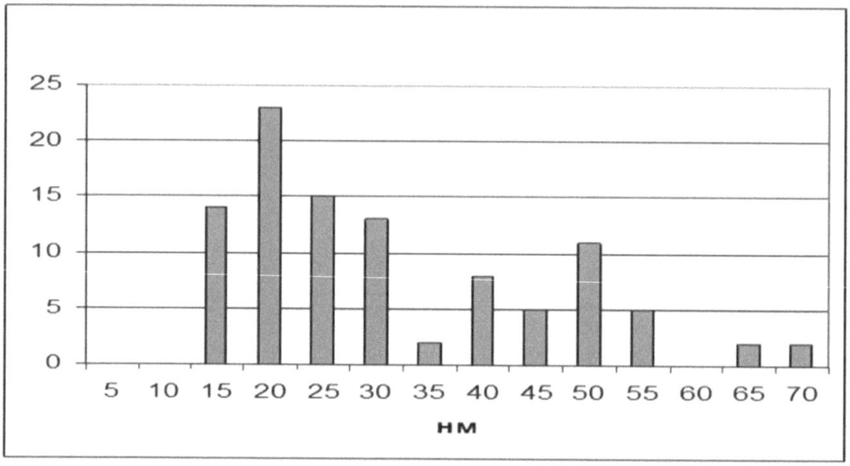

FIGURE 3.10 Distribution of copper–carbon mesocomposites in alcohol.

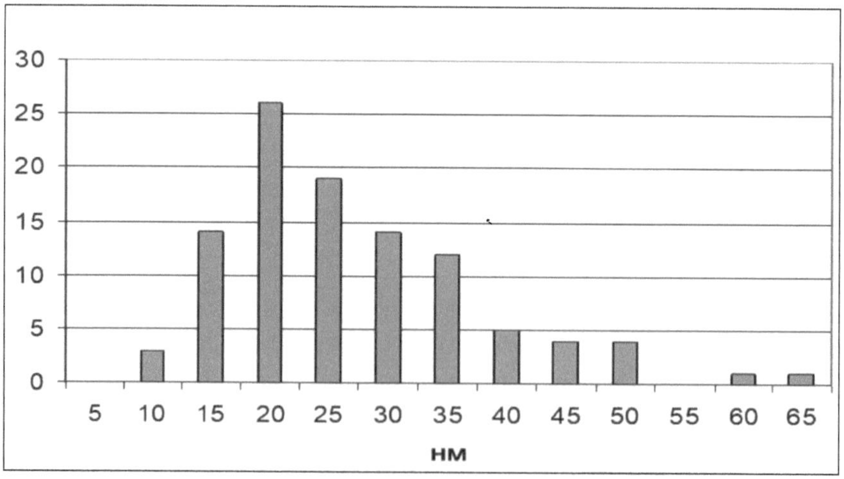

FIGURE 3.11 Distribution of copper–carbon mesocomposites in water.

Assuming that the nanocomposites obtained can be considered as oscillators transferring their oscillations onto the medium molecules, we can determine to what extent the IR spectrum of liquid medium will change, e.g., PEPA applied as a hardener in some polymeric compositions, when small and super small quantities of MC into it are introduced.

The introduction of a modifier based on Me–C MC into the composition results in medium structuring, decreasing the number of defects and improving the material physical and mechanical characteristics [5, 6]. The availability of metal compounds in MCs can provide the final material with additional characteristics, such as magnetic susceptibility and electric conductivity. Data of EPR spectra investigations for copper–carbon and nickel–carbon MCs are given in Figure 3.12 and Table 3.7.

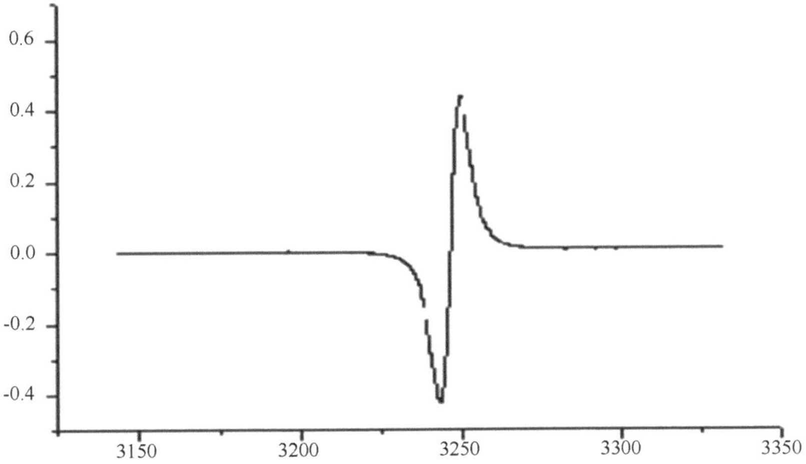

FIGURE 3.12 EPR spectrum of copper–carbon mesocomposite.

EPR spectrum of copper/carbon MC is presented as a singlet spectrum in which the distance between points of incline maximum ΔH_{pp} = 6.8 Hz, g-factor equals to g-factor of biphenyl picryl hydrazyl (g = 2.0036), and the unpaired electrons number corresponds to value 1.2×10^{17} spin/g (Figure 3.11). In the comparison with this spectrum, the EPR spectrum for nickel–carbon MC has ΔH_{pp} corresponding to 2,400 Hz, g – 2.46 and the unpaired electrons number – 10^{22} spin/g (Table 3.7).

It is possible, the spectra difference explains the Carbon shape difference for these nanocomposites. Thus, the Me–C MC are stable radicals with the migration of unpaired electrons from metal to carbon shell and back. These mesoparticles can stimulate the transport of electrons within media.

TABLE 3.7 Data of EPR Spectra for Copper/Carbon and Nickel/Carbon Nanocomposites

Type of Metal/Carbon Mesocomposite	g-Factor	Number of Unpaired Electrons (spin/g)
Copper/carbon mesocomposite	2.0036	1.2×10^{17} spin/g
Nickel/carbon mesocomposite	2.46	10^{22} spin/g

3.4 CONCLUSION

In this chapter, the possibilities of developing new ideas on base of Mesoscopic Physics principles for self-organization processes during red-ox synthesis within nanoreactors of polymeric matrices are discussed on the examples of Me–C MC. It is proposed to consider the obtaining of Me–C MC in nanoreactors of matrices as the self-organization process similar to the formation ordered phases analogous to crystallization. The perspectives of investigations are looked through in an opportunity of thin regulation of processes and the entering of corrective amendments during processes stages.

According to the analysis of Me–C MC characteristics, which are determined by their sizes and content, their activities are stipulated the correspondent dipole moments and vibration energies.

The introduction of phosphorus in MCs by the method similar to red-ox synthesis in mesoreactors leads to the growth of dipole moments and metal magnetic moments for corresponded MCs. It's shown that the MCs vibration energies depend on their average masses.

It's noted that the specific surface of Me–C MC particles changes in dependence on the nature of MC in other order than the correspondent order of the vibration energies. Therefore, the energetic characteristics of MCs are more important for the activity determination in comparison with their size characteristics.

KEYWORDS

- **atomic force microscopy**
- **Kolmogorov-Avrami equations**
- **mesoreactors**
- **metal-carbon mesocomposite**
- **polyethylene polyamine**
- **polyvinyl alcohol**
- **x-ray photoelectron spectroscopy**

REFERENCES

1. Imri, (2009). *Introduction in Mesoscopic Physics* (p. 304). M.: Physmatlit.
2. Moskalets, M. V., (2010). *Fundamentals of Mesoscopic Physics*. Khar'kov: NTU KhPI.
3. Morozov, A. D., (2002). *Introduction in Fractal Theory* (p. 160). M.-Izhevsk: ICT.
4. Kolmogorov, A. N., & Fomin, S. V., (2009). *Introductory Real Analysis* (p. 403). USA, Portland: Prentice Hall.
5. Wunderlikh, B., (1979). *Physics of Macromolecules* (Vol. 2, p. 422). M.: Mir.
6. Brown, J. F. Jr., & White, D. M., (1960). *J. Am. Chem. Soc., 82*, 5671.
7. Morgan, P. W., (1970). Condensation polymers by interfacial and solution methods (translated) L.; *Chemistry*, 448.
8. Buchachenko, A. L., (2003). *Nanochemistry–Direct Way to High Technologies* (Vol. 72, No 5, pp. 419–437). Uspechi Chimii.
9. Kodolov, V. I., & Khokhriakov, N. V., (2009). *Chemical Physics of Formation and Transformation Processes of Nanostructures and Nanosystems* (V. 1, 2, pp. 361, 415). Izhevsk: Publ. IzhSACA.
10. Shabanova, I. N., Kodolov, V. I., Terebova, N. S., & Trineeva, V. V., (2012). *X-Ray Photoelectron Spectroscopy Investigations of Metal/Carbon Nanosystems and Nanostructured Materials* (p. 252). M.: Izhevsk: Publ. Udmurt university.
11. Kodolov, V. I., Khokhriakov, N. V., Trineeva, V. V., & Blagodatskikh, I. I., (2008). Nanostructure activity and its display in nanoreactors of polymeric matrices and in active media. *Chemical Physics and Mesoscopy, 10*(4), 448–460.
12. Kodolov, V. I., (2009). The addition to previous paper. *Chemical Physics and Mesoscopy, 11*(1), 134–136.
13. Trineeva, V. V., Vakhrushina, M. A., Bulatov, D. I., & Kodolov, V. I., (2012). The obtaining of metal/carbon nanocomposites and investigation of their structure phenomena. *Nanotechnics, 4*, 50–55.
14. Trineeva, V. V., Lyakkhovich, A. M., & Kodolov, V. I., (2009). Forecasting of the formation processes of carbon metal containing nanostructures using the method of atomic force microscopy. *Nanotechnics, 4*(20), 87–90.

15. Kodolov, V. I., Blagodatskikh, I. I., Lyakhovich, A. M., et al., (2007). Investigation of the formation processes of metal containing carbon nanostructures in nanoreactors of polyvinyl alcohol at early stages. *Chemical Physics and Mesoscopy, 9*(4), 422–429.
16. Kodolov, V. I., Khokhriakov, N. V., & Kuznetsov, A. P., (2006). To the issue of the mechanism of the influence of nanostructures on structurally changing media at the formation of "intellectual" composites. *Nanotechnics, 3*(7), 27–35.
17. Kodolov, V. I., Khokhriakov, N. V., Trineeva, V. V., & Blagodatskikh, I. I., (2010). Problems of nanostructure activity estimation, nanostructures directed production and application. *Nanomaterials Yearbook–2009: From Nanostructures, Nanomaterials and Nanotechnologies to Nanoindustry* (pp. 1–18). N.Y.: Nova Science Publishers, Inc.
18. Fedorov, V. B., Khakimova, D. K., Shipkov, N. N., & Avdeenko, M. A., (1974). *To Thermodynamics of Carbon Materials* (Vol. 219. No. 3. pp. 596–599). Doklady AS USSR.
19. Fedorov, V. B., Khakimova, D. K., Shorshorov, M. H., et al., (1975). *To Kinetics of Graphitation* (Vol. 222. No. 2. pp. 399–402). Doklady AS USSR.
20. Khokhriakov, N. V., & Kodolov, V. I., (2005). Quantum-chemical modeling of nanostructure formation. *Nanotechnics, 2*, 108–112.
21. Lipanov, A. M., Kodolov, V. I., Khokhriakov, N. V., et al., (2005). Challenges in creating nanoreactors for the synthesis of metal nanoparticles in carbon shells. *Alternative Energy and Ecology, 2*(22), 58–63.
22. Kodolov, V. I., Didik, A. A., Volkov, Yu. A., & Volkova, E. G., (2004). Low-temperature synthesis of copper nanoparticles in carbon shells. HEIs' news. *Chemistry and Chemical Engineering, 47*(1), 27–30.
23. Serkov, A. T., (1975). Theory of chemical fiber formation (p. 548). edited by – M.: Himiya.
24. Palm, V. A., (1967). *Basics of Quantitative Theory of Organic Reactions* (p. 356). L: Himiya.
25. Kodolov, V. I., (1965). On modeling possibility in organic chemistry. *Organic Reactivity* (Vol. 2. No. 4. pp. 11–18). Tartu: TSU.
26. Chashkin, M. A., Kodolov, V. I., Zakharov, A. I., et al., (2011). Metal/carbon nanocomposites for epoxy compositions: Quantum-chemical investigations and experimental modeling. *Polymer Research Journal, 5*(1), 5–19.
27. Kodolov, V. I., & Trineeva, V. V., (2013). Fundamental definitions for domain of nanostructures and metal/carbon nanocomposites. In: *Nanostructure, Nanosystems and Nanostructured Materials: Theory, Production and Development* (pp. 1–42). Toronto-New Jersey: Apple Academic Press.
28. Kodolov, V. I., Trineeva, V. V., Blagodatskikh, I. I., Vasil'chenko, Yu. M., Vakhrushina, M. A., & Bondar, A. Yu., (2013). The nanostructures obtaining and the synthesis of metal/carbon nanocomposites in nanoreactors. In: *Nanostructure, Nanosystems and Nanostructured Materials: Theory, Production and Development* (pp. 101–145). Toronto-New Jersey: Apple Academic Press.
29. Shabanova, I. N., & Terebova, N. S., (2012). Dependence of the value of the atomic magnetic moment of d metals on the chemical structure of nanoforms. In: *Problems of Nanochemistry for the Creation of New Materials* (pp. 123–131). Torun, Poland: IEPMD.
30. Kodolov, V. I., & Trineeva, V. V., (2012). Perspectives of idea development about nanosystems self-organization in polymeric matrixes. In: *"The Problems of Nanochemistry for the Creation of New Materials* (pp. 75–100)/Torun. Poland: IEPMD.

CHAPTER 4

Mechanochemical Modification of Metal-Carbon Mesocomposites

V. I. KODOLOV,[1,2] V. V. TRINEEVA,[1,3] R. V. MUSTAKIMOV,[1,4] N. S. TEREBOVA,[1,5] T. M. MAKHNEVA,[1,3] and I. N. SHABANOVA[1,5]

[1]*Basic Research–Research Center of Chemical Physics and Mesoscopic, Udmurt Federal Research Center, Russian Academy of Sciences, Izhevsk, Russia*

[2]*M.T. Kalashnikov Izhevsk State Technical University, Izhevsk, Russia*

[3]*Institute of Mechanics, Udmurt Federal Research Center, Russian Academy of Sciences, Izhevsk, Russia*

[4]*Izhevsk Electromechanical Plant "KUPOL," Izhevsk, Russia*

[5]*Physicotechnical Institute, Udmurt Federal Research Center, Russian Academy of Sciences, Izhevsk, Russia*

ABSTRACT

The conditions of copper (nickel) carbon mesocomposites (MCs) modification by such substances as silica, ammonium polyphosphate (APPh), ammonium sulfite, and metal oxides are given. The mechanochemical modification mechanism of correspondent nanocomposites is proposed. The results of modified nanocomposite production with the mechanism confirmation are discussed. The following methods investigate the structures and properties of modified nanocomposites: X-ray photoelectron spectroscopy (XPS), radiography, IR spectroscopy, and transition electron microscopy (TEM). The structure changes of modified nanocomposites and the increasing metal

Modern Magnetic Materials: Properties and Applications. Iuliana Stoica, Ann Rose Abraham, & A. K. Haghi (Eds.)

atomic magnetic moments are presented. In the phosphorus-containing Cu/C nanocomposites, the Copper atomic magnetic moment is increased from 1.3 to 3 Bohr Magnetons, and the oxidation state of the Phosphorus atom is decreased from 5 to zero, according to XPS.

4.1 INTRODUCTION

Some possible methods of nanostructures (nanosized systems) modification are known [1–3]. The substances interaction processes with nanostructures can be occurred at the formation of covalent or coordinative bonds, and also possibly at dispersion interaction. These processes are realized when the gaseous or liquid substances act on nanostructures or, as in our case, when the interaction between solid mesocomposite (MC) and active components takes place. In other words, in our case the mechanochemical modification is carried out with inclusion Red-Ox process. For the investigation and the far production of MCs modified the following MCs and substances in pairs are used: cooper carbon, nickel–carbon, iron carbon MCs, and also silica, silica gel, ammonium polyphosphate (APPh), ammonium sulfite. The investigations of initial MCs show that metal-carbon mesocomposite (Me–C MC) represent to the metal clusters associated with carbon fibers or the metal clusters within carbon shell consisting of 3–4 layers of carbon fibers. Each from carbon fiber includes polyacetylene and carbine fragments with delocalized electrons on fragments joints. In other words, the Me–C MC obtained by the mechanochemical method have active surface, which contains double bonds and delocalized electrons [4].

4.2 METHOD OF METAL/CARBON NANOCOMPOSITES MECHANOCHEMICAL MODIFICATION

The representation about Me–C MC is based on the results of such investigations as IR or Raman spectroscopy, X-ray photoelectron spectroscopy (XPS), electron paramagnetic resonance (EPR), transition electron microscopy (TEM), atomic force microscopy (AFM). The image of copper–carbon MC structure shows on Figure 4.1.

The presence of active double bonds and delocalized electrons in the carbon shell of metal/carbon nanocomposites gives the possibility for their modification by means of red-ox and addition processes. Therefore, the substances interactions, in which elements (Si and P) have the highest oxidation state (+4 or +5), with such electron containing compounds as metal/carbon nanocomposites, is evident. The plain scheme of reduction process is shown below:

$$\text{НК} (-\delta, \text{или} \sum e) + P^{+5} \rightarrow \text{НК} - P^{+3} (P^{+2}, P^{0})$$

$$\text{НК} (-\delta, \text{или} \sum e) + Si^{+4} \rightarrow \text{НК} - Si^{0} (Si^{+2})$$

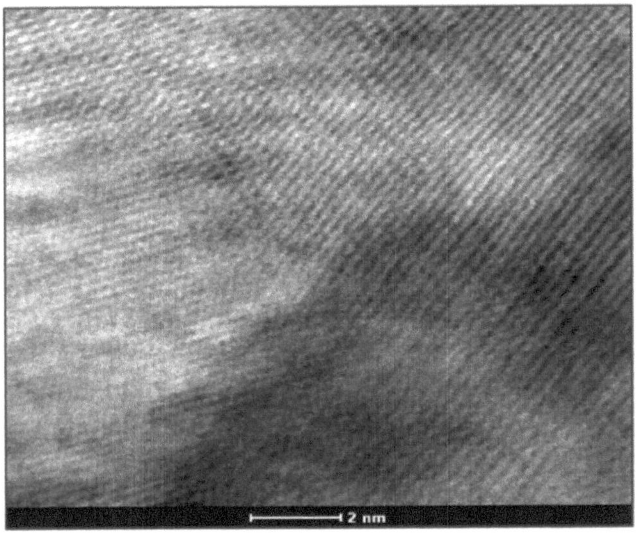

FIGURE 4.1 TEM microphotograph for copper–carbon mesocomposite.

Mechanism based on the chemical mesoscopics notions for electron transport across mesoparticle (metal cluster) from one to other bank (meso-reactor walls) is proposed (Figure 4.2).

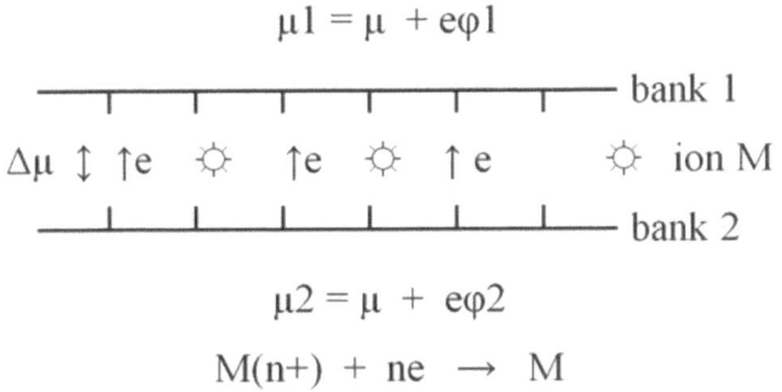

FIGURE 4.2 Scheme of electron transport and the red-ox reactions within nanoreactors.

The chemical activity of nanosized granule obtained is determined by their interaction with polar substances containing p elements with the positive oxidation state, for example, Si^{+4}, P^{+5} and S^{+6}.

The Me–C MC modification process consists in the grinding of copper–carbon or nickel–carbon MCs with the corresponding substances containing P or Si. As the mechanochemical process result – the xerogel is formed which is dried at 80°C.

4.3 INVESTIGATION OF METAL-CARBON MESOCOMPOSITES (ME–C MCS) PROPERTIES AND CHARACTERISTICS

The early created method [5] is used for the copper–carbon MC modification mechanism investigation. The inner levels spectra for C1s, O1s, Si2p, P2p, S2p Cu3s are studied to investigate the mechanism of chemical bonds formation between metal or carbon atoms and silicon or phosphorus atoms. Cu3s spectra of nanosized systems modified by Si, P, and S containing compounds are presented in Figure 4.3.

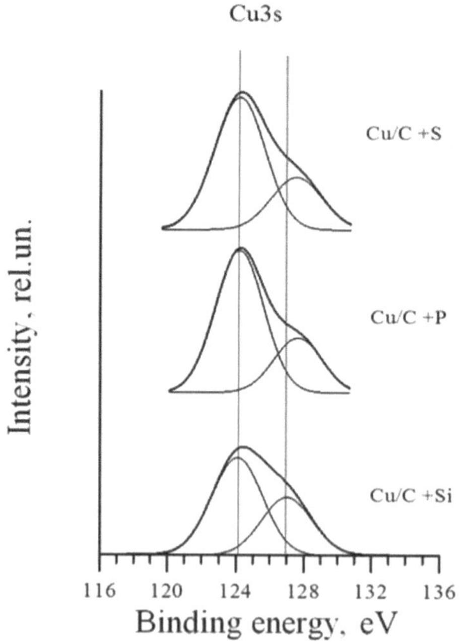

FIGURE 4.3 X-ray photoelectron spectra of copper/carbon nanocomposites modified by p-element containing substances.

As early as in chapter [1], it's shown that the Copper atomic magnetic moment in copper–carbon MC is appeared. At the modification the atomic magnetic moments of Metal/Carbon nanocomposites are increased in the comparison with non-modified nanocomposites (Table 4.1). The analysis of Table 4.1 data shows the influence of modifier nature on the magnetic characteristics as well as the influence of its layer thickness on the nanosized granule surface. At the same time, it's shown that atomic magnetic moment grows in modifier (elements) row Si, P, S at the identical conditions. It's established that at the modification all modifiers are changed because of elements reduction.

TABLE 4.1 The Parameters of the Multiple Splitting of the 3s Spectra of the Metal/Carbon Nanocomposites Modified with Phosphorus-, Silicon-, and Nickel-Containing Compounds

Sample	I_2/I_1	Δ (eV)	μ_{Ni}, μ_B	μ_{Cu}, μ_B	μ_{Fe}, μ_B
Ni3*s* (crystal)	0.15	4.3	0.5	–	–
Ni3*s* (non-modified)	0.32	3.0	1.8	–	–
Ni3*s* (NC/Si s – 1)	0.80	3.8	4.0	–	–
Ni3*s* (NC/P s – 1)	0.60	3.6	3.0	–	–
Ni3*s* (NC/S s – 1)	0.56	4.0	2.8	–	–
Cu3*s* (non-modified)	0.20	3.6	–	1.3	–
Cu3*s* (NC/Si s – 1)	0.60	3.0	–	3.0	–
Cu3*s* (NC/P s – 2)	0.85	3.5	–	4.2	–
Cu3*s* (NC/P s – 1)	0.42	3.6	–	2.0	–
Cu3*s* (NC/S s – 1)	0.40	3.6	–	1.8	–
Cu3*s* (NC/Ni s – 1)	0.4 Cu/0.4 Ni	3 Cu/2 Ni	–	2.0 Cu/2.3 Ni	–
Fe3*s* (non-modified)	0.42	3.9	–	–	2.2
Fe3*s* (NC/P s – 1)	0.50	4.0	–	–	2.5
Fe3*s* (NC/P s – 2)	0.60	4.0	–	–	3.2

Note: I_2/I_1 is the relation of the maxima intensities of the lines of the multiple splitting; and Δ is the energy distance between the maxima of the multiple splitting in Me3s-spectra.

According to P2p spectrum (Figure 4.4) Phosphorus in Phosphorus containing copper–carbon MC changes the oxidation state from +5 to zero. The binding energy P2p changes from 135 eV, corresponding to PO_4 group, to 129 eV for P^0. The process flows on 90%. It's possible the interaction Copper and Phosphorus in this case.

C1s spectrum of this MC (Figure 4.5(b)) is distinguished from C1s spectrum of non-modified MC (Figure 4.5(a)): the correspondent form C–H on 15% smaller than in spectrum of non-modified MC. In turn, the relation of intensities for sp^2 and sp^3 hybridization is increased that can be linked with the nanosized granule increasing and approaching its form to roundish.

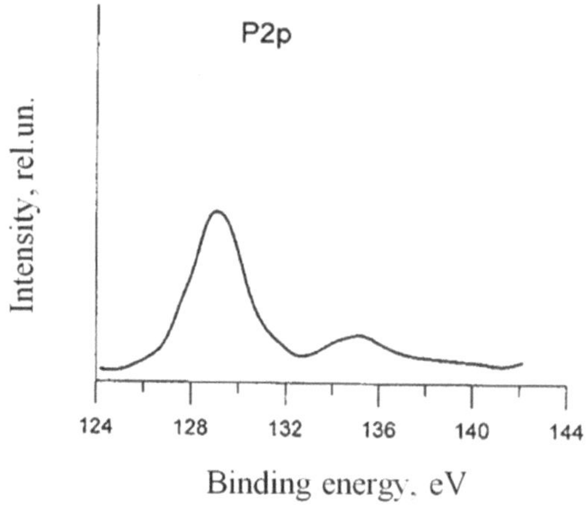

FIGURE 4.4 P2p spectrum of Cu C mesocomposite modified by ammonium polyphosphate at the relation 1:1.

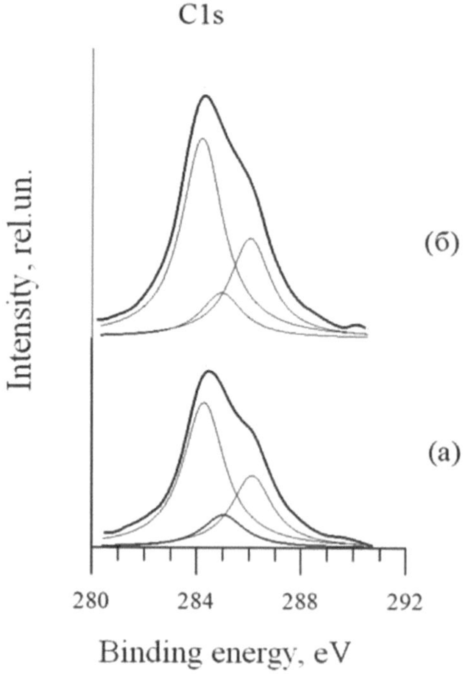

FIGURE 4.5 C1s spectra: (a) C1s spectrum of copper–carbon mesocomposite; and (b) C1s spectrum of modified phosphorus containing copper–carbon mesocomposite.

Analogous investigations are carried out [4, 5] for copper– and nickel–carbon MCs modified by Silicon containing substances. Spectra Si2p and C1s for Cu–C MC modified by Silica at the relation MC/Silica = 1 shown in Figure 4.6.

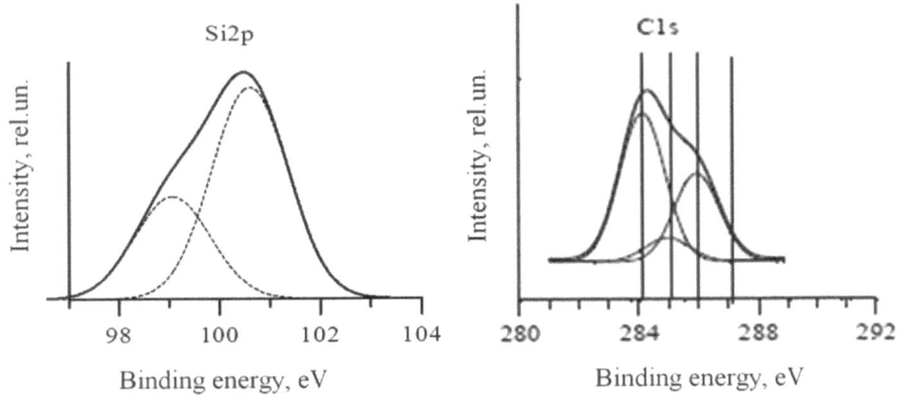

FIGURE 4.6 X-ray photoelectron spectra for modified Cu C (by silica) mesocomposite: (a) Si2p spectrum; and (b) C1s spectrum.

According to Si2p spectrum the relation of spectrum form intensities shows the red-ox process development on 51.4%.

C–H intensity in spectrum of modified Si containing MC on 65% smaller the correspondent value in spectrum of initial MC.

The thickness Si containing shell for Cu C (Si) MC in comparison with a shell of modified P containing MC is higher in 4 times.

For confirmation and development of notions about the modification process and the obtaining modified magnetic Me–C MC, the investigations with application of the following methods: radiography, IR spectroscopy, and TEM—are carried out.

It should be taken that in all cases of Me–C MC modification by Si, P, S containing substances the results corresponding to reduction of these elements are appeared.

For example, the X-ray pattern analysis of Phosphorus containing Cu C MC shows the presence of peaks for groups Cu–C–P at θ equaled to 43°.

Modified nanocomposites are studied by IR spectroscopy. For Phosphorus containing MCs the highest intensity of spectra lines corresponds to P–O–C group vibration which is formed at the interaction of APPh with the MC shell (Table 4.2).

TABLE 4.2 The Characteristics of Lines for IR Spectra of Samples Cu C MC ↔ P (1:1); Cu C MC ↔ P (1:0.5); and Cu C MC ↔ P (1:1.5)

No. of Lines	Spectrum of Samples	Characteristics of Spectrum Lines				Corresponding Groups
		Wave Number (cm−1)	Intensity	Half Width	D	
1.	Cu/C↔P (1:0.5)	911	0.010519	42,324	2.49×10^{-4}	P–O–P
	Cu/C↔P (1:1)	904	0.011797	51,674	2.28×10^{-4}	P–O–P
	Cu/C↔P (1:1.5)	895	0.024485	57,447	4.26×10^{-4}	P–O–P
2.	Cu/C↔P (1:0.5)	1,073	0.054722	1,49,260	3.67×10^{-4}	P–O–C
	Cu/C↔P (1:1)	1,072	0.039904	1,35,310	2.95×10^{-4}	P–O–C
	Cu/C↔P (1:1.5)	1,073	0.019365	71,075	2.72×10^{-4}	P–O–C
3.	Cu/C↔P (1:0.5)	1,254	0.005739	50,479	1.14×10^{-4}	P=O
	Cu/C↔P (1:1)	1,254	0.0090368	48,373	1.87×10^{-4}	P=O
	Cu/C↔P (1:1.5)	1,253	0.022251	53,025	4.20×10^{-4}	P=O

The investigation of copper–carbon MC modified by Silica with using IR spectroscopy shows the increasing of medium self-organization index D (Figure 4.7) which is calculated on formula:

$$D = I/(a/2)$$

where; 'I' is the line intensity; and a/2 is the half width of line.

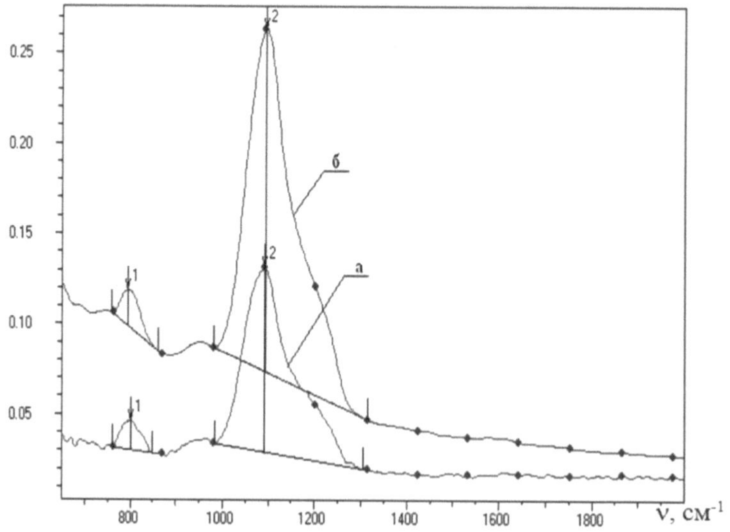

FIGURE 4.7 IR spectra of samples: (a) silica; and (b) silica + Cu C mesocomposite.

According to data of Table 4.3, the self-organization is increased in field 980–1,300 cm^{-1} (52.11%). The line in this field corresponds to the vibration of Si–O and, probably, Si–C bonds.

TABLE 4.3 Characteristics of IR Spectra Lines for Samples Silica and System Silica with Cu–C Mesocomposite

No. of Lines	Sample	Characteristics of Spectra Lines					Corresponding to Lines of Spectrum
		Start (cm^{-1})	Finish (cm^{-1})	Intensity	Half Width	D	
1.	$(SiO_2)_n$	759.98	848.7	0.016287	53.711	3.032×10^{-4}	Si–O–H
	$(SiO_2)_n$ + Cu/C HK	763.83	854.49	0.022069	47.481	4.648×10^{-4}	Si–O–H Si–C
2.	$(SiO_2)_n$	983.73	1303.9	0.10186	102.66	9.922×10^{-4}	Si–O Si–O–Si
	$(SiO_2)_n$ + Cu/C HK	983.73	1292.3	0.19596	102.2	19.174×10^{-4}	Si–O Si–O–Si Si–C

The investigation of modified Me–C MC by means of TEM of energy resolution shows that the shell from carbon fibers on mesogranule surface is well preserved. For instance, the TEM image of Phosphorus containing Cu C MC is presented in Figure 4.8.

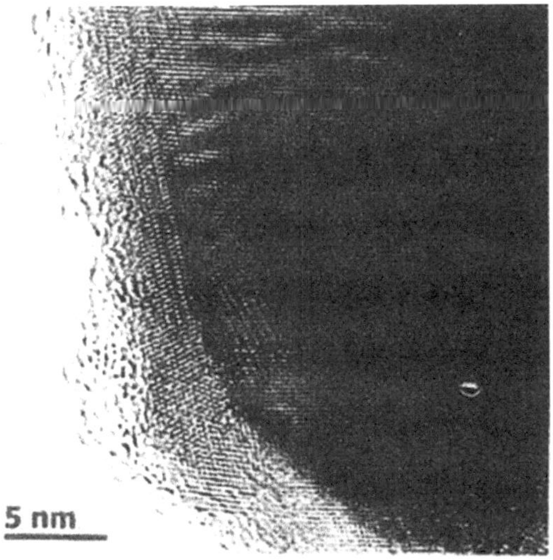

FIGURE 4.8 TEM image of phosphorus containing copper–carbon mesocomposite.

In contrast to Figure 4.1, in this case there are bulges and strands of fibers. At the same time the possible phases fields are found Carbon – 70%; phases Cu–C–P–O and C–Cu–P–O in which the content of Copper equals to 15%, and Phosphorus – 7%.

The results obtained cause natural question about the reason of magnetic characteristics growth for metal atoms in the clusters within the MCs meso-granule. Answer on this question may be found in the following hypothesis:

"The hypothesis is concluded in the formation of delocalized (unpaired) electrons because of the electrons shift on high energetic levels at the action electromagnetic (EM) radiation (field). It's possible the appearance of this radiation may be conditioned by the annihilation of quants of positive and negative charges which are appeared in the Red-Ox process at mechano-chemical modification of Me–C MC. Certainly, this hypothesis is necessary to confirm by corresponding experiments."

4.4 CONCLUSION

For the first time the metal/carbon nanocomposites which contain silicon, phosphorus, or sulfur, are produced within active media by means of mechanochemical modification. In this case the change of element oxidation states as well as the increasing of metal (Copper, Nickel, Iron) atomic magnetic moment takes place. At the same time above elements and functional groups with them are appeared in carbon shell of nanocomposites. These facts open a new era for further investigations and development of metal/carbon nanocomposites application fields.

KEYWORDS

- **atomic magnetic moment**
- **IR spectroscopy**
- **mechanochemical modification**
- **metal/carbon nanocomposites**
- **radiography**
- **red-ox synthesis**
- **X-ray photoelectron spectroscopy**

REFERENCES

1. Pat. RF 2393110, publ., (2010). *Applicant and Patent Holder "IEMZ-KUPOL" Method for Preparation of Carbon Metal-Containing Nanostructures.*
2. Kodolov, V. I., & Trineeva, V. V., (2017). New scientific trend-chemical mesoscopics. *Chemical Physics & Mesoscopy, 19*(3), 454–465.
3. Shabanova, I. N., Kodolov, V. I., Terebova, N. S., & Trineeva, V. V., (2012). *X Ray Photoelectron Spectroscopy for Investigation of Metal Carbon Mesoscopic Systems and Nanostructured Materials* (p. 252). M. - Izhevsk: Publisher "Udmurt University."
4. Kopylova, A. A., & Kodolov, V. I., (2014). Investigation of the copper carbon mesocomposite action with silicon atoms from silica compounds. *Chemical Physics & Mesoscopy, 16*(4), 556–560.
5. Kopylova, A. A., Zaitseva, E. A., & Kodolov, V. I., (2015). The functionalization of copper/carbon mesocomposite by silicon atoms. *Abstracts of Fifth International Conference from Nanostructures, Nanomaterials and Nanotechnologies to Nanoindustry* (pp. 93–95). Izhevsk.
6. Mustakimov, R. V., (2015). *The Elaboration of Intumescent Fireproof Coating Modified by Phosphorus Containing Copper Carbon Mesocomposite.* Magister Thesis, IzhSTU, Izhevsk.
7. Pigalev, S. A., & Kodolov, V. I., (2015). Quantum chemical modeling of copper/carbon nanocomposite functionalization by Sulphur containing compounds. *Abstracts of Fifth International Conference from Nanostructures, Nanomaterials and Nanotechnologies to Nanoindustry* (p. 153). Izhevsk.

CHAPTER 5

Investigation of Copper/Carbon Nanocomposites Modified with Phosphorus-Containing Groups as Inhibiting Additives in Mineral Oil

I. N. SHABANOVA,[1] S. M. RESHETNIKOV,[1,2] V. I. KODOLOV,[3] N. S. TEREBOVA,[1] R. V. MUSTAKIMOV,[3] F. F. CHAUSOV,[1] and S. G. BYSTROV[1]

[1]*Udmurt Federal Research Center, Izhevsk, Russian Federation*

[2]*Udmurt State University, Izhevsk, Russian Federation*

[3]*M.T. Kalashnikov State Technical University, Izhevsk, Russian Federation*

ABSTRACT

The mechanism responsible for a protective anticorrosive preservative layer composed of mineral oil with various amounts of a phosphorus-containing copper/carbon nanocomposite was studied by the XPS method from room temperature to 700°C. It was shown that a durable corrosion-protective layer was formed due to the donor-acceptor bonds between iron and phosphorous atoms. No changes in the nature of the contact between the protective layer and the steel substrate were found upon heating in a spectrometer chamber (*in vacuo*) with simultaneous acquisition of X-ray electron spectra. Optical microscopy surface analysis revealed that, at room temperature, the nanocomposites in mineral oil formed separate conglomerates on the steel surface. The conglomerates disintegrated into individual nanoparticles (NPs) at temperatures of 100–200°C, and NPs distributed over the entire surface. The interactions between the nanocomposite and iron led to the formation

Modern Magnetic Materials: Properties and Applications. Iuliana Stoica, Ann Rose Abraham, & A. K. Haghi (Eds.)

of a durable protective coating that hindered the attack of corrosive surface environments. The corrosion protection of alloy-free steel was studied by a Monicor corrosion meter. A sulfate–chloride solution simulating moisture condensate during atmospheric corrosion was used as the corrosive medium. The study provided evidence that the incorporation of copper/carbon nanocomposites modified with phosphorus-containing groups could suppress atmospheric corrosion by 70–95%.

5.1 INTRODUCTION

Mineral oils with various additives are known to be utilized for application of protective coatings on metals in order to prevent their corrosion damage [1]. The additives based on nitrated mineral oils are most effective [2–4]. A mineral-oil based lubricant containing nitrated mineral oil, synthetic fatty acid soaps, alkyl sulfates, alkyl sulfonates, and butyl ethers is described in Ref. [4]. The high anti-corrosion properties of the protective lubricant are attributed to a combination of complex additives of very diverse chemical nature. A drawback of the lubricant is its complex composition, which makes it challenging to prepare.

A series of studies [5–8] provided evidence of high protective properties shown by nanosized layers, which are formed by additives capable of chemosorption followed by chemical transformations, under conditions simulating atmospheric corrosion. In fact, under these conditions, complex nanocomposites largely suppressing corrosion are synthesized on the surface.

Nanocomposites are known to exhibit high reactivity due to the high surface energy of nanoparticles (NPs) [9]. A phosphorus-containing copper/carbon nanocomposite is one such example of a highly active nanocomposite; its synthesis and properties were reported in Ref. [10]. An attempt was made to utilize this nanocomposite as a component of a mixed amine-based inhibitor to protect steel surface from corrosion in a neutral aqueous environment [11]. It should be noted that the increase in the protection efficiency was not as high as expected. Apparently, this is due to the fact that a steel surface in contact with an aqueous solution of neutral electrolyte contains iron oxides or hydroxides that prevent the formation of strong binding between the nanocomposite elements and the surface being protected.

We assumed that a protective coating containing a phosphorus-containing copper/carbon nanocomposite could be obtained in a hydrocarbon environment created by mineral oil applied to a steel surface. In this case, the contact between the steel surface and the aqueous solution of the inhibitor was excluded.

Based on this assumption, the purpose of this work was to study interatomic interactions and protective properties of a coating obtained with the use of the high chemical and adsorption activity of a copper/carbon nanocomposite modified with phosphorous-containing groups that was added to the mineral oil base as an inhibitor.

For this purpose, a copper/carbon nanocomposite modified with phosphorous-containing groups was prepared by the mechanochemical method [10].

5.2 EXPERIMENTAL METHODS

The copper/carbon nanocomposite modified with phosphorous-containing groups [10] and I–20 industrial oil [GOST 20799-88] were used.

The nanocomposite is a nanostructure including a copper cluster. The copper clusters with an average size of 15 nm are enclosed in carbon shells containing polyacetylene and carbyne fragments with unpaired electrons at their joints. According to the electron diffraction and X-ray photoelectron spectroscopy (XPS) data, the phosphorus atoms are located between carbon fibers in P–C bonds [12].

X-ray photoelectron spectroscopic measurements were carried out on an X-ray electron magnetic spectrometer with a resolution of 10–4 and an instrument luminosity of 0.085% with excitation by the AlKα line at 1486.5 eV [13]. The advantages of X-ray photoelectron magnetic spectrometers are due to separation of the magnetic electron analyzer from the vacuum chamber of the spectrometer. Hence, a sample can be heated in the chamber with simultaneous recording of XPS spectra.

An OLYMPUS SZ-STB2O optic microscope was used to obtain the optical images of the samples. The pure I–20 industrial oil and I–20 industrial oil-based formulations containing 1, 5, 30, 50, or 75 wt.% of the copper/ carbon nanocomposite modified with phosphorous-containing groups were tested as anti-corrosion preservative lubricants. The copper/carbon nanocomposite modified with phosphorous-containing groups was mixed with the base mineral oil and applied to a sample after grinding in a porcelain mortar.

Steel samples with 20×20×20 mm dimensions were cleaned with P-600 sandpaper, then covered with two layers of Zapon lacquer, leaving 400 mm^2 as the working surface. Afterward, the samples were immersed in the test protective (preservative) lubricants for 5–7 min. Where, according to experimental procedure, the coated samples were heated in a muffle furnace, the non-working surfaces were insulted with Zapon lacquer after cooling the samples. The working surface exposed in a corrosion meter cell had an area of 400 mm^2.

The protective properties of the formulations were examined by evaluating the corrosion rates of steel 08 samples. The corrosion rate was determined from the polarization resistance measured on a Monicor corrosimeter. Solution No. 7 (NaCl – 30 mg/L, Na_2SO_4 – 70 mg/L in accordance with GOST 9.502-82), which simulates a moisture condensate under weathering conditions of industrial cities, was employed as the corrosive medium. The corrosion rates were expressed in relative units as the fraction of the corrosion rate relative to an uncoated reference sample whose corrosion rate was taken as unity.

The samples were mounted on the corrosion meter sensor. After immersing the sensor with samples into the model corrosive medium, the corrosion rates were measured for two hours. Within this period of time, the corrosion rates of all samples reached constant values. These values were then used in the evaluation of the protective (inhibitory) effect. Measurements and data processing were carried out according to GOST 9.514-99.

5.3 RESULTS AND DISCUSSION

We examined the chemical composition of the surface protective layers produced on steel upon adsorption of the preservation oil formulations containing 5 to 75% nanocomposite prepared in this way. The X-ray photoelectron spectra of core levels such as C1s, O1s, Fe2p, P2p, Cu2p, and Cu3s were studied.

The XPS study of the steel substrate surface onto which the protective coatings were subsequently applied showed that after cleaning from contamination by heating in the spectrometer chamber from 50 to 100°C, oxides were removed from the steel surface and the Fe2p spectrum showed a Fepure peak (707 eV) (*see* Figure 5.1, spectrum 1).

In addition to the maximum in the range of pure iron, the Fe2p spectrum of the oil formulations applied on the steel substrate before heating showed two weak peaks attributed to Fe–O surface oxides at 709–711 eV (*see* Figure 5.1, spectrum 2). When the sample was heated from ~50 to –100°C and up to 700°C, the position of the main peak at 707 eV in the Fe2p spectrum corresponded to pure iron or the Fe–P covalent (donor-acceptor) bond (*see* Figure 5.1, spectrum 3).

In addition to the main P–Fe peak at 130 eV in the P2p spectrum of the non-heated sample, a weak maximum at high bond energies corresponding to the P–O bond was recorded (Figure 5.2, spectrum 1) that corresponded to the fragments adsorbed on the surface. When the sample was heated to 50–70°C, this maximum disappeared. However, an intense P–Fe maximum with a bond energy of 130 eV persisted (Figure 5.2, spectrum 2). The P–Fe bond persisted even when the sample was heated to 700°C or higher *in vacuo*.

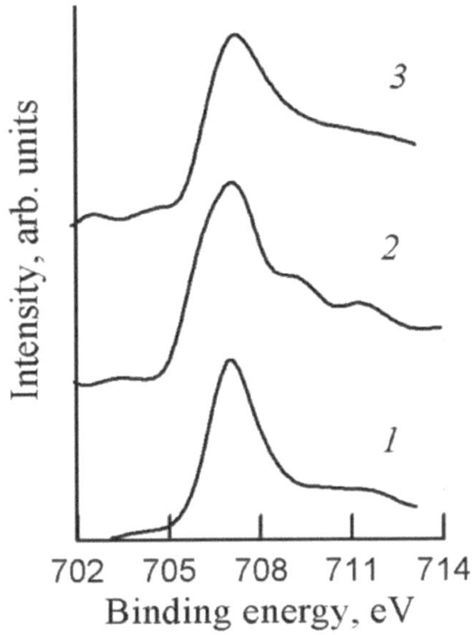

FIGURE 5.1 Fe2p spectra of uncoated steel surface upon heating (1); steel surface with the protective coating without heating (2); and steel surface with the protective coating upon heating (3).

No copper was found in the surface layers of the protective coating. The peak position in the C1s spectrum (285 eV) corresponded to the C–H bond in the hydrocarbons, but after the sample was heated, the contrast ratio of the spectrum decreased by half. No formation of Fe–C bonds was found.

Thus, the phosphorus atoms are bonded to carbon (C–P bond) in the copper/carbon nanocomposite; the P–O bond that is quite strong under ordinary conditions is not formed even at room temperature. The reason lies in the high reductive activity of the copper/carbon nanocomposite modified with phosphorous-containing groups, including the highly active carbyne and polyacetylene moieties, combined with the Red-Ox catalytic activity of copper/carbon moieties in which copper as the transition metal with a highly labile 3D-subshell serves as a common Red-Ox catalyst, and a developed carbon matrix acts as an electron density transfer carrier. The C–P bond is broken on contact of the copper/carbon nanocomposite modified with phosphorous-containing groups with steel surface, and the phosphorus atoms form a strong covalent (donor–acceptor) bond with iron. Therefore, the copper/carbon nanocomposite modified with phosphorous-containing groups is involved in the creation of a highly efficient anticorrosive protective coating both as a depot of reactive phosphorus atoms enclosed in the carbon matrix and as a transport structure that ensures the delivery of reactive forms of phosphorus to the steel surface to form an iron phosphide layer.

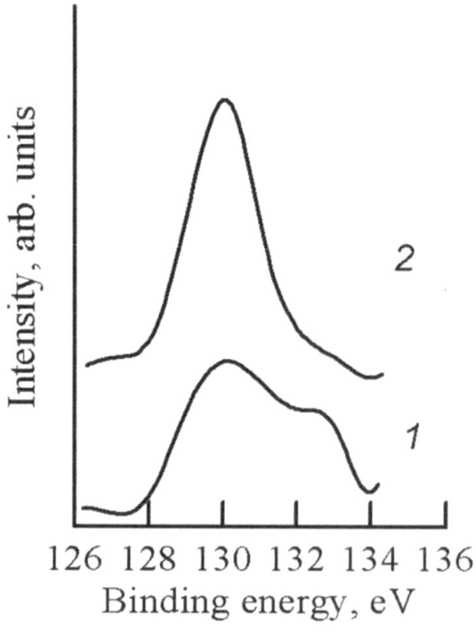

FIGURE 5.2 P2p spectra of the steel surfaces with a protective coating before (1); and after heating (2).

It is crucial that due to the structure of the copper/carbon nanocomposite modified with phosphorous-containing groups determined by the technology of its preparation, copper NPs are encapsulated in the carbon matrix, which prevents their direct contact with the steel surface. This eliminates the stimulating effect of NPs of metal copper on the corrosion of the steel surface.

An OLYMPUS SZ-STB2 optical microscope was used to obtain the surface images of the samples at room temperature and after heating to 50–100°C. This study revealed that at room temperature, the nanocomposites in the mineral oil formed separate conglomerates on the steel surface

(Figure 5.3). When the samples were heated to 50–100°C, these conglomerates disintegrated, and the nanocomposite was evenly distributed in the mineral oil over the entire coating surface (Figure 5.4).

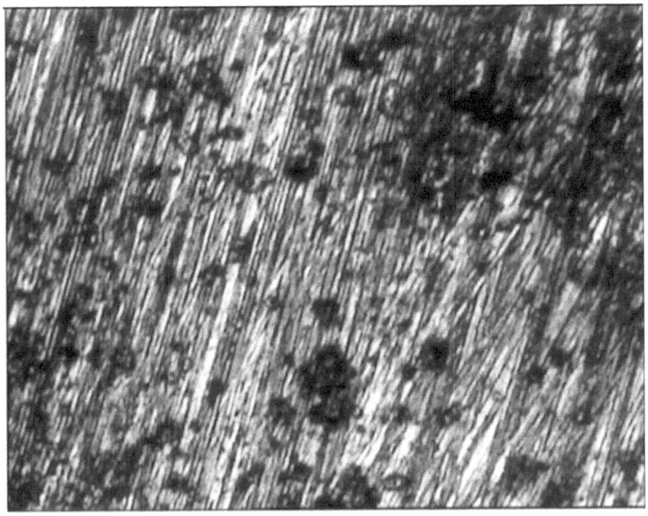

FIGURE 5.3 Optical images of the protective coating at room temperature. The eyepiece readings were 3.5, and the frame size was 740×580 mkm.

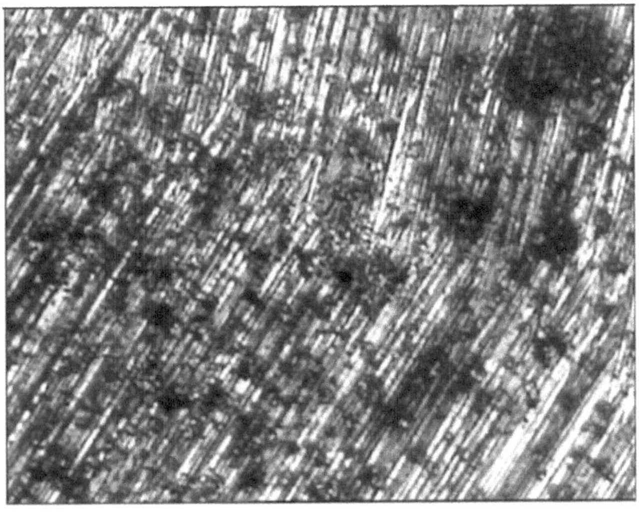

FIGURE 5.4 Optical images of the protective coating after heating up to 50–100°C. The eyepiece readings were 3.5, and the frame size was 740×580 mkm.

The protective properties of the formulations were examined using the Monicor corrosion meter. The Monicor corrosion meter enabled the measurement of relative corrosion rates. The corrosion rate in the background solution was taken as 1.

The samples for testing were prepared by dipping into a suspension of the nanocomposite in I–20 industrial oil. After that, some samples were subjected to thermochemical treatment, i.e., heated for 20 min in a muffle furnace at 100°C and 200°C in air. After cooling, the corrosion rates of the samples were measured by the Monicor corrosion meter. Table 5.1 summarizes the test results.

TABLE 5.1 The Corrosion Rates in the Model Corrosion Medium, Arb. Units

Composition of the Protective Coating (wt.%, Nanocomposite in I–20 Mineral Oil)	The Corrosion Rate, Arb. Units, at Activation Temperature (°C)		
	Room Temperature	100	200
Reference (Uncoated Steel)	10	0.90	0.90
0	0.91	0.88	0.85
1	0.88	0.72	0.70
5	0.88	0.70	0.65
30	0.72	0.12	0.10
50	0.68	0.08	0.06
75	0.66	0.06	0.04

Note. *Legend*: Reference (without a coating); 0 – I–20 oil without the NC additive; numbers 1, 5, 30, 50, and 75 correspond to the NC concentration (wt.%) in I–20 oil.

It was found in the experiments that heating of a sample coated with a mixture of the oil with the nanocomposites studied to 100–200°C increased its chemical activity. As a result, a protective coating was formed on the surface that reduced the corrosion rate by 70–95% depending on the nanocomposite concentration.

Thus, a formulation with a high nanocomposite content becomes efficient in forming a durable surface layer only after thermal activation.

5.4 CONCLUSION

The protective coating produced by thermal treatment of a suspension of a copper/carbon nanocomposite modified with phosphorous-containing groups in a hydrocarbon medium consisting of a mineral (petroleum) oil applied to a steel surface was studied.

The X-ray photoelectron spectra of the adsorbed layers on the steel surface were recorded. The interatomic interactions in these layers were studied.

It was shown that the formation of a protective layer upon adsorption of the inhibited mineral-oil formulation occurs due to the formation of covalent (donor–acceptor) bonds between iron and phosphorus atoms of the nanocomposite.

Corrosion tests revealed that the thermal activation of protective surface layers on steel surface from preservative oil formulations comprising 5 to 75% copper/carbon nanocomposite modified with phosphorous-containing groups increased the chemical activity of the nanocomposite. The resultant efficient protective layer on steel surface reduces the corrosion rate by 70–95% depending on the nanocomposite concentration.

ACKNOWLEDGMENTS

Thus, study was carried out within the framework of a state task (state registration no. AAA-A17-117022250040-0). The investigations were performed using the facilities of the Center for collective usage "Center of physical and physicochemical methods of analysis, study of properties and characteristics of surfaces, nanostructures, materials, and products" of UdmFTIS UB RAS supported by the Ministry of Science and Higher Education of the Russian Federation under the Federal Target Program "Study and Development in Priority Areas of Development of the Scientific and Technological Complex of Russia for 2014–2020" (unique project identifier: RFMEFI62119X0035).

KEYWORDS

- **anticorrosive layer**
- **copper/carbon nanocomposite**
- **corrosion meter**
- **optical microscopy**
- **X-ray electron spectroscopy**
- **X-ray photoelectron spectroscopic**

REFERENCES

1. Bogdanova, T. I., & Shekhter, Yu. N., (1984). *Inhibited Oil Compositions for Corrosion Protection* (p. 284). Moscow, Khimiya, (in Russian).
2. Shekhter, Yu. N., Legezin, N. E., Murav'eva, S. A., & Muryzyova, N. O., (1998). Corrosiological principles of the protection of inner metallic surfaces with inhibitors and inhibited formulations. *Prot. Met., 33*(3), 239–246 (in Russian).
3. Shekhter, Yu. N., Murav'eva, S. A., Puzevich, N. L., Podchinov, V. M., & Egorov, S. A., (1998). The use of the corrosion inhibitor AKOR-1B for improving the protective properties of lubricants. *Prot. Met., 34*(3), 322–324 (in Russian).
4. Shekhter, Yu. N., Bogdanova, T. I., Bakaleynikov, V. M., Zubareva, M. A., Shkaruba, E. V., Aliev, A. E., & Ustalov, A. V., (1994). *Protective Lubricant*. RF Patent 2017798 C1, 15.08. (In Russian).
5. Avdeev, Ya. G., Kuznetsov, D. S., Tyurina, M. V., & Chekulaev, M. A., (2015). Protection of nickel-chromium steel in sulfuric acid solution by a substituted triazole. *Int. J. Corros. Scale Inhib., 4*(2), 146–161. doi: 10.17675/2305-6894-2015-4-1-146-161.
6. Gladkikh, N. A., Maleeva, M. A., Maksaeva, L. B., Petrunin, M. A., Rybkina, A. A., Yurasova, T. A., Marshakov, A. I., & Zalavutdinov, R. Kh., (2018). Localized dissolution of carbon steel used for pipelines under constant cathodic polarization conditions. Initial stages of defect formation. *Int. J. Corros. Scale Inhib., 7*(4), 683–696. doi: 10.17675/2305-6894-2018-7-4-14.
7. Grafov, O. Yu., Kazansky, L. P., Dubinskaya, S. V., & Kuznetsov, Yu. I., (2019). Adsorption of depocolin and inhibition of copper dissolution in aqueous solutions. *Int. J. Corros. Scale Inhib., 8*(3), 549–559. doi: 10.17675/2305-6894-2019-8-3-6.
8. Shabanova, I. N., Reshetnikov, S. M., Naimushina, E. A., & Terebova, N. S., (2020). XPS investigation of adsorption protective layers based on industrial inhibited oil. *Int. J. Corros. Scale Inhib., 9*(2), 903–911. doi: 10.17675/2305-6894-2020-9-3-6.
9. Suzdalev, I. P., (2009). *Nanotechnologies: Physicochemistry of Nanoclusters and Nanomaterials* (p. 592). Moscow, Librokom, (in Russian).
10. Mustakimov, R. V., (2019). *Method for Modification of Metal/Carbon Nanostructures with Ammonium Polyphosphate*. RF Patent No. 2694092 C1, 09.07.2019, Bul. No. 19 (in Russian).
11. Pletnev, M. A., Ovechkina, O. A., Buldakova, N. C., Novikova, N. V., & Miller, V. K., (2014). Effect of metal-carbon nanocomposites on protective effect of corrosion inhibitors. *Intell. Syst. Prod.*, (1), 150–152 (in Russian).
12. Kolodov, V. I., Trineeva, V. V., Terebova, N. S., Shabanova, I. N., Makhneva, T. M., Mustakimov, R. V., & Kopylova, A. A., (2018). The changes in the electron structure and magnetic characteristics of modified copper/carbon nanocomposites. *Chem. Phys. Mesoscopy, 20*(1), 72–79 (in Russian).
13. Shabanova, I. N., Trapeznikov, V. A., Dobysheva, L. V., Varganov, D. V., Karpov, V. G., Kovner, L. G., Klyushnikov, O. I., et al., (1986). Advanced automated x-ray photoelectron magnetic spectrometers: Spectrometer with technological attachments and manipulators, melt spectrometer. *Izv. AN SSSR, 50*(9), 1677–1682 (in Russian).

CHAPTER 6

Investigation of the Formation of the Electron Structure of Metallocarbonic Nanoforms

I. N. SHABANOVA,[1] V. I. KODOLOV,[2] and N. S. TEREBOVA[1]

[1]*Physical–Technical Institute, Ural Branch, Russian Academy of Sciences, Izhevsk, Russia*

[2]*M.T. Kalashnikov State Technical University, Izhevsk, Russia*

ABSTRACT

Model samples of nanostructures are synthesized with the use of transition metals, with a different number of *d* electrons and *sp* elements of groups II and III with a different number of valence *p* electrons, as modifiers. The activity of nanostructure synthesis is investigated depending on the addition of *sp* elements. A procedure is developed for determining the atomic magnetic moment of the *d* metals and the change in the distances between atoms of the metal and the *sp* element via the parameters of the X-ray photoelectron spectrum of transition metals, which allows the degree of hybridization of valence electrons in the chemical bond of adjacent atoms (Me–X) to be determined. As a result of the investigation, the laws of nanoform growth are found, which facilitate the development of new directions in the synthesis of nanostructures with unique properties.

6.1 INTRODUCTION

The nanotechnological approach lies in the purposeful control of the properties of materials at the molecular and supramolecular levels, i.e., obtaining

Modern Magnetic Materials: Properties and Applications. Iuliana Stoica, Ann Rose Abraham, & A. K. Haghi (Eds.)

nanostructures bottom-up from molecules to supramolecular structures. In spite of the enormous number of theoretical and experimental works in the study of carbon nanosystems, until now, there has been no single pattern allowing the structure and properties of new forms to be explained.

For understanding the nature of quantum mechanical effects, there is a need to apply experimental methods investigating the chemical structure of nanoparticles (NPs) at the atomic level. One of the principal problems is the development of diagnostic methods allowing one to control the intermediate and final results of nanomaterial creation. In this context, the development of X-ray photoelectron spectroscopy (XPS) has become very significant.

The aim of the work is to determine the laws of the effect of filling the *d* and *p* atom shells of *d* metal and *sp* elements of groups II and III on the chemical structure and form of the created metallocarbonic nanoforms in polymer nanoreactors by XPS. The basis of the investigations into the formation of nanostructures with a defined form and their properties is the concept of the study of the interatomic interaction of the initial components and the formation of a hybridized chemical bond between *d* electrons of metal atoms and *p* electrons of *sp*-element atoms (carbon).

For studying the interatomic interaction of the initial components and the formation of the hybridized chemical bond of *d–p* electrons, a systematic investigation was carried out for the dependence of nanoform growth on filling of the *d* shell of the modifiers and the component content; the electron structure of the systems of *d* metals with *sp* elements of groups II and III; the effect of additions of *sp* elements (N, Si, P) into a polymer matrix on the activity of the synthesis of met allocarbonic nanostructures; and the dependence of the nanostructure form on the nanoreactor structure. The atomic magnetic moment of the *d* metals in the nanostructures was estimated.

6.2 EXPERIMENTAL METHODS

For determination of the form of metallocarbonic structures and the chemical bond of their components atoms, XPS was applied. An important feature of the method is its nondestructive character; this cannot be said of surface-analysis methods connected to ion or electron bombardment.

The investigations were carried out on an X-ray electron magnetic spectrometer with a resolution of 10^{-4} in vacuum with a residual pressure of 10^{-8}–10^{-10} Pa. The device's luminosity was 0.085% upon excitation of the Al*K*α line. The given spectrometer has advantages in comparison with electrostatic X-ray photoelectron spectrometers; these lie in the constancy of the luminosity and resolution capability without dependence on the electron energy, high spectral contrast, and the possibility of external influences on the sample during measurements [1–4].

The subjects of investigation were metallocarbonic nanostructures obtained by low-temperatures synthesis (to $T = 400°C$) in nanoreactors which were extended cavities formed by macromolecules in gels of polymeric materials. *d*-metals were used as metal modifiers [5]. The samples were finely dispersed powders. The powder was rubbed into a corrugated copper substrate into a thin layer with a thickness of about 1 μm. As nanosystem references, we used single- and multi-walled nanotubes, amorphous carbon, etc. The formation of *d*-metal-containing carbon nanostructures occurs in the course of the oxidation–reduction reaction in which the *d*-metal compounds are oxidizers, and hydrocarbon and amine groups are reducing agents. During the process the interaction region releases chlorine, hydrogen, and oxygen; carbonization occurs with the formation of corresponding nanostructures. Transition-metal nanoclusters deposited by the galvanic method into nanoporous silicon were also studied. The obtained results were compared with data of transmission electron microscopy.

6.3 RESULTS AND DISCUSSION

Systematic investigations of the electron structure of nanosystems were carried out with synthesized model samples of nanostructures using transition metals with a different number of *d* electrons and *sp* elements of groups II and III with a different number of valence *p* electrons as modifiers. For the development of XPS to study the chemical bond, *sp* hybridization of valence electrons and the vicinity of carbon atoms within the nanostructures, there is a need to interpret the C1*s* spectrum, i.e., to decompose it into its components. For that, graphite C–C (sp^2), diamond C–C (sp^3), and hydrocarbons C–H were used as standards.

Polymer systems with unsaturated bonds may possibly demonstrate the appearance of a shake-up satellite at a distance of 7 eV from the main line with an intensity that is 10% that of the main line. Such a structure unambiguously corresponds to two transitions with the excitation of an electron, situated at one of two filled orbitals and at an unfilled orbital. Consequently, the presence of these satellites in the C1*s* spectrum is an indicator of the presence of C–H groups in the material under investigation.

The satellite in the graphite spectrum is at a distance of 22 eV from the principal maximum, and it has a relative intensity of 10%. The satellite is caused by plasma losses [6]. The X-ray photoelectron study of diamond reveals a satellite with a binding energy of 313 eV at a distance of 27 eV and with a relative intensity of 10% of the principal maximum. Spectral calibration was carried out by the position and intensity of the satellite lines. The spectral width was determined from reference spectra. The component decomposition of the spectra was performed by the least squares method. The accuracy in the determination of the peak position is 0.1 eV. Subtraction of the background intensity was carried out by a standard procedure proposed by Shirley [7]. The error in the determination of the electron spectrum contrast in this case was no more than 5%.

In order to identify the form of the nanostructures of samples investigated, model nanoforms with well-known structures were studied, namely, single and multilayer nanotubes and amorphous carbon. The C1*s* spectrum of single-layer nanotubes shows the presence of C–C components with the sp^2 and sp^3 hybridizations of valence electrons with a binding energy of 284.3 and 286.1 eV and an intensity ratio of 2: 1. A similar situation is observed in the C1*s* spectrum of multilayer nanotubes. The C1*s* spectrum of a fullerene demonstrates the preferential presence of a component characteristic of the C–C bond with the sp^3 hybridization of valence electrons. For single and multilayer nanotubes and fullerenes with large diameters, one can explain the significant increase in the sp^2 hybridization of valence electrons of carbon atoms in comparison with the sp^3 hybridization. Consequently, the ratio of sp^2 and sp^3 hybridized valence electrons of carbon changes depending on the size and form of the nanostructures obtained. Thus, the study of standards of NPs with well-known structures enables identification of the structure under investigation by the form of its C1*s* spectra, which is of importance for the development of a new technology for creating nanostructures.

It was found by means of XPS and electron microscopy that the relative content of C–C and Me–C bonds and the hybridization type of the *s* and *p* electrons of carbon atoms allow one to control the formation and growth of nanostructures. The increase in the intensity of the peaks which is evidence of an increase in the content of these bonds in the spectrum correlates, according to the data of electron microscopy, with the growth of metal-carbon nanostructures (Figure 6.1).

The study of the formation of metal-carbon nanostructures and the effect of the type and concentration of the complexing element and metal modifiers on the yield of carbon nanotubes (CNTs) according to the C1*s* and Me2*p* spectra showed the following. The application of salts of transition *d* metals as modifiers results in the formation of the C–C bond with the sp^2 and sp^3 hybridizations of valence electrons and the C–Me bond. The content ratio of the C–C (Me) and C–H bonds depends on the modifier content. The number of C–C and C–Me bonds relative to the C–H bond also depends on the filling of the *d* shell of the metal modifier. With the filling of the *d* shell upon the transfer from Mn to Co and Ni, the relative number of C–C and C–Me bonds increases. Because Mn atoms have a lapping of the *d*–*p*-wave functions of the Mn and metalloid (C, Cl) atoms occurs, a significant amount of hydrocarbons remains. The formation of the strong Mn atom–C atom bond inhibits graphitization. Upon substituting manganese with nickel with a *d*-shell, close to complete filling, which has a strongly localized *d*-wave function in the vicinity of its atom, weaker bonds of Ni atoms with the metalloid (Cl) form. Therefore, hydrocarbon decomposition because of the formation of H–Cl bonds proceeds easier than in the case of

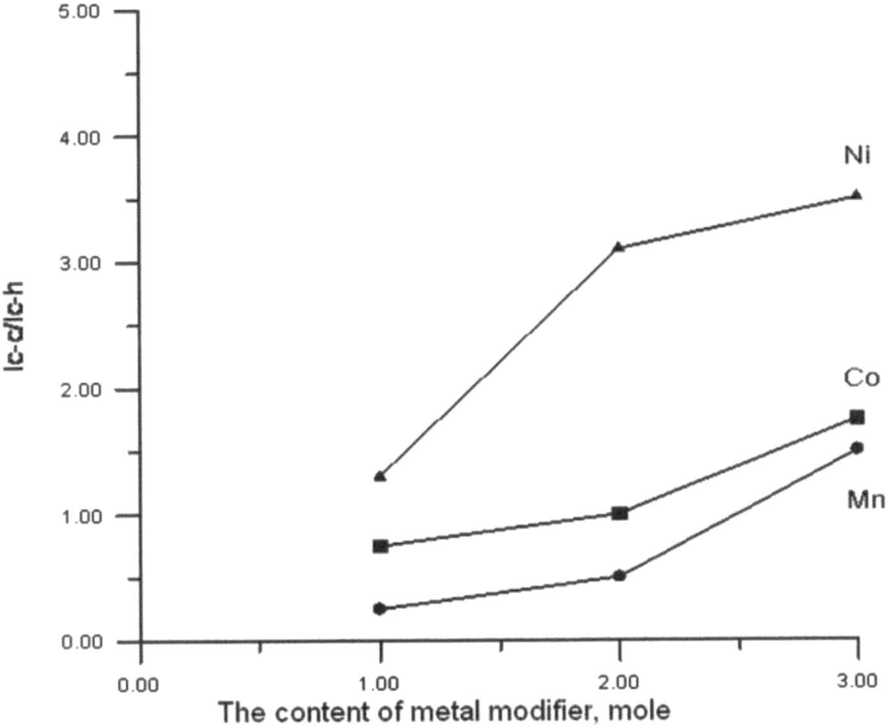

FIGURE 6.1 Ratio of the number of C–C and C–H chemical bonds as a function of the content of the metal modifier and filling of the metal *d* shell.

Mn. Comparison of the results of XPS with the data of transmission electron spectroscopy showed that multiple carbon accumulations form around Ni atoms, the sp^3 hybridization of valence electrons of carbon forms parallel with the sp^2 hybridization, and the graphene lattice and then, the nanotube form. In the case of cobalt, an intermediate situation occurs, namely, an insignificant portion of hydrocarbons remains, and C–C bonds with the sp^2 and sp^3 hybridization of valence electrons of carbon appear in a ratio less than for Ni. The quantum-chemical calculations given in Ref. [8] are in good agreement with the experimental results of XPS. Data obtained by the C1*s* spectra are represented in Figure 6.2.

It is found that the structure of the nanoforms being formed depends on the sizes and form of the nanoreactors [9]. The formation of nanoreactors from gels of complex composition, including polyethylene polyamine (PEPA) and acetyl acetone (AA), results in the acceleration of metal reduction and graphitization.

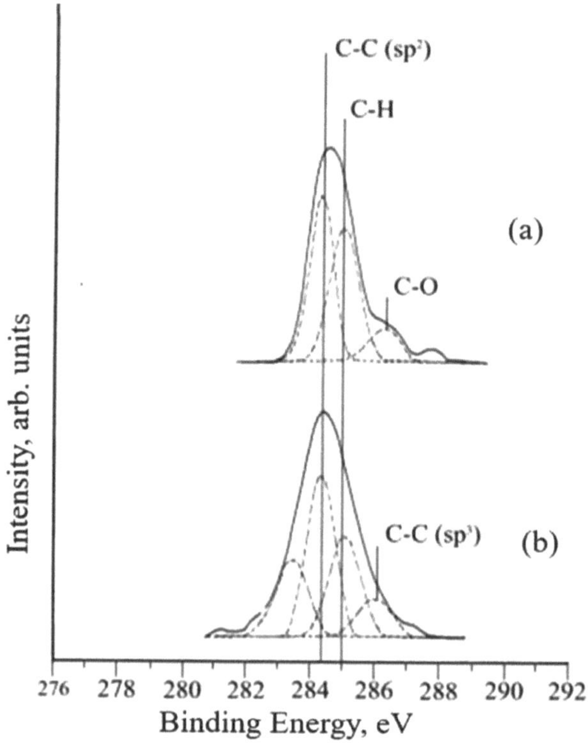

FIGURE 6.2 The X-ray photoelectron C1*s* spectra of the simple and complex gels of PVA and PEPA.

In contrast to PVA gels, the content of reduced nickel coordinated on C=C bonds increases within complex PEPA gels. As the nickel content in the nanoreactor increases, the formation of nanotubes is observed.

The effect of *sp* element (N, P, Si) additions into the polymer matrix on the activity of nanostructure growth was investigated. The most effect is shown by nitrogen forming a strong covalent bond with hydrogen and releasing carbon atoms and by phosphor interacting easily with oxygen. This is evidenced by the example of the synthesis of nanostructures in nanoreactors when Fe_2O_3 is used as the modifier. In this case reduction is difficult, and the yield of the nanoproduct is small. The reduction activity rises when, during synthesis, $(NH_4)_5PO_4$ is added in addition to the polymer and iron oxide. In this case, iron reduces with the formation of nanostructures.

For investigation of the formation of nanostructures with a certain form, there is a need to study the interatomic interaction of the initial components, the possibility of the formation of a hybridized chemical bond of *d* electrons of metal atoms with *p* electrons of atoms of *sp* elements. For that, a systematic investigation was carried out for the 3*s* spectra in order to reveal a correlation between the parameters of the multiplet splitting of the 3*s* spectra and the number of uncompensated *d* electrons of metal atoms and also the change in the neighborhood of the atom.

For most transition metal-ligand systems, the model connecting the parameters of the X-ray photoelectron 3*s* spectra with the change in the spin state on the metal atom correctly predicts a tendency toward a change in the atomic structure [10]. The relative intensity of the maxima of 3*s*-spectrum multiplets correlates with the number of unpaired *d*-electrons of atoms in the systems of 3*d* metals. The interval between the maxima of multiplets yields information on the exchange interaction of the 3*s* and 3*d* shells which depends on changes occurring in the 3*d* shell (localization or hybridization), i.e., we obtain information on the change in the spacing between adjacent atoms. The form of the valence bands, which is the energy distribution of the density of *d* states, determines their localization connected to interatomic interaction yielding information on the neighborhood of the atom. The presence of changes in the shape of the 3*s* spectra and the valence bands yields information on structure variation in the immediate vicinity of the 3D-metal atoms [11].

The given model was used in the investigation of metal–carbon nanostructures obtained in a nanoreactor of PVA with the use of superdispersed Fe, Co, Ni, and Cu powders as modifiers. The O1*s* spectrum in all the studied samples is low in intensity and typical for adsorbed oxygen. The C1*s* spectrum for all the investigated samples of nanostructures contains the Me–C and C–C components which are evidence of the formation of metal-carbon nanostructures [12].

The study of special features of the structure of the Me3*s* spectra for the samples under investigation revealed a change in the relative intensities of the multiplet-splitting maxima and the interval between them in comparison with the 3*s* spectra of pure metals (Table 6.1).

TABLE 6.1 Parameters of Multiplet Splitting for the 3*s* Spectra in the Reference Samples and Formed Nanostructures

Sample	I_2/I_1	Δ (eV)	μCo μ_B	μNi μ_B	μCu, μ_B	μFe, μ_B
Ni3s_{body}	0.15	4.3	–	0.5	–	–
Ni3s_{nano} (C)	0.32	3.0	–	1.8	–	–
Ni3s_{nano} (Si)	0.4	4.3	–	2.0	–	–
Co3s_{body}	0.29	4.6	1.6	–	–	–
Co3s_{nano} (C)	0.48	5.4	2.3	–	–	–
Cu3s_{nano} (C)	0.24	3.5	–	–	1.5	–
Fe3s_{body}	0.48	3.9	–	–	–	2.2
Fe3s_{nano} (PVA + $(NH_4)_5PO_4$)	0.5	4.0	–	–	–	2.5

Note: I_2/I_1 is the ratio of the line intensities of multiplet splitting; Δ is the energy range between the maxima of multiplet splitting in the 3*s* spectra of the pure metals Fe, Co, and Ni and nanostructures of Fe, Ni, Co, and Cu.

The obtained results indicate an increase in the number of uncompensated *d* electrons in metal atoms in the nanostructures with Co and Ni and their appearance in Cu atoms. The increase in the number of uncompensated *d* electrons can be explained by the participation of *d* electrons of metal atoms in the hybridized chemical bond with p electrons of carbon atoms. Consequently, the atomic magnetic moment in Co and Ni increases in comparison with that in pure metals in the nanostructures and appears on Cu atoms. The difference in the intervals of multiplet splitting in the 3*s* spectra of Fe, Co, and Ni in comparison with pure metals are evidence of the difference in the spacings between carbon and metal atoms in the nanostructures. This can be explained by the appearance of different nanoforms depending on the modifiers used.

The decrease in the interval between the multiplets in Ni nanostructures in comparison with pure Ni indicates the approaching of Ni and C atoms at significantly smaller spacings than between the Ni–Ni and Ni–C atoms in metal carbides [13] due to carbon atom implantation into the metal particle. This is evidence of the formation of a strong hybridized Ni–C bond. The increase in the interval between the multiplets in the Co and Fe nanostructures in comparison with pure metals indicates an increase in the spacing between the Co and C atoms in the nanostructures. This spacing has the same order as that in Me–C carbides.

Based on XPS data and analysis of the computational models for nanostructure formation, we proposed mechanisms for the growth of metal-carbon nanostructures. An explanation of the formation of multilayer nanotubes is proposed in the mechanism of turning a graphite sheet into a scroll. The broken bonds are compensated by interaction with metal (Co, Fe, Mn) atoms [14].

The single-layer nanotubes in the presence of modifiers (salts or metal oxides) grow on metal (Ni, Cu) particles with carbon atoms implanted into them.

These nanotubes set down (are adsorbed) at the particle surface in the form of islands and form a fullerene like structure with *sp*3 hybridization (Figure 6.3) which is a center for the formation of nanostructures with dimensions and a form determined by the metal modifier. In the case of the application of Ni and Cu as modifiers, single-layer nanotubes grow on their particles. The *d* electrons in Pt and Ag metals are more strongly localized than in Ni and Cu, and the bonds between the *d* and *p* electrons of the metal and carbon are significantly weaker. Therefore, the growth of fullerene-like structures on Pt, Pd, and Ag metal particles is accompanied by their separation. This mechanism is verified by the data of published works [15, 16].

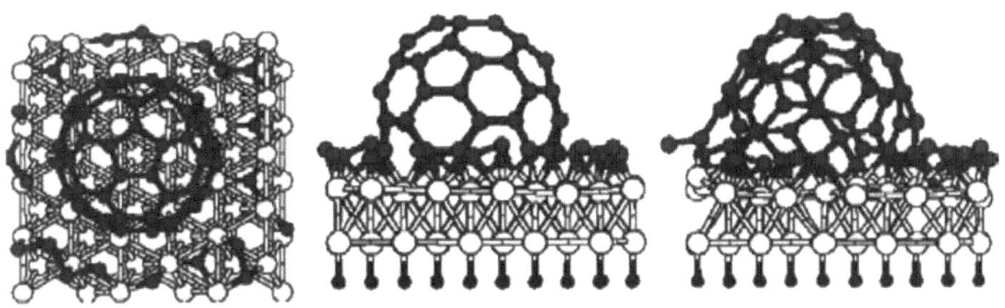

FIGURE 6.3 Integration of carbon adatoms in the base of the fullerene half-molecule at the nickel surface.

The table lists data of the X-ray photoelectron study of Ni nanostructures obtained by galvanic deposition into nanoporous silicon [17]. The analysis of the Ni3*s* spectrum for the given sample shows that the mechanism for the formation of such nanostructures is different than that for the formation of metal-carbon nanostructures. The interval of the multiplet splitting of the Ni3*s* spectrum for nanostructures with silicon remains the same as for crystalline nickel. This is evidence that Si atoms replace Ni atoms in the lattice, but do not implant into it as occurs in the case with metal-carbon nanostructures. The ratio of the line intensities of the multiplet splitting in the given sample is 0.4, which corresponds to ~2.0 μB, i.e., the number of uncompensated *d* electrons of Ni increases in these nanostructures, and, consequently, the atomic magnetic moment on the Ni atom grows. The work of Shabanova et al. [18] describes the formation of a strongly hybridized covalent bond in Me–Si systems in comparison with Me–C, Me–P, etc. The formation of a strongly hybridized covalent Ni–Si bond results in an increase in uncompensated *d* electrons of Ni and, consequently, in an increase in the atomic magnetic moment on the Ni atoms, which is greater than for metal-carbon nanostructures.

6.4 CONCLUSIONS

Thus, it is shown that investigations of the formation of nanostructures with a defined form and also of their properties are based on the concept of studying the interatomic interaction of the initial components and the formation of the hybridized chemical bond of *d* electrons of metal atoms with *p* electrons of atoms of *sp* elements. The laws of metal reduction and the synthesis of various nanoforms in the nanoreactors were studied depending on the content, modifier type, and nanoreactor structure.

An increase in the amount of metal-carbon nanostructures directly depends on the degree of filling of the 3*d* electron shell of the metals. The same mixture content shows the growth of C–C and Me–C bonds among Mn–Co–Ni. The dependence of the formation activity and form of the nanostructures on the nanoreactor structure and the effect of additions of *sp* elements (N, P) to the polymer matrix on the synthesis activity for metal-carbon nanostructures were studied. It is shown that the *sp* hybridization of carbon atoms and the filling and type of *d* shell determine the form of metal-carbon nanostructures.

KEYWORDS

- **acetylacetone**
- **catalysts**
- **low-energy synthesis**
- **metal-carbon nanosystems**
- **metallocarbonic structures**
- **polyethylene polyamine**
- **satellite structure of C1s spectra**
- **X-ray photoelectron spectroscopy**

REFERENCES

1. Shabanova, I. N., Varganov, D. V., Dobysheva, L. V., et al., (1986). Izv. Akad. Nauk SSSR, Ser. Fiz. 50, 1677.
2. Vasil'ev, L. S., Lomaeva, S. F., & Shabanova, I. N., (1993). *Poverkhnost*, (11), 94.
3. Bragin, V. G., Shabanova, I. N., Kulyabina, O. A., Kholin, N. A., & Trapeznikov, V. A., (1982). *Poverkhnost*, (11), 105.
4. Shabanova, I. N., Samoilovich, S. S., & Zhuravlev, V. A., (1982). *Poverkhnost*, (2), 129.
5. Kodolov, V. I., & Khokhryakov, N. V., (2009). *Chemical Physics of Processes of Formation and Transformations of Nanostructures and Nanosystems* (Vol. 1, 2). IzhGSKhA, Izhevsk, in Russian.
6. Pines, D., (1963/1965). *Elementary Excitations in Solids* (Benjamin, New York; Mir, Moscow).
7. Briggs, D., & Seah, M. P., (1983/1987). *Practical Surface Analysis by Auger_ and X_ray Photoelectron Spectroscopy*. Wiley, New York; Mir, Moscow.
8. Kodolov, V. I., Shabanova, I. N., Szargan, R., Kuznetsov, A. P., Nicolaeva, O. A., Makarova, L. G., Khokhryakov, N. V., et al., (2001). *J. Surf. Interface Anal., 32*, 10.
9. Kodolova, V. I., Khokhryakov, N. V., Tyneeva, V. V., & Blagodatskikh, I. I., (2008). *Khim. Fiz. Mezoskop., 10*, 448.
10. Lomova, N. V., Shabanova, I. N., Terebova, N. S., et al., (2005). *Izv. Akad. Nauk, Ser. Fiz., 69*, 1015.
11. Lomova, N. V., Shabanova, I. N., Kholzakov, A. V., et al., (2008). *Inorg. Mater., 44*, 818.
12. Makarova, L. G., Shabanova, I. N., Kodolov, V. I., et al., (2004). *Electron Spectrosc. Relat. Phenom., 137–140*, 239.
13. Slavinskii, M. P., (1952). *Physicochemical Properties of Elements* (Nauchno_Tekh. Izd. Liter. Chern. Tsvetn. Metallurgii, Moscow) [in Russian].
14. Makarova, L. G., Shabanova, I. N., Kodolov, V. I., et al., (2008). *Bull. Russ. Acad. Sci.: Phys., 72*, 459.

15. Pazhetnov, E. M., Koshcheev, S. V., & Boronin, A. I., (2003). *Kinet. Catal., 44*, 414.
16. Tontegode, A. Y., (1991). *Prog. Surf. Sci., 38*, 201.
17. Kashkarov, V. M., Len'shin, A. S., & Agapov, B. L., (2009). *Tech. Phys. Lett., 35*, 827.
18. Shabanova, I. N., Mitrokhin, Yu. S., & Terebova, N. S., (2004). *Electron Spectrosc. Relat. Phenom., 137–140,* 565.

CHAPTER 7

Dependence of the Value of the Atomic Magnetic Moment of *d*-Metals on the Chemical Structure of Nanoforms

I. N. SHABANOVA,[1] N. S. TEREBOVA,[1] and V. I. KODOLOV[2]

[1]*Udmurt Federal Research Center, Ural Branch, Russian Academy of Sciences, Izhevsk, Russia*

[2]*Kalashnikov Izhevsk State Technical University, Izhevsk, Russia*

ABSTRACT

X-ray photoelectron spectroscopy (XPS) was used to study the dependence of the 3D-metal spin state in nanoforms on the nanoform chemical structure. For studying the nanoform chemical structure, a model developed by us was used, which allowed us to find out the interrelation between the parameters of the X-ray electron 3s spectra of transition metals and the number of uncompensated electrons in metal atoms. It is shown that the nanoform chemical structure is determined by the methods used for their preparation and by the nanoreactor form. By changing the structure of nanoforms, it is possible, to a certain extent, to control the magnetic characteristics of nanoform-based materials.

7.1 INTRODUCTION

In the creation of the effective nanoscale materials for wide application and the implementation of nano-technological ideas, it is important to investigate the principles of self-organization of a substance and the creation of

Modern Magnetic Materials: Properties and Applications. Iuliana Stoica, Ann Rose Abraham, & A. K. Haghi (Eds.)

nanoscale materials from molecules to per molecular structures. By controlling the sizes and forms of nanostructures, one can impart quite new and often unprecedented properties to an object. To develop new approaches to the nanostructure preparation, experimental studies of the regularities of the formation of nanoparticles (NPs) in nanoreactors are needed.

In Ref. [1], a model was developed and used for revealing the interrelation between the parameters of the XPS 3s spectra and the change of the spin state on the atoms of d metals, which was helpful in the determination of the nanoform atomic magnetic moment. It was shown that the atomic magnetic moment of 3D-transition metals in carbon metal-containing nanostructures was larger than that in massive samples.

One of the most powerful direct methods for studying electronic structure, chemical bond, nearest surrounding of atoms of a substance is X-ray photoelectron spectroscopy (XPS). The choice of an electron magnetic spectrometer is due to a number of its advantageous features in comparison with electrostatic spectrometers, which are the persistence of luminosity and resolution capacity independent on the electron energy, and high contrast of spectra. In addition, the constructive separation of the energy analyzer of a magnetic type from the spectrometer vacuum chamber allows various actions upon a sample in vacuum directly during spectra taking. Thus, sample heating and cooling or the mechanical cleaning of the sample surface from impurities do not deteriorate the spectrometer resolution.

The objective of the present work is the XPS study of the dependence of the spin state of a 3D-metal contained in nanoforms on the chemical structure of the nanoforms, which greatly depends on methods used for the nanoform preparation, and the nanoreactor structure.

7.2 EXPERIMENTAL METHODS

The XPS spectra were obtained on an X-ray electron magnetic spectrometer constructed in the Physicotechnical Institute, Ural Branch of Russian Academy of Sciences [2]. The chemical structure of nanoforms was studied; the following methods were used for their production:

i. Low-temperature synthesis (in the polyvinyl alcohol nanoreactor (PVA)), wherein superdispersed powders of Fe, Co, Ni, and Cu were used as modifiers. Multilayer nanotubes were produced.
ii. Pyrolysis (700C) of injected ferrocene and nickel acetylacetonate solutions in benzene-ethanol mixture. The multilayer carbon nanotubes

(CNTs) produced were cleaned from amorphous carbon impurities by annealing in the air at the temperature 330°C for two hours.

iii. Galvanic (electrochemical) deposition of transition-metal nanoclusters in nanoporous silicon.

The samples were evaluated by the method of transmission electron microscopy.

In the present chapter, the dependence of the value of the atomic magnetic moment of 3D-metals on the size and form of nanostructures was investigated. The investigations were conducted:

a. For multilayer carbon metal-containing nanotubes produced by low-temperature synthesis and pyrolysis;
b. For multilayer carbon metal-containing nanotubes prepared in nanoreactors with different composition and structure;
c. Metal-silicon nanocomposites produced by the galvanic (electrochemical) deposition of a d-metal (Ni) into the pours of silicon.

In contrast to conventional methods presenting the information integrated over a sample, the XPS spectra give the information about the local characteristics of the structure of a substance. It is known that the parameters of the XPS 3s spectra of transition metals correlate with the magnetic moment of metal atoms [3]. The XPS 3s spectra of reference metals are characterized by the presence of two maxima, which indicates a multiplet splitting in the spectra [1]. The intensity ratio of the multiplet maxima (I_2/I_1) correlates with the number of uncompensated d electrons of the atoms of 3D-metals; and the distance between the multiplet maxima (Δ) characterizes the exchange interaction between 3*s*- and 3D-shells, and, thus, the localization of 3D electrons in the neighborhood of their own atom, which, in its turn, determines a change in the distance between neighboring atoms [4]. Consequently, a change in the 3s spectra parameters presents the information about a change in the chemical structure of nanoforms.

7.3 RESULTS AND DISCUSSION

Changes in the structure of carbon metal-containing nanoforms prepared by different methods of synthesis are investigated. Figures 7.1(a) and (b) show Me3s spectra and Table 7.1 presents the parameters of the Me3s spectra obtained from the studied samples. Multilayer carbon metal-containing nanotubes were produced by low-temperature synthesis and pyrolysis. In the case of pyrolysis and low-temperature synthesis, multilayer nanotubes grown on the d-metal particles are formed [5, 6].

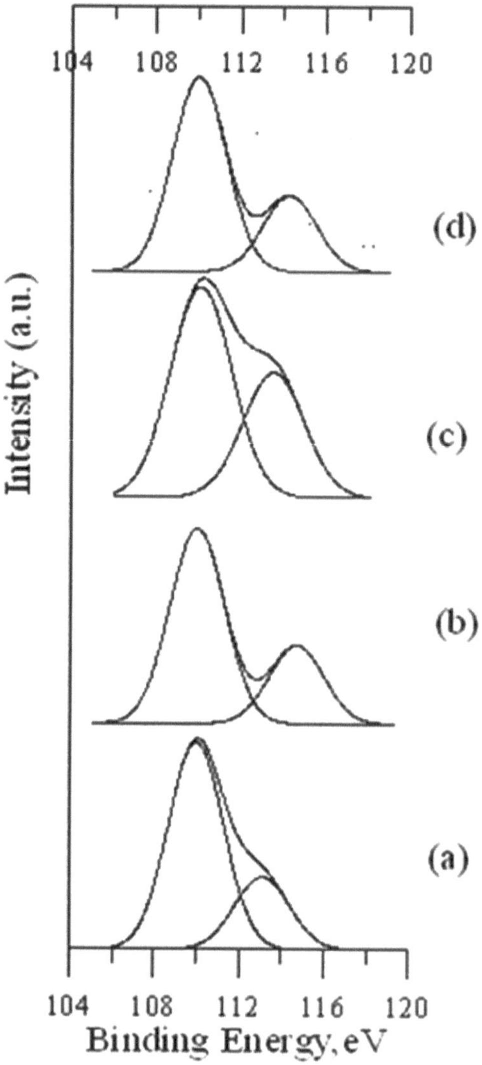

FIGURE 7.1 X-ray photoelectron Ni3s spectra of nanostructures produced by different methods: (a) low-temperature synthesis (400°C) in the PVA + NiO nanoreactor; (b) pyrolysis (700°C) of the injected solutions of ferrocene and nickel acetylacetonate in the benzene-ethanol mixture; (c) low-temperature synthesis (400°C) in the PVA + NiO reactor with addition of $(NH_4)_5PO_4$; and (d) galvanic (electrochemical) deposition of Ni into nanoporous silicon.

TABLE 7.1 The Parameters of the Multiplet Splitting of the 3s Spectra of Massive Samples and Fe-, Ni-, and Cu-Containing Nanostructures Produced by Different Methods

Samples	I_2/I_1	Δ (eV)	μ_{Ni}, μ_B	μ_{Cu}, μ_B	$\mu_{Fe}\ \mu_B$
$Ni3s_{nano}$ (C) low temperature	0.32	3.0	1.8	–	–
$Ni3s_{nano}$ (C) pyrolysis	0.40	4.7	2.0	–	–
$Ni3s_{nano}$ (PVA + $NH_4)_5PO_4$	0.60	3.6	3.0	–	–
$Ni3s_{nano}$ (Si)	0.40	4.3	2.0	–	–
$Fe3s_{massive}$	0.42	3.9	–	–	2.2
$Fe3s_{nano}$ (PVA +$NH_4)_5PO_4$)	0.50	4.0	–	–	2.5
$Cu3s_{nano}$ (C)	0.24	3.5	–	1.5	–
$Cu3s_{nano}$ (PVA + $NH_4)_5PO_4$	0.40	3.6	–	2.0	–

It is shown that the nanotubes produced by pyrolysis have a considerably smaller diameter (2–6 nm) than that of the multilayer nanotubes produced by low-temperature synthesis (17 nm). Table 7.1 presents the multiplet splitting parameters and the atomic magnetic moments for nickel in the nanostructures produced by the above methods. When the size of the nanostructures prepared by pyrolysis decreases, the ratio of maxima intensities of the multiplet splitting lines (I_2/I_1) increases, i.e., the magnetic moment of a nickel atom increases. It can be explained by the approachment of nickel and carbon atoms and an increase in the number of uncompensated d electrons due to the overlapping of their wave functions.

Further, the influence of the nanoreactor composition and structure on the chemical structure of the nanoforms produced by low-temperature synthesis has been studied (Figure 7.1(c) and (d); Table 7.1). In particular, the influence of the addition of sp-elements into the polymer matrix on the activity of the nanostructure growth has been shown. Most influence is produced by the presence of nitrogen and phosphorus. Nitrogen forms a strong covalent bond with hydrogen and doing so it releases carbon atoms; phosphorus readily interacts with oxygen reducing metal atoms. It is observed in the synthesis of nanostructures in nanoreactors, when d-metal oxides are used as d-metal compounds. The process of the Cu, Ni, and Co reduction goes without any difficulties. In the case of Fe_2O_3, the iron reduction process is complicated and the yield of a nanoproduct is insignificant. In this case, the activity of the reduction process increases when in addition to polymer and iron oxide, $(NH_4)_5PO_4$ is used during the synthesis. It results in the reduction of iron and the formation of carbon metal-containing nanostructures. In the Fe-containing nanostructures, the change of the atomic magnetic moment is insignificant in comparison with the Fe massive sample (Table 7.1).

Consequently, in this case the formation of nanostructures is not active. The number of d electrons participating in the bond with carbon p electrons and the number of uncompensated d electrons change to a very little degree (within the experimental error). Such behavior can be explained by a greater degree of ionicity in the chemical bond of the atoms of these nanostructures, which is characteristic of Me–C carbides.

Let us consider a change in the atomic magnetic moment of carbon Ni- and Cu–Containing nanotubes without additions of nitrogen and phosphorus and, vice versa, with the nitrogen and phosphorus additions into a polymer matrix (a nanoreactor). In both cases (for Ni and Cu–Containing nanotubes), the growth of the Ni and Cu atomic magnetic moments is observed, i.e., the growth of the number of uncompensated d electrons of the metals, which is indicated by the increase of the ratio of the maxima intensities of the 3s spectra multiplets (I_2/I_1) (*see* Table 7.1 and Figure 7.1). It indicates the change of the nanoform structure at adding phosphorous and nitrogen into the polymer matrix, and first of all, as mentioned above, a decrease in the sizes of the nanotubes. The decreasing distance between atoms leads to an increase in the number of uncompensated d electrons, which is confirmed by an increase in the distance between the maxima (Δ) of the multiplet splitting in the nickel and copper 3s spectra. The results received for the dependence of the atomic magnetic moment of nanoforms on the nanoreactor structure are in agreement with the data for the nanostructure growth with the increasing number of electrons on the d shell of a metal (Fe, Ni, and Cu).

Further, silicon metal-containing nanocomposites prepared by galvanic (electrochemical) deposition of a 3D metal into nanoporous silicon [7] have been studied. The mechanism of the formation of such nanostructures differs from that of the formation of the carbon metal-containing nanostructures. Nanocomposites are formed only on the basis of Ni, since the Ni–Si bond is the strongest [8]. The distance between the maxima of the multiplet splitting of the Ni3s spectrum of silicon nanostructures is the same as that for crystalline nickel, which indicates that Si atoms replace Ni atoms in the lattice, while in the case of carbon metal-containing nanostructures they incorporate into the lattice [9]. It is evidenced by the Δ values showing that the distance between Ni and C atoms is smaller than that between Ni and Si atoms. However, despite this fact the ratio of the maxima intensities of the multiplet splitting in the studied sample is 0.4, which corresponds to $\sim 2.0\mu_B$, and it is $\sim 1.8\mu_B$ for Ni–C, which means that in the silicon nanostructures the number of uncompensated d electrons of Ni increases. In Ref. [8], the formation of a much stronger hybridized covalent bond in the Me–Si systems is shown in comparison with the Me–C systems. The formation of the strongly hybridized covalent Ni–Si bond leads to an increase in the number of uncompensated d electrons of Ni, and, consequently, to a greater increase in the atomic magnetic moment on Ni atoms in comparison with that in carbon metal-containing nanostructures.

Thus, by changing the composition and structure of nanoreactors or the method of nanostructure production, it is possible, to a certain extent, to control the magnetic properties of materials based on nanostructures.

7.4 CONCLUSION

The nanostructures formed as the result of the interaction of transition metal oxides with polyvinyl alcohol (PVA) are not of the same type. The differences are due to different filling of the d shell of the transition metals participating in their formation.

An increase in the atomic magnetic moment of d metals in nanostructures is shown to be dependent on the decrease of the nanotube diameter which is associated with the decrease of the distance between the atoms of a metal and carbon, the increase of the overlapping of the wave functions of their d and p electrons and the increase of the number of the metal uncompensated d electrons.

In nanostructures, the degree of the chemical bond covalence of the component atoms increases with the d-shell filling growth from Fe to Cu, and consequently, the number of uncompensated d electrons increases to a greater extent.

A model was developed showing the dependence of the parameters of the multiplet splitting of the 3s spectra on the spin state or atomic magnetic moments; it was used in the carbon metal-containing nanoform investigation for studying the hybridized bond of ‘d’ electrons of a metal atom with ‘p’ electrons of an sp-element and a change in the number of uncompensated d electrons:

1. The comparative study of the influence of carbon and other *sp* elements (N, P, Si) on the activity of the synthesis of carbon metal-containing nanostructures, their form and properties has been conducted. The addition of *sp* elements (P, N) into a polymer matrix results in the acceleration of the reduction of the Fe, Co, Ni, and Cu metals and the increased activity of the nanostructure synthesis. Nanoreactors are mainly intended for facilitating the formation of

the activated complex and decreasing the activation energy of the main reaction between the components.

2. The study of the interatomic interaction in carbon metal-containing and silicon metal-containing nanostructures shows that it has the same features in both cases, i.e., the formation of the hybridized chemical bond of d and p atoms of components depends on d and *sp* shell filling, and there is an increase in the atomic magnetic moment on a metal atom in nanostructures in comparison with massive samples.
3. A number of properties of nanostructures depend on the method of production and their structure. In the cases of low-temperature synthesis and pyrolysis on the basis of polymers and d metals, there is a strong interatomic interaction, and carbon metal-containing nanotubes with different diameters are readily formed. In nanocomposites prepared by the electrochemical deposition of a metal into nanoporous silicon, nanoforms are formed only on the basis of Ni, since the Ni–Si bond is the strongest.

Thus, for manufacturing magnetic materials with a super-high atomic magnetic moment, both the iron- and cobalt-containing and nickel- and copper-containing nanostructures with minimal sizes can be used.

KEYWORDS

- **atomic magnetic moment of *d*-metals**
- **chemical structure of nanoforms**
- **electronic structure**
- **interatomic interaction**
- **nanoforms**
- **spin state**
- **X-ray photoelectron spectroscopy**

REFERENCES

1. Shabanova, I. N., & Terebova, N. S., (2010). Application of the x-ray photoelectron spectroscopy method for studying the magnetic moment of 3d metals in carbon-metal nanostructures. *Surface and Interface Analysis, 42*(6, 7), 846–849.

2. Trapeznikov, V. A., Shabanova, I. N., & Zhuravlev, V. A., (2004). The use of the x-ray photoelectron spectroscopy for studying inverse β-decay. *Electron Spectroscopy and Related Phenomena, 137–140,* 731–734.
3. Briggs, D., & Sikh, M. P., (1987). *The Surface Analysis by Auger- and X-Ray Photoelectron Spectroscopy* (p. 600). M.: Mir.
4. Lomova, N. V., & Shabanova, I. N., (2004). The study of the electronic structure and magnetic properties of invar alloys based on transition metals *J. of Electr. Spectr. and Rel. Phen., 137–140,* 511–517.
5. Kirikova, M. N., (2009). *Physicochemical Properties of Functionalized Multi-Walled Carbon Nanotubes* (p. 24). After abstract of dissertations for competitive academic degrees, PhD, Mostov.
6. Kodolov, V. I., Blagodatskikh, I. I., Volkova, E. G., Terebova, N. S., & Makarova, L. G., (2008). *Method for Production of Carbon Metal-Containing Nanostructures by Interaction of Organic Substances and D-Metal Salts*. Patent No. 2323876.
7. Kashkarov, V. M., Lenshin, A. S., Agapov, B. L., Turishchev, S. Yu., & Domashevskaya, E. P., (2009). Electron structure of iron and cobalt nanocomposites on the basis of porous silicon. *Phys. Status Solidi C., 6*(7), 1656–1660.
8. Shabanova, I. N., Mitrokhin, Yu. S., & Terebova, N. S., (2004). Experimental and theoretical study of electronic structures of Ni-X(X = Al, Si, P) systems. *Electron Spectroscopy and Related Phenomena, 137–140,* 565–568.
9. Shabanova, I. N., & Terebova, N. S., (2011). X-ray photoelectron investigation of the regularities of the growth of carbon metal-containing nanoforms in nanoreactors. *Journal of Structural Chemistry.*

CHAPTER 8

Some Aspects of Magnetic Metal Carbon Mesoscopic Composites With Regulated Magnetic Characteristics

V. I. KODOLOV,[1,2] V. V. KODOLOVA-CHUKHONTSEVA,[1,3] YU. V. PERSHIN,[1] R. V. MUSTAKIMOV,[1,2] I. N. SHABANOVA,[1,4] and N. S. TEREBOVA[1,4]

[1]*Basic Research–High Educational Centre of Chemical Physics & Mesoscopics, Izhevsk, Russia*

[2]*Kalashnikov Izhevsk State Technical University, Izhevsk, Russia*

[3]*Great Peter St Petersburg Polytechnic University, St Petersburg, Russia*

[4]*Udmurt Federal Research Centre, Ural Division, Russian Academy of Sciences, Izhevsk, Russia*

ABSTRACT

This book chapter reviews some important aspects of the metal-carbon mesocomposites in detail. We have also focused on the chemical mesoscopic characteristics. The process takes place at the quantization of the charge considering specific phase consistency along with the chemical bond formation. This is because of real involvement and the reduction-oxidation process destruction that could possibly take place. Meanwhile, the process is analyzed by the destruction origin. In such conditions, we have to note that, in fact, the metal atomic magnetic moment growth will be in the reliance on electrons number that could be evidently engaged in oxidation decrement approach (i.e., red-ox process). Metal carbon production of mesoscopic composites could be clearly carried out with the application of mechanical–chemical interaction.

Modern Magnetic Materials: Properties and Applications. Iuliana Stoica, Ann Rose Abraham, & A. K. Haghi (Eds.)

8.1 INTRODUCTION

The mesoscopic proposition could be used to analyze the chemical particles reactivity when the charge quantization along with phase coherence is an important factor. These incidences are the main cause for the intervention and feasible destruction aspects. This transformation is worked-out by means of the reflection of mesoscopic systems interaction along with the macroscopic systems. Based on the studies presented in Refs. [1, 2], the mesoscopic system (ms) frequently depends on the types of applied samples. Here, the macroscopic systems are called reservoirs, contacts, or banks. Therefore, these banks or reservoirs are expected to be the main sources and/or drains for particles and energy. These reservoirs are large to transfer the energy between ms particles. So in any condition, the electron systems within reservoirs are expected to appear in equilibrium condition. This is characterized by a certain amount of temperature (T) and chemical potential (μ).

The restrictions for processes are explained clearly by the limits of Chemical Mesoscopics. These could be summarized as:

1. The mesoscopic particle;
2. The size of phase consistency is placed in limits up to 1000 nm;
3. As the circumstances such as spectrum quantization, charge quantization, and interference, are appeared.

8.2 MESOSCOPIC REACTORS (NANOREACTORS)

Mesoscopic reactors are very particular nanostructures. These reactors could be provided as a nanosized cavity in polymeric matrices or could be designed as the space bounded part.

Mesoscopic one-dimensional nanoreactors are classified as [3,4]: clearance between probe and surface, crystal canals, complexes, crystal solvates, macromolecules, micelles, vesicles, or pores.

Mesoscopic two-dimensional nanoreactors (mesoscopic reactors) are classified as double electrical layers, monomolecular layers on the surface, membranes, interface layers (boundaries), or adsorption layers.

Different types of mesoscopic reactors exist with various conditions that could take place. Meanwhile, the formation of activated complexes is observed within the reactors. It should be pointed out that the formations of nanostructures within mesoscopic reactors are observed by the nature of reactants. These, of course, participate in synthesis and by the energetic and geometric characteristics of mesoscopic reactors.

According to our analysis of modern scientific data and based on our studies, the following form in the nanoreactors could be noted:

1. The main abnormality is the decline of parallel collateral processes and the process direction to the special product side.
2. The low energetic costs and the high rates of processes.
3. The dependence of nanostructures properties obtained by energetic and geometric characteristics of nanoreactors.

The basic parameters and equations needed the matrices. First of all, the coordination number of elements or elements group is needed. And then, the growth and form of nanostructures could be observed using Kolmogorov–Avrami equations [5–7]. Expansion of the nanostructures self-organization process is approximated by means of the Kolmogorov-Avrami equations [8, 9].

In this equation, the Red Ox potentials are considered. As the metal reduction happens in the process of Metal Carbon mesocomposite formation, we can show;

$$W = 1 - k \exp\left[-\tau^{n} \exp(zF\Delta\varphi/RT)\right],$$

where; the parameter (k) is the proportionality coefficient, the parameter (τ) is the duration of the process, the parameter (n) is the fractal dimension, the parameter (z) is the number of electrons participating in the process, the parameter ($\Delta\varphi$) is the difference of potentials on the border "mesoreactor wall–reaction mixture," the parameter (F) is the Faraday number, the parameter (R) is the universal gas constant, and the parameter (T) is the temperature.

During the calculating process, our process duration takes within half-hour intervals. In this regard, for such calculations, we considered the following:

$$n = 2 \text{ (two-dimensional growth)},$$

where; potential of Red ox process during the metal reduction = 0.34V, temperature = 473K, Faraday number = 26.81 A×hr/mol, and universal gas constant (R) = 2.31 W×hr/mol×degree.

However, our calculations are done for the thermal stage duration of the synthesis of Copper Carbon mesocomposites. The results are shown in Table 8.1.

It should be noted that Table 8.1 results practically agree with experimental data.

TABLE 8.1 Results Obtained

Duration (hrs)	0.5	1.0	1.5	2.0	2.5
Content of product, %	22.5	63.8	89.4	98.3	99.8

8.3 MESOSCOPIC REACTORS FOR METAL CARBON MESOSCOPIC COMPOSITES SYNTHESIS

Metal carbon mesoscopic composites are designed using metal oxides (e.g., CuO) and polyvinyl alcohol (PVA). In the dependence of metal oxidation state (for Cu^{+1}, CN is equaled to 2; for Cu^{+2} – 4). The coordination number (CN) of copper could be equal to 2 or 4. The reagents relation (CuO/PVA) can be equated to 4. At the first stage (at the grinding of reagents), when the reagents relation is given as 1:4 (1 part – metal oxide, for example, copper oxide, and four parts – PVA), the decline of metal-containing phase sizes is observed using reduction-oxidation (redox) process. With the metal-containing cluster, the carbon formation from PVA chains is characterized by four chain fragments. The reduction of metals with the change of electron structure is expected [10–13], and the difference of banks of chemicals are obtained (Figure 8.1)

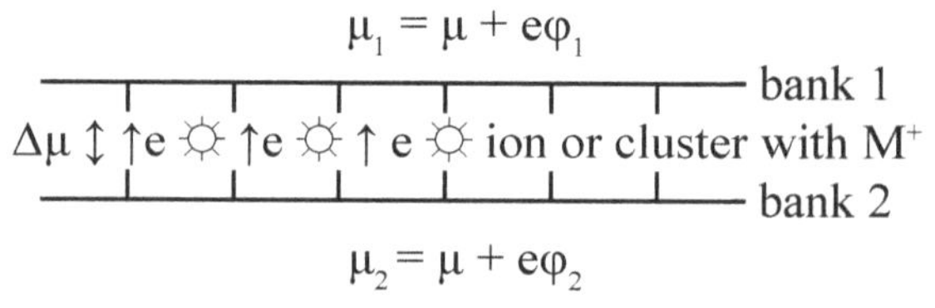

FIGURE 8.1 Nanostructures formation stages.

In this particular case study, metal (e.g., copper) within the cluster is considered with a positive charge. Meanwhile, the annihilation process stimulates the formation of (d) electrons flow from metal to the carbon shell of the mesoscopic particle formed, which is presented as metal (e.g., copper) and carbon mesoparticle (or mesogranul).

This process is observed by the growth of metal atomic magnetic moment. Such scheme of copper carbon mesoscopic composites formation is accepted by: TEM (transition electron microscopy), (X-ray) photoelectron spectroscopy, roentgenograph or X-ray image, and also EPR spectroscopy (the electron paramagnetic resonance) or ESR spectroscopy (electron spin resonance).

It should be noted that mesoscopic composites have magnetic characteristics. Considering the delocalized electrons on the carbon shell, our mesoscopic composites are active at the modification by the mechanic chemical methods.

The structures of metal-carbon mesoscopic composites with active carbon shells are defined by means of complex methods, including x-ray photoelectron spectroscopy, transition electron microscopy with high permission, electron microdiffraction, and also EPR spectroscopy (please refer to Figures 8.2–8.5 and Tables 8.1–8.3).

The image of copper/carbon mesoscopic composite structure observed by transition electron microscopy (TEM). This is shown in Figure 8.2 as the TEM microphotograph for copper/carbon mesocomposite.

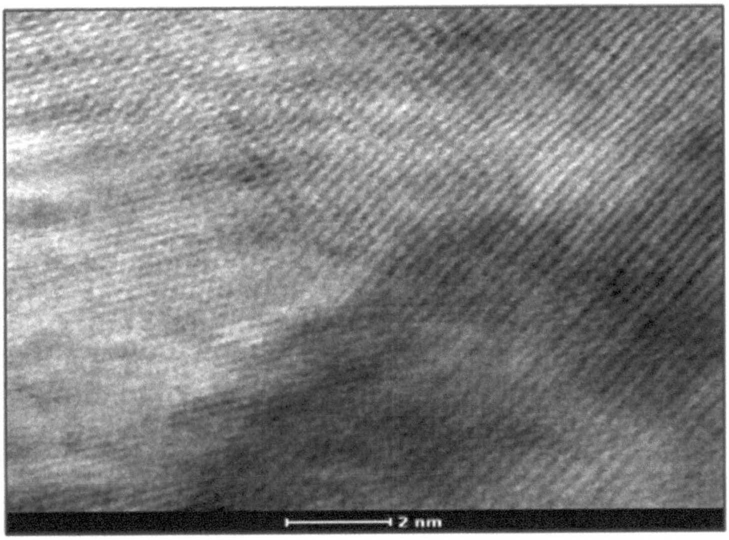

FIGURE 8.2 Transition electron microscopy (TEM microphotograph).

Data from transition electron microscopy with high solution indicates that the carbon fiber contains Carbon atoms because of the correspondence of the fiber diameter and carbon atom diameter or C–H group diameter value.

In Figure 8.2, the carbon structure shells for clusters of copper are presented as carbon fibers. This fact is confirmed by electron microdiffraction results. The carbon fiber formation is generated by the realization of the reduction-oxidation process with the appearance of declined copper and the carbonization of polymeric hydrocarbon chains.

The declined copper formation at the red ox process is confirmed in Figure 8.3 [The diffract gram of copper carbon mesoscopic composites obtained from copper oxide within matrices for different marks of PVA Cu/C NC (red) – PVA mark BF-14, NC (blue) – BF-17, NC (dark) – BF-24].

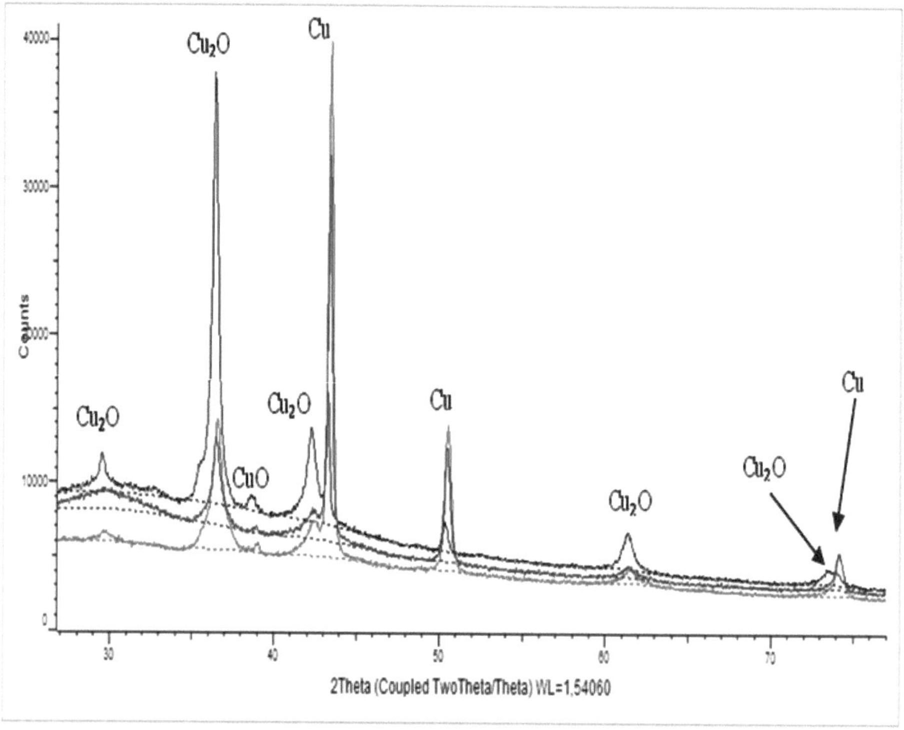

FIGURE 8.3 The diffract gram of copper carbon mesoscopic composites.

Table 8.2 shows the dependence of process results from the nature of the polymeric matrix or marks of PVA.

TABLE 8.2 Influence of PVA

Marks of PVA	The Line color on diffract grams of Copper Carbon mesoscopic composite	Process completion, %
PVA mark BF-17	Blue	77
PVA mark BF-14	Red	82
PVA mark BF-24	Dark	72

The correspondent marks of PVA or polyvinyl alcohol contain different quantities of acetic groups. The big number for PVA designates the most quantity of acetic groups in the polymer. Meanwhile, the process completion using this PVA mark is less in comparison with PVA mark B-14. Therefore the PVA mark B-14 is used for Copper Carbon mesoscopic composite synthesis. The composition of the metal-containing phase in this mesoscopic particle is shown in Table 8.3.

TABLE 8.3 Composition of Metal-Containing Phases in Copper Carbon Mesocomposite

Phase	Cu/C mesocomposite
CuO	1.17%
Cu_2O	5.19%
Cu	93.64%

Therefore, the decline of the oxidation process observed by the copper formation takes place at the interaction of copper oxide and PVA. Meantime, in this process, the carbon fibers are formed.

In correspondence with C1s spectra (please refer to Figure 8.4), the carbon fibers contain the carbine and polyacetylene fragments.

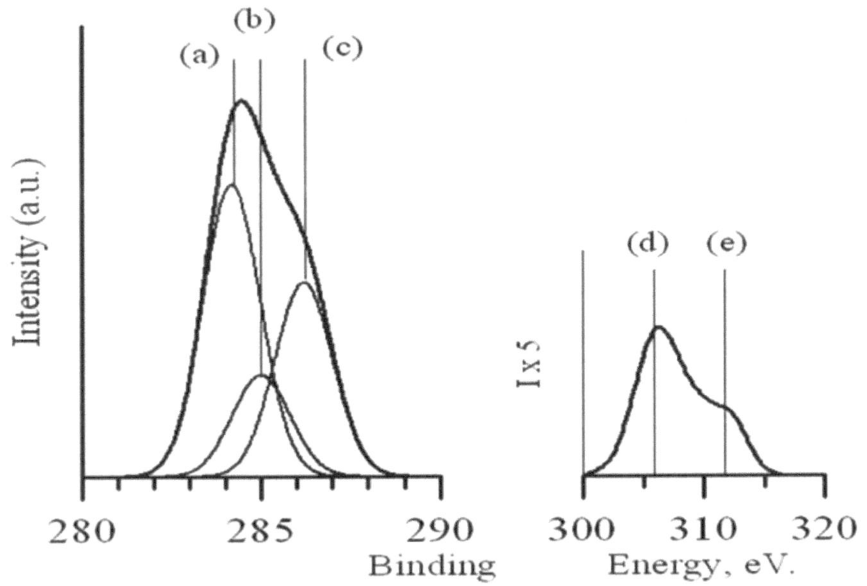

FIGURE 8.4 C1s spectrum of Cu/C mesoscopic composite.

Here is our observation:

- In C1s spectra, there are lines corresponding to the C1s energy for CH groups.
- Peak at 285 eV (C–H bond) can be considered poly acetylene fragment of carbon fiber. In addition, the peak at 281–282 eV is considered to be a Carbine fragment.
- Based on C1s spectra three types of satellites (sp, sp^2, and sp^3 hybridization) with different relation between them are noted. These satellites also take place in the C1s spectrum.
- The intensities relation as I_{sp^2}/I_{sp^3} for Cu/C mesocomposite corresponds to 1.7. The connection of these fragments is only possible at the unpaired electrons formation on the joints of connections. Therefore the investigations of EPR spectra are carried out.
- The results of these investigations are shown in Figure 8.5 and Table 8.4.

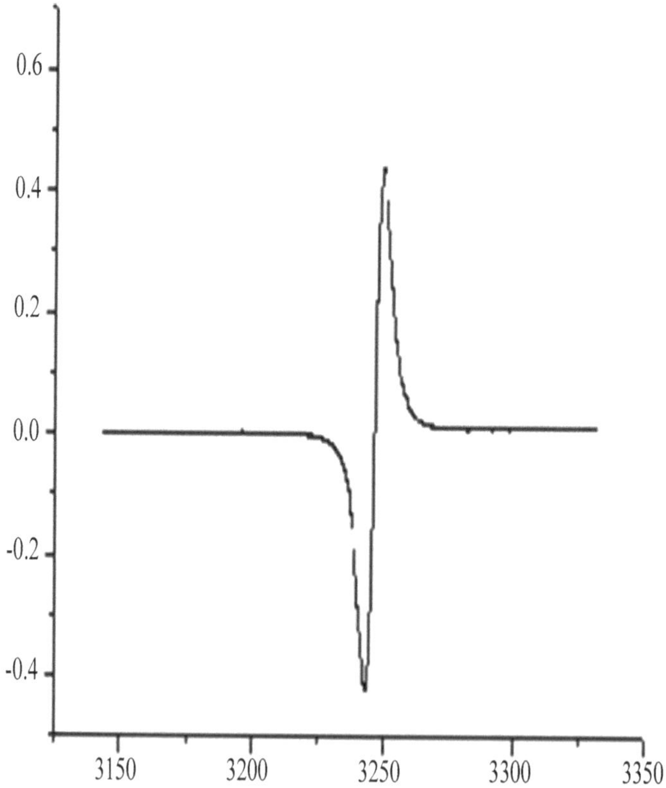

FIGURE 8.5 EPR spectrum of Cu-C mesoscopic composite carbon shell.

The coordination processes lead to changes in metal electron structure with unpaired electrons formation. That is established by an increase of metal atomic magnetic moment and also the appearance of unpaired electrons on the carbon shell surface (please refer to Table 8.4).

TABLE 8.4 Empirical Results for EPR Data and Atomic Magnetic Moments (μ_B) for Copper/Carbon Mesoscopic Composite

Copper carbon mesoscopic composite	g-factor	Number of unpaired electrons, spin/g	Atomic magnetic moment, μ_B copper carbon mesoscopic composite/massive sample
Copper carbon mesoscopic composite	2.0036	1.2×10^{14}	1.3

Based on Ref. [15], the energetic characteristics of copper carbon mesoscopic composite can be shown in Table 8.5.

TABLE 8.5 Energetic Features of Copper /Carbon Mesoscopic Composite

Relation of Copper and Carbon content, %	50/50
The density of Copper Carbon mesocomposite, g/cm^3	1.71
Summary mass of Copper Carbon mesocomposite, au	36.75
The middle size of Copper Carbon mesocomposite, d, nm	25
The specific surface of Copper Carbon mesocomposite, m^2/g	160
Frequency of skeleton vibration of Copper/Carbon mesoscopic composite, s^{-1}	4×10^{11}
Middle vibration energy of Copper Carbon mesoscopic composite, erg	1.6×10^{13}

Source: Adapted from Ref. [15]

Based on our observation, the mesoscopic particles needed to compare to the energetic parameters, for example, on middle vibration energy or the quant's radiation energy for mesoscopic particles.

8.4 REACTIVITY OF COPPER CARBON MESOSCOPIC COMPOSITE AT THE REACTIONS WITH OXIDIZERS

The existence of active double bonds and delocalized electrons in the carbon shell of metal-carbon mesoscopic composites gives a chance for their moderation by means of Red Ox and other processes. For this reason, in such substances interactions, in which elements (Si and P) have the highest oxidation state (+4 or +5) are evident. Such electron-containing compounds as metal/carbon nanocomposites. The scheme of the reduction process is shown below.

$$HK\ (-\delta, \text{or} \sum e) + P^{+5} \rightarrow HK - P^{+3}\ (P^{+2}, P^{0})$$

$$HK\ (-\delta, \text{or} \sum e) + Si^{+4} \rightarrow HK - Si^{0}\ (Si^{+2})$$

Such a mechanism is based on the chemical mesoscopic notions for electron transport across positively charged chemical particles. Therefore the proposed scheme is controlled by x-ray photoelectron spectroscopy (P2p and Si2p spectra).

Based on the P2p spectrum (please refer to Figure 8.6) Phosphorus containing Copper/Carbon mesoscopic composite changes the oxidation state from +5 to zero. The binding energy P2p changes from 135 eV, corresponding to the PO_4 group, to 129 eV for P^0 (The process flows on 90%). In such cases, we can expect that the interaction between Copper and Phosphorus occurs [16]. Meanwhile, from the X-ray pattern analysis of phosphorus-containing Cu-C mesocomposite, we can easily see the presence of peaks for groups Cu-C-P at θ equaled to 43°.

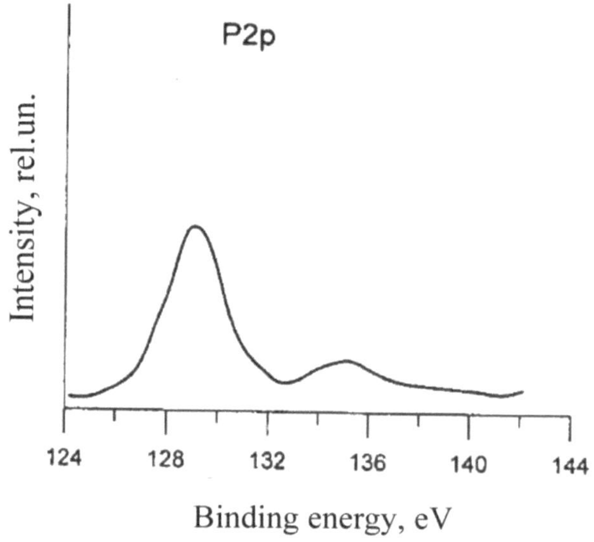

FIGURE 8.6 P2p spectrum of Cu/C mesocomposite modified by ammonium polyphosphate at the relation 1:1.

C1s spectrum of this mesoscopic composite (please refer to Figure 8.7b) is eminent from the C1s spectrum of non-modified mesocomposite (please refer to Figure 7a) [Figue 8.7(a) shows the C1s spectrum of Copper/Carbon mesocomposite, and Figure 8.7(b) shows C1s spectrum of modified P containing Copper/Carbon mesoscopic composite].

The result shows the C–H on 15% smaller than in the spectrum of non-modified mesocomposite. Hence, the relation of intensities for sp^2 and sp^3 hybridization is augmented. Then this can be linked with the mesoscopic granule increasing and approaching its form to roundish.

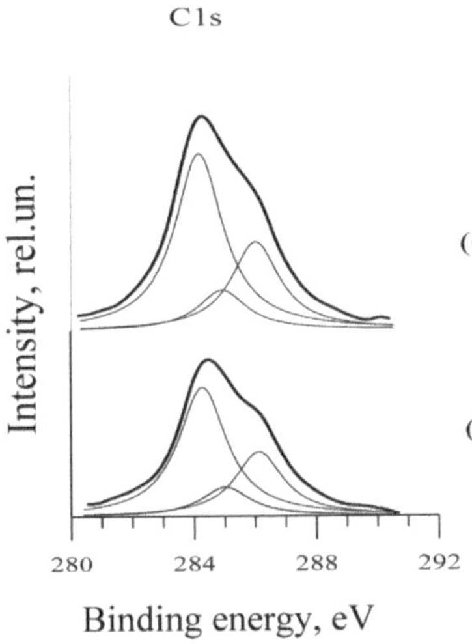

FIGURE 8.7 C1s spectra: (a) C1s spectrum of Copper/Carbon mesocomposite; (b) C1s spectrum of modified P containing Copper/Carbon mesoscopic composite.

Similar studies are carried out in references [17, 18] for Copper/Carbon mesoscopic composite modified by Silicon containing substances. Spectra Si2p and C1s for Cu/C nanocomposite (MC) modified by Silica at the relation MC/Silica = 1 leads to Figure 8.8.

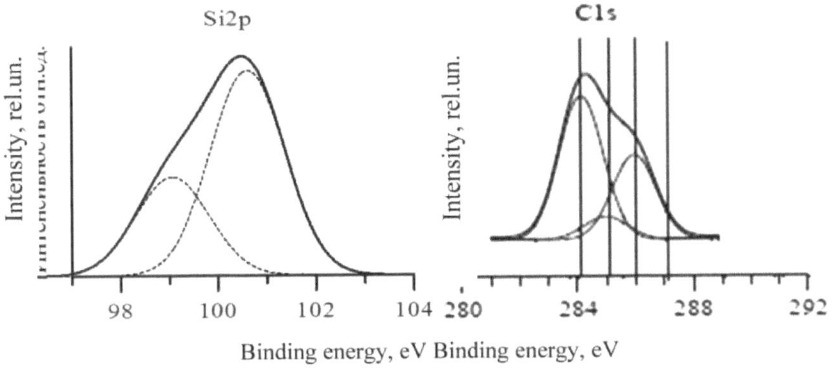

FIGURE 8.8 X-ray photoelectron spectra for modified Cu/C (Si) mesocomposite: (a) Si2p spectrum; (b) C1s spectrum.

Based on the Si2p spectrum, the relation of spectrum form intensities indicated that the reduction-oxidation process developed at 51.4%. C–H intensity in the spectrum of modified Si-containing mesocomposite on 65% smaller than the correspondent value in the spectrum of initial mesocomposite. The thickness of the Si-containing shell for Cu/C (Si) mesocomposite in comparison with a shell of modified P-containing mesoscopic composite is higher than four times.

The studies of modified metal/mesoscopic carbon composites by means of transition electron microscopy (TEM) of energy resolution indicated that the shell from carbon fibers on mesoscopic granules surface is well preserved. As an example, the TEM image of phosphorus-containing Cu-C mesocomposite cab be seen in Figure 8.9.

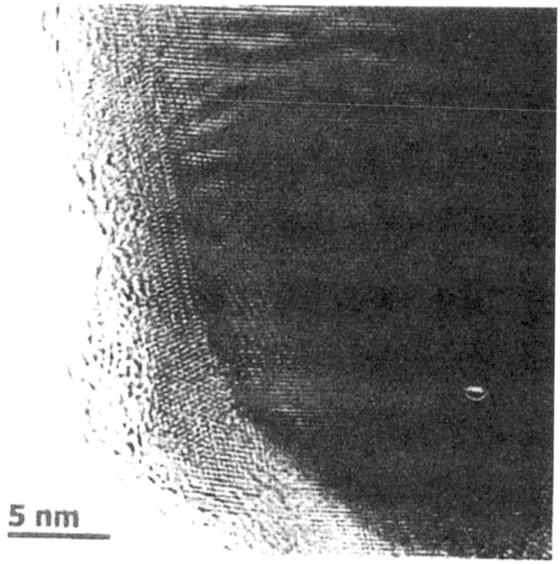

FIGURE 8.9 Transmission electron microscopy image of phosphorus-containing copper carbon mesoscopic composite.

At the reduction process, the Copper atomic magnetic moment growth is observed (please refer to Table 8.6).

This is accompanied by an increase in unpaired electron values (Table 8.7).

The observed results cause natural questions regarding the reason for magnetic characteristics growth for metal atoms in the clusters within the nanocomposite granules. The correct response to this question may be found in the following hypothesis:

- The hypothesis is concluded in the formation of delocalized (unpaired) electrons because the electrons shift on high energetic levels at the action electromagnetic radiation (field).
- It's possible the appearance of this radiation may be conditioned by the annihilation of quants of positive and negative charges, which are appeared in the redox process at mechanochemical modification of metal/carbon nanocomposites.
- As a matter of fact, this hypothesis is necessary to confirm by corresponding experiments.

It's worth mentioning that these processes are realized by mechanochemical methods, which can be a perspective for mesoscopic composite production.

TABLE 8.6 The Values of Copper Atomic Magnetic Moment in the Interaction Products for Systems Cu-C NC – APPh (or SiO_2)

Systems Cu/C NC – substances	μ_{cu}
Cu/C NC – Silica	3.0
Cu/C NC – APPh	2.0
Cu/C NC - APPh, relation 1:0,5	4.2

TABLE 8.7 The Unpaired Electron Values (from EPR spectra) for Systems "Cu/C NC–silica" and "Cu/C NC – APPh" (relation 1:1) in Comparison with Mesoparticle Cu/C NC

Substance	Quantity of unpaired electrons, spin/g
Cu-C mesocomposite	1.2×10^{17}
system «Cu/C NC – SiO_2»	3.4×10^{19}
system «Cu/C NC – APPh»	2.8×10^{18}

8.5 MECHANOCHEMICAL METHOD

The magnetic mesoparticles obtaining the possibility to the regulation of metal magnetic atomic moment is based on the combination of third variants of action on mesosystems formed on the boundary of reagents phases because of sign variable loadings (Figure 8.10).

The modification of different materials by metal-carbon mesoscopic composites is also studied in Refs. [20–23]. The theoretical and empirical methods indicate that the mesoscopic composites' electromagnetic field action on the medium molecules ends up in the formation of nanostructured fragments in the material composition (Figure 8.11).

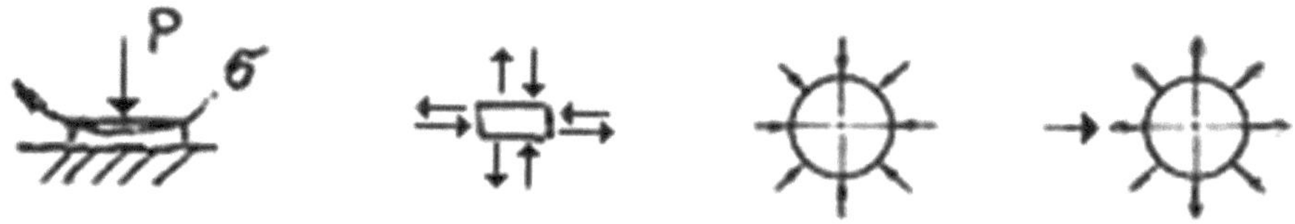

FIGURE 8.10 Mechanic loadings on reactive mesoscopic systems (Left–the combination of pressure with displacement loadings, Middle–mechanical pulsation loadings, Right–the pressure increasing with next momentary decreasing).

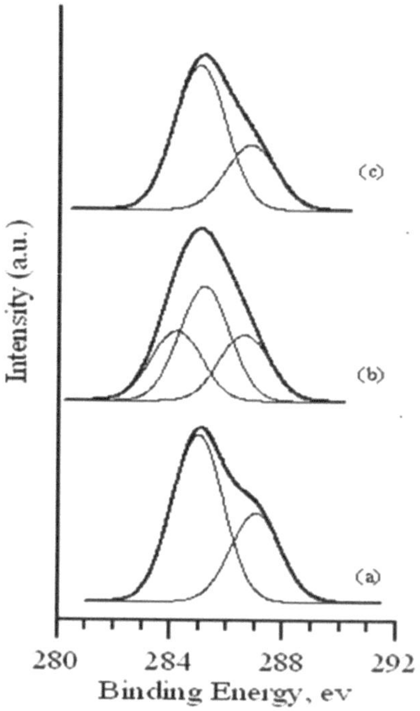

FIGURE 8.11 C1s spectra for polyvinyl alcohol modified by: (a) Cu-C mesocomposite in comparison with initial mesoscopic composite; (b) 0.001%; (c) 0.0001%.

8.6 CONCLUSION

The process studied in this book chapter, in general, is expected to happen among microscopic particles of metal oxides along with the macromolecules of polymers in the active medium presence.

Thereafter, and during the production process of the mesoscopic composites, changeable loadings are applied. It should be noted that these loadings became available at the grinding with successive pressing through the existing pores in the stream of inert heat gas. This stream moves through the nozzle into the vapor phase, which carries a protective polymer solution.

KEYWORDS

- **Chemical Mesoscopics**
- **Electron Transport on High Energetic Levels**
- **Metal Atomic Magnetic Moment**
- **Metal Carbon Mesoscopic Composite**
- **Quantization Spectrum**
- **Reduction Oxidation Processes**

REFERENCES

1. Imri, I. (2009). Introduction in Mesoscopic Physics. *Physmatlit.* 304p.
2. Moskalets, M. V. (2010). *Fundamentals of Mesoscopic Physics*. Khar'kov: NTU KhPI, 186p.
3. Kodolov, V. I., & Khokhriakov, N. V., (2009). *Chemical Physics of Formation and Transformation Processes of Nanostructures and Nanosystems*–Izhevsk: Publ. IzhSACA,– Vol. 1, 361 p; Vol. 2, 415p.
4. Kodolov, V. I., & Trineeva, V. V. (2017). New scientific trend–Chemical Mesoscopics. *Chemical Physics & Mesoscopics, 19*(3), 454–465.
5. Morozov, A. D. (2002). *Introduction in Fractal Theory*. M.-Izhevsk: ICT, 160p.
6. Kolmogorov, A. N., & Fomin, S. V. (2009). *Introductory Real Analysis*. USA, Portland: Prentice Hall, 403p.
7. Wunderlikh, B. (1979). *Physics of Macromolecules*. Vol. 2, M.: Mir, 422p.
8. Kodolov, V. I., Khokhriakov, N. V., Trineeva, V. V., Blagodatskikh, I. I. (2008). Nanostructure activity and its display in nanoreactors of polymeric matrices and in active media. *Chemical Physics and Mesoscopy, 10*(4), 448–460.
9. Kodolov, V. I. (2009). The addition to previous paper – *Chemical Physics and Mesoscopy, 11*(1), 134–136.
10. Trineeva, V. V., Vakhrushina, M. A., Bulatov, D. I., & Kodolov, V. I. (2012). The obtaining of metal/carbon nanocomposites and investigation of their structure phenomena. *Nanotechnics, 4*, 50–55.
11. Trineeva, V. V., Lyakhovich, A. M., & Kodolov, V. I. (2009). Forecasting of the formation processes of carbon metal containing nanostructures using the method of atomic force microscopy. *Nanotechnics, 4*(20), 87–90.

12. Kodolov, V. I., Blagodatskikh, I. I., Lyakhovich A.M. et al. (2007). Investigation of the formation processes of metal containing carbon nanostructures in nanoreactors of polyvinyl alcohol at early stages. *Chemical Physics and Mesoscopy, 9*(4), 422–429.
13. Kodolov, V. I., Trineeva, V. V., Blagodatskikh, I. I., Vasil'chenko Yu. M., Vakhrushina, M. A., & Bondar, A. Yu. (2013). The nanostructures obtaining and the synthesis of metal/carbon nanocomposites in nanoreactors. In *Nanostructure, Nanosystems and Nanostructured Materials: Theory, Production and Development*. Toronto-New Jersey: Apple Academic Press, pp. 101–145.
14. Kodolov, V. I., & Trineeva, V. V. (2013). Fundamental definitions for domain of nanostructures and metal/carbon nanocomposites. In *Nanostructure, Nanosystems and Nanostructured Materials: Theory, Production, and Development*. Toronto-New Jersey: Apple Academic Press, pp. 1–42.
15. Kodolov, V. I., & Trineeva, V. V. (2015). Energetic characteristics of Metal/Carbon nanocomposites. *JCDNM, 7*(2), 223–228.
16. Kodolov, V. I., Trineeva, V. V., Kopylova, A. A. et al. (2017). Mechanochemical modification of metal/carbon nanocomposites. *Chemical Physics & Mesoscopics, 19*(4), 569–580.
17. Shabanova, I. N., Kodolov, V. I., Terebova, N. S., & Trineeva, V. V. (2012). *X-Ray Electron Spectroscopy in Investigations of Metal/Carbon Nanosystems and Nanostructured Materials*. M.-Izhevsk: Publ. "Udmurt University," 252p.
18. Kodolov, V. I., Trineeva, V. V., Terebova, N. S. et al. (2018). Changes of electron structure and magnetic characteristics of modified copper/carbon nanocomposites. *Chemical Physics & Mesoscopics, 20*(1), 72–79.
19. Kopylova, A. A., & Kodolov, V. V. (2014). Investigation of coper/carbon nanocomposite interaction with silicium atoms from silicon comp ounds. *Chemical Physics & Mesoscopics, 16*(4), 556–560.
20. Shabanova, I. N., & Terebova, N. S., (2012). Dependence of the value of the atomic magnetic moment of d metals on the chemical structure of nanoforms. In *The Problems of Nanochemistry for the Creation of New Materials*, Torun, Poland: IEPMD, 123–131.
21. Kodolov, V. I., Khokhriakov, N. V., & Kuznetsov, A. P. (2006). To the issue of the mechanism of the influence of nanostructures on structurally changing media at the formation of "intellectual" composites. *Nanotechnics, 3*(7), 27–35.
22. Kodolov, V. I., Khokhriakov, N. V., Trineeva, V. V., & Blagodatskikh, I. I. (2010). Problems of Nanostructure Activity Estimation, Nanostructures Directed Production and Application. *Nanomaterials Yearbook–2009: From Nanostructures, Nanomaterials and Nanotechnologies to Nanoindustry*. N. Y.: Nova Science Publishers, Inc., 1–18.
23. Chashkin, M. A., Kodolov, V. I., Zakharov, A. I., et al. (2011). Metal/carbon nanocomposites for epoxy compositions: quantum-chemical investigations and experimental modeling. *Polymer Research Journal, 5*(1), 5–19.

CHAPTER 9

Metal-Carbon Mesocomposites Application Possibilities as Magnetic Mesoscopic Materials

V. I. KODOLOV,[1,2] V. V. KODOLOVA-CHUKHONTZEVA,[1,3]
YU. V. PERSHIN,[1,2] and R. V. MUSTAKIMOV[1,2]

[1]*Basic Research–High Educational Center of Chemical Physics and Mesoscopic, UD, RAS, Izhevsk, Russia*

[2]*M.T. Kalashnikov Izhevsk State University, Izhevsk, Russia*

[3]*Peter Great St. Petersburg Polytechnic University, St. Petersburg, Russia*

ABSTRACT

The chapter is dedicated to the consideration of the metal-carbon mesocomposites (Me–C MCs) application possibilities for the different practical trends. Each of these trends is determined by correspondent peculiarities of content and structure of mesoscopic composites. The main peculiarities of these nanosized particles are the following: (i) the presence of unpaired electrons on the carbon shell; (ii) the structure of the carbon shell from poly acetylene and carbine fragments; and (iii) the atomic magnetic moment of inner metal is equaled to more than 1–3 μB. The creation of reactive mesoscopic materials with regulated magnetic characteristics that can be used as modifiers of materials properties, catalysts for different processes, and effective inhibitors of oxidation processes, including corrosion processes, sorbents, and plant growth stimulators as magnetic medicine transport within organisms, is very topical. The present investigation has a fundamental character. It's based on the ideas concerning the change of Me–C MC reactivity. The using is possible

Modern Magnetic Materials: Properties and Applications. Iuliana Stoica, Ann Rose Abraham, & A. K. Haghi (Eds.)

as Me–C MC both, and they have modified analogously. The application examples are presented for the property's improvement of concretes, polymer compounds, and plastics and for the increasing positive qualities of such substances as glues, corrosion inhibitors, fire retardants, stimulators of plant growth, and magnetic medicine transport.

9.1 INTRODUCTION

The manuscript is presented as the review of a series of papers, manuscripts, and patents on the obtaining, investigations, and applications of uncials magnetic mesoparticles which are mesoscopic metal-carbon composites [1–41]. The correspondent structure and content of these mesocomposites (MCs) are caused by the special conditions of their production. In as much as the application of mesoparticles is determined by the properties of these particles which are defined by their structure and content. At the beginning the initial metal-carbon mesocomposite (Me–C MC) electron structure and correspondent magnetic characteristics (metal atomic magnetic moments, and the spin quantities on carbon shell) are considered. The activity of metal-carbon mesoscopic composites is caused by the structure and composition of corresponding MCs, which contain the delocalized electrons and double bonds on the surface of the carbon shell. Therefore, the initial mesoscopic composites are early participated in reactions, especially radical processes and reduction oxidation processes. This activity may be used in modification processes MCs accompanied by the magnetic characteristics changes in modified mesoscopic composites. That development of possibilities processes opens a new era for further investigations and development of metal-carbon mesoscopic composites application fields. The expansion of Me–C MC application chances takes place. Seven trends of MCs application are presented in succeeding sections.

9.2 THE GLUES AND ADHESIVES ON THE BASE OF MAGNETIC METAL-CARBON MESOSCOPIC COMPOSITES

The presence of unpaired electrons and double bonds in the carbon shell of Me–C MC guarantees the additional conditions for the adhesion increasing especially at the metal materials connection. In these cases, the positive meaning has the presence of above MCs magnetic properties. For example, the introduction of nickel–carbon MC in the composition of Silver containing current conductive glue leads to the improvement of adhesion and electric conductivity (Figures 9.1 and 9.2; Table 9.1).

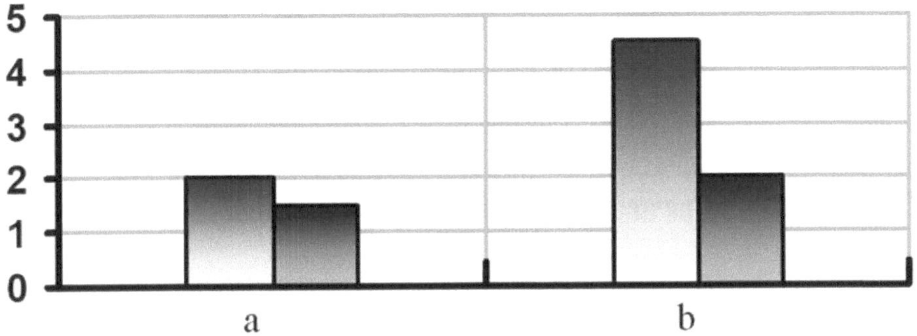

FIGURE 9.1 The adhesion durability on shift for Ag containing glue (pink) and paste (blue), initial (a) and modified by nickel–carbon mesocomposite.

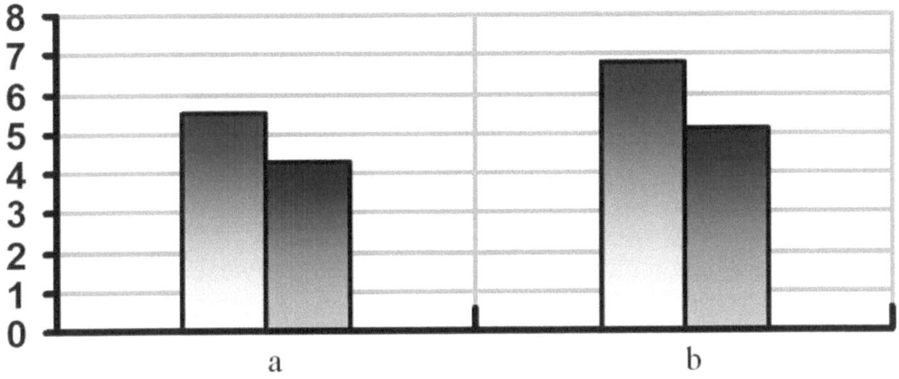

FIGURE 9.2 The adhesion durability on separation for Ag containing glue (pink) and paste (blue), initial (a) and modified by nickel–carbon mesocomposite.

TABLE 9.1 The Measuring Results of the Current Conductive Paste (Glue) Electro-Resistance, Modified by Nickel–Carbon Mesocomposite

Characteristics	Current Conductive Paste	Current Conductive Glue
Specific volume electro-resistance, Om·cm (initial sample)	2.4×10^{-4}	3.6×10^{-4}
Specific volume electro-resistance, Om·cm (modified sample)	2.2×10^{-5}	3.3×10^{-5}

The specific volume electro-resistances for current conductive pastes in comparison with current conductive glues are presented in Table 9.1. From the comparison of the specific volume resistances of pastes and glues it's shown that the pastes on the electroconductivity better than the correspondent glues.

Thus, the Me–C MCs can be applied for the improvement of the current conductive materials characteristics.

Analogous results are obtained at the modification of cold hardened epoxy resins (ERs) by the Me–C MC. According to the investigation on the modification results of cold hardened ERs the following conclusion may be made:

> "The test for defining the adhesive strength and thermal stability correlate with the data of quantum-chemical calculations and indicate the formation of a new phase facilitating the growth of cross-links number in polymer grid when the concentration of Cu–C MC goes up. The optimal concentration for elevating the modified ERs adhesion equals 0.003% from ER weight. At this concentration the strength growth is 26.8%. At the same time, the optimal quantity of Cu–C MC for elevating the modified industrial epoxy materials adhesion equals 0.005% that leads to the strength growth equals 60.7%. From the concentration range studied, the concentration 0.05% from ER weight is optimal to reach a high thermal stability. At this concentration the temperature of thermal destruction beginning increases up to 195°C."

The modification of hot hardened ERs by means of copper–carbon MCs is carried out with the application of the finely dispersed suspension based on isomethyl tetra phthalic anhydrate or based on toluene. After testing the samples of four different schemes, the increase in the strength at detachment σ_{det} up to 50% and shear τ_{sh} up to 80% takes place, the concentration of copper–carbon MC introduced corresponds to 0.0001–0.0003%.

The application of these materials as adhesives for the gluing of metals and vulcanite is realized on the schemes "metal$_1$–adhesive$_1$–vulcanite–adhesive$_2$–metal$_2$." To define the adhesive tear and shear strengths the above proposed scheme was used (Figures 9.3 and 9.4). The investigations carried out revealed that the modification of the conventional recipe of the glue 51-K-45 results not only in increasing the glue adhesive characteristics but also in changing the decomposition character from adhesive-cohesive to cohesive one.

The availability of metal compounds in MCs can provide the final material with additional characteristics, such as magnetic susceptibility and electric conductivity.

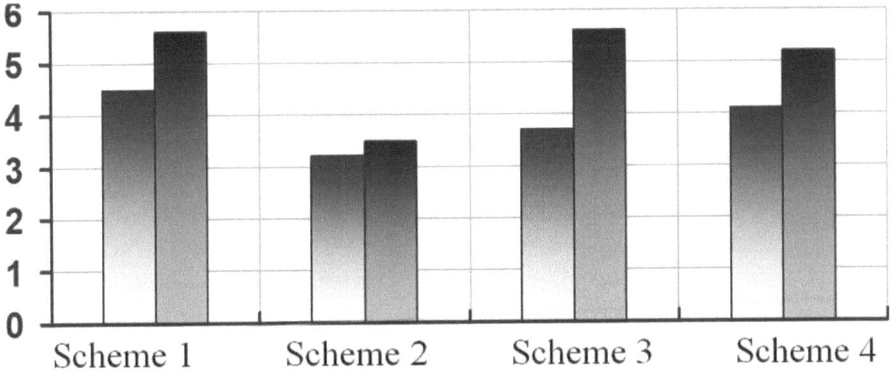

FIGURE 9.3 Relative changes of adhesive tear strengths of epoxy glues modified by copper–carbon mesocomposites (content of MC – 0.0001%).

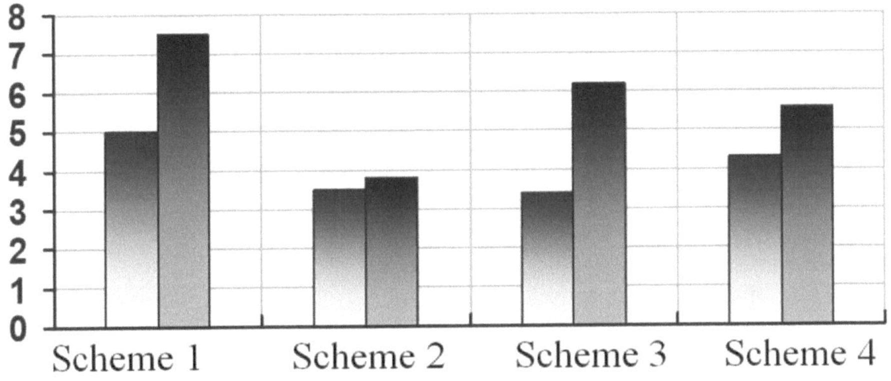

FIGURE 9.4 Relative changes of adhesive shear strengths of epoxy glues modified by copper–carbon mesocomposites (content of MC–0.0001%).

The modification of different materials by minute quantities of Me–C MC allows improving their technical characteristics, decreasing material consumption and extending their application.

9.3 MAGNETIC TRANSPORT OF MEDICINES WITHIN ORGANISM FROM METAL-CARBON MESOCOMPOSITES (ME–C MCS)

At last time the great attention is spared to the creation of medicine remedies with address direction for the action on the organs which are needed in correspondent healing. Usually, the transport of therapeutically active substances to the correspondent organ is carried out by means of the magnetic mesoparticles containing the linker connected with medicine substance. In other words, the remedies of medicine carriage in organisms are constructed on the following scheme: transport magnetic mesoparticle-linker-medicine [42]. In above patents the iron containing nanosized particles, for example, Fe_3O_4, are used as magnetic mesoparticle, and the organic substances connected with this mesoparticle by covalent or coordinative bonds as linker. In this case it's possible the certain difficulties on the undoing of medicine because of the bio active substances big interaction with some functional groups. The medicine release from magnetic mesoparticles with linker occurs at the variable magnetic fields. The best linker can be phosphorus organic compounds which, as it's known, are easily destructed in water media. Therefore, its proposed [34] to accomplish the modification of metal containing magnetic mesoparticle by ammonium polyphosphate (APPh). At the same time, the copper–carbon MC [37] is proposed as a magnetic mesoparticle since the Copper has bactericides and anti-microbes' properties and increases the organism's protective forces. For that reason, the copper–carbon mesoscopic composite modified by APPh chooses as the investigations object. The modification process is carried out by mechanical chemical method in processing which the Phosphorus reduction and the formation of linker which consists of phosphorus with the following oxidation states such as 0, +3, and +5 takes place.

The obtaining of remedy for medicine delivery to definite organ is carried out on the following scheme:

> "The phosphorus containing copper–carbon mesoscopic composites ($mc_{(P)}$) is applied as magnetic mesoparticles. Then the medicines (M) are linked with $mc_{(P)}$ at the relation ($mc_{(P)}/M$) correspondent to 1:(0.02–0.5) by means of mechanic chemical method with the using the mechanical mortar. The $mc_{(P)}$ applied are obtained by the interaction of copper–carbon mesocomposite (Cu–C MC) and ammonium polyphosphate (APPh) at the relation 1:0.5 for the obtaining of high atomic magnetic moment of copper. The therapeutically active substances such as adenosine tri phosphorus acid (the relation $mc_{(P)}/M$ = 1:0.02), ascorbic acid ($mc_{(P)}/M$ = 1:0.2) and Urotropine ($mc_{(P)}/M$ = 1:0.5) is linked with phosphorus containing MC across the phosphorus containing liker."

The investigations of phosphorus containing copper–carbon MC and its analogous are realized with the application of methods complex from which the basic methods are X-ray photoelectron spectroscopy (XPS), transition electron microscopy (TEM) of high solution (permeation) (TEM), electron micro-diffraction, electron paramagnetic resonance (EPR), X-ray diffraction measuring.

According to these investigations, the copper atomic magnetic moment growth is established at the copper–carbon MC modification by APPh and therapeutically active substances. The atomic magnetic moment of copper is obtained more ($\mu = 4.5\ \mu_Б$) in the comparison with iron containing nanostructures ($\mu = 2.5\ \mu_Б$). Hence the proposed magnetic mesoparticles can be interested as the remedies of medicine carriage in organisms by means of drive magnetic field.

The production of magnetic mesoscopic particles with connected therapeutically active substances is realized by mechanic chemical method on the next scheme:

- The reduction-oxidation synthesis of copper–carbon MC from CuO and PVA at the reagents relation equaled to 1 mol:4 mol [31].
- The copper–carbon MC modification by APPh at the reagent's relation [33, 39].
- The phosphorus containing copper–carbon MC modification by therapeutically active substances, such, as adenosine tri phosphoric acid (relation–1:0.02), ascorbic acid (relation–1:0.2) and Urotropine (relation–1:0.5) (Pat. 2018143197).

The mechanic chemical synthesis is carried out with the use of mechanical mortar by the joint grinding of reagents at the energetic expenses approximately equaled to 260–270 kJ/mol. After mechanic chemical process the mesoscopic product obtained is dried in the closed crucible at 400°C for the first stage, and at 150°C for second and third stages. Then the product obtained is endured in vacuum at 100–150°C for 3 minutes.

The results of mechanic chemical process with thermochemical finishing are estimated with the application of the following methods: XPS, EPR, TEM with high permission.

The transition analytic electron microscope FET Tecnai G2F20 with prefix EDAX is used for the investigation of copper–carbon mesoscopic composite structure and phase content. High permission corresponds to 2 nm in 1.5 cm.

X-ray photoelectron spectroscopic investigations are realized by X-ray photoelectron magnetic spectrometer with the permission 10^{-4} at the activation AlK_α line 1,486 eV in vacuum 10^{-8}–10^{-10} Torre. On the basis of Van Fleck theory, the model for metal atomic magnetic moments calculation is proposed.

EPR investigations are carried out by means of EPR spectrometer E-3 of firm "Varian." The peak correspondent to binding energy 132.5 eV is ascribed [43] bond C=P that can be possible at the appearance of interference phenomenon because of the direct electromagnetic (EM) field which arises at annihilation in red-ox process. The unpaired electrons presence on the carbon shell of MC can promote the decreasing of free radical activity in the defeat part of the organism. The high atomic magnetic moment of copper (more than 4 μ_B) can be used for the address carriage of medicine.

Thus, the transport remedy included the copper–carbon MC and linker containing phosphorus for the connection of therapeutically active substances is proposed.

The investigations of medicines modified by the above magnetic MCs are carried out with the application of XPS. C1s spectra for these medicines are presented in Figure 9.5.

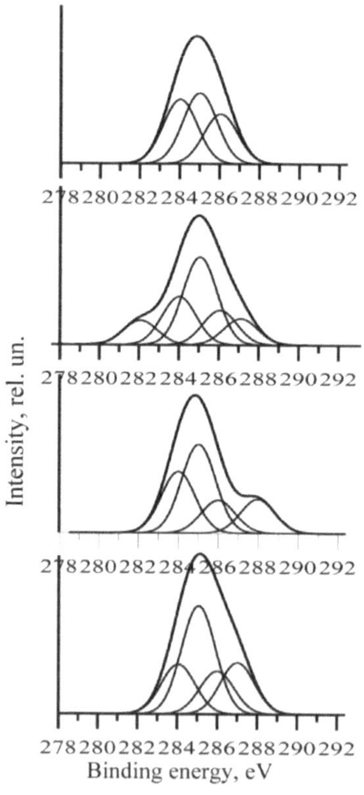

FIGURE 9.5 C1s spectra phosphorus containing copper–carbon mesocomposite (a); and its analogous modified adenosine tri-phosphorus acid (relation 1:0.02) (b); ascorbic acid (relation 1:0.2) (c); and urotropine (relation 1:0.5) (d).

The discussion of the electron structures changes for the modified bioactive substances takes place below in the following order: (i) ATP; (ii) AA; and (iii) U.

After the modification of phosphorus-containing copper–carbon MC, obtained at the relation Cu–C MC/APP equaled to 2, by adenosine-tri-phosphorus acid (relation 1:0.02) the following changes of the mesoparticle surface structure takes place:

- In C1s spectrum the maximum for C–H bond is considerably increased;
- In this spectrum the appearance new content at 283 eV correspondent to *sp* hybridization, and the maximum for sp^2 (C–C) and also sp^3 (C–C) hybridization are decreased;
- In C1s spectrum new maximum (286.7 eV) for C–N bond is observed;
- In Cu2p and P2p spectra the peaks for Cu–P and Cu–P–O are found.

It's necessary to note that the magnetic moment determined on Cu3s spectrum did not exchange and is equaled to 4.5 μ_B.

Above changes in spectra can be explained by the carbon phosphorus shell deformation at the interaction of adenosine tri phosphorus acid with transport remedy.

In the X-ray photoelectron spectra of mesoparticles obtained by means of the phosphorus containing MC modification by ascorbic acid at the reagent's relation 1:0.2 the following changes are discovered:

- In C1s spectrum in comparison on the spectrum of initial MC the new maximum (E_b = 288 eV) for C–OH group is present;
- In Cu2p and P2p spectra the maximums for Cu–P, Cu–O and P–O bonds are observed.

The magnetic moment determined on Cu3s spectrum also as in above example is not changed and equaled to 4.5 μ_B.

The changes in X-ray photoelectron spectra of phosphorus containing Cu–C MC modified by Urotropine at the relation 1:0.5 also take place. The following peculiarities in spectra are given:

- In C1s spectrum the C–H maximum increasing is found;
- In the same spectrum the maximum for C–N bond (E_b = 287 eV) is appeared, and the maximums attributed to sp^2 and sp^3 hybridization have near intensity;
- In P2p spectrum the maximums at 135 eV (P^{+5}) and 129 eV (P^0) take place;
- In N1s spectrum the covalent bond C–N is determined.

In this case the copper atomic magnetic moment is near to 4.5 μ_B. It's interesting to note that the carbine component (sp hybridization) is observed at the MC quantity increasing.

Thus, the above said examples show on the connection of therapeutically active substances with transport remedy (phosphorus containing copper–carbon MC) at the conservation of magnetic properties. Latter is necessary for the application of the obtained substances in practice.

9.4 STIMULATORS FOR PLANTS GROWTH WITH ADDITIVES OF MAGNETIC METAL-CARBON MESOCOMPOSITES (ME–C MCS)

The metal-carbon mesoscopic composites owing to structure and properties can be effective for the increasing and improvement of agricultural production. These substances can be initiators or inhibitors for the definite vital processes in different fields of agriculture. In this chapter, as examples the stimulation of plants growth under the action of Me–C MC are considered. As examples of plants the lilies are discussed. In this case the bulbs of lilies came in peat; they were packed in film bags, from a dense film that had 18 holes 1 cm in diameter for air intake. Packages with lily bulbs were stored in container boxes at a temperature of –1°C, which allows them to remain at rest. Samples of this peat were sampled, and agrochemical analysis was carried out [35] using a volumetric analysis of the preparation of aqueous extract (with recalculation of the results of the analysis for volume) and a sample with recalculation of the results of the analysis for dry weight (Table 9.2).

The moisture content of the peat corresponds to the normative indices. According to the degree of acidity, peat refers to a very acidic. The specific electrical conductivity of peat does not exceed 1.0 mS/cm. Peat has a low content of nitrogen, phosphorus, and potassium in terms of the degree of supply of food elements. Such properties of this peat contribute to the long storage and transportation of bulbs of long lilies and do not allow them to germinate.

For the growth of plants on top peat, the optimum moisture content is within 78–85% of the mass. The moisture content in the peat mixture is normal.

A very important factor for the development of roots and the absorption of nutrients by them is the acidity of the soil. High acidity of soil leads to insufficient intake of such elements as phosphorus, magnesium, and iron. When growing Oriental hybrids of lilies, which include our variety, the pH should be between 5.0 and 6.5.

TABLE 9.2 Agrochemical Characteristics of Peat During Storage of Lily Bulbs Imported from Holland

Index	Result	The Value of the Horse Peat Sphagnum
Moisture of peat (%)	60.0	59.46
pH (suspension)	4.0	5.5–6.1
Specific conductivity of the suspension (mS/cm), at 25°C	0.093	–
The Content of Mobile Forms of Batteries		
Nitrate nitrogen (N-NO_3)	0.57	26.3
	0.1	29.0
Nitrogen ammonium (N-NH_4)	7.36	135.6
	1.5	149.0
P_2O_5	1.64	291.0
	0.3	320.0
K_2O	34.56	353.0
	7.0	388.0
Ca^{2+} + Mg^{2+} (mmol/g)	2.98	–
	0.6	

Note: Numerator (for dry matter) – mg/100 g; denominator – mg/l.

Lilies are sensitive to soil salinity. With a high content of salts in the soil, the roots of the lilies become hard, brittle, and acquire a yellow-brown color. Soil salinity is determined by the specific electrical conductivity. Lilies require low EU levels.

The content of mobile forms of nutrients is low, but the level of plant nutrition is regulated by a computer program in accordance with the needs of lily bulbs. Thanks to the inherent peat strength of buffer and high sorption capacity, mineral fertilizers are not washed out and stored in an accessible form for plants; at the same time, the danger of creating an increased concentration of salts harmful to plants is reduced.

High productivity and quality of flowers largely depend on the full provision of plants with trace elements, so when planting lily bulbs, an analysis was carried out for the content of trace elements in the peat mix (Table 9.3).

According to the content of trace elements, peat mixes contain a very low content of such elements as zinc, copper, iron, manganese, and boron [8]. By processing bulbs of lilies, copper–carbon MC, we contribute to replenishing available forms of copper in lily plants. In the period of growth and development, the level of nutrition of lily plants is regulated by a computer program in accordance with the needs of macro- and microelements.

TABLE 9.3 Chemical Analysis of Peat Mixture for Micronutrients Content

Microelements (mg/kg Dry Weight)	The Actual Value of the Test Results
Zinc	15.60
Copper	3.81
Iron	126.60
Manganese	57.30
Boron	8.27

For two years, the reaction of the lily to the treatment of copper–carbon MC was studied. The biometric parameters of the lily were determined as the height of the flower bud in the bud budding phase (Table 9.4), the number of buds, the diameter of the open flower, and the height of the stem when cutting plants.

When processing with MCs, there was a significant increase in the height of the flowering shoot of lilies from 4.9 to 8.4 cm (at HSR = 1.7) (Table 9.4). The highest plants were Santander lily plants when treated with a 0.01% MC.

TABLE 9.4 Height of Flowering Shoot (cm) (Average for 2015–2016, the Phase of the Beginning of Budding)

Hybrids of Lilies (Factor A)	Concentration of Mesocomposites (Factor B)					Average by Factor A HSR_{05} A = 1.7
	Without Processing (k)	Water (к)	0.01 (%)	0.02 (%)	0.05 (%)	
Siberia (к)	79.6	71.9	87.7	87.9	87.2	82.8
Santander	77.9	81.6	89.1	82.2	82.3	82.6
Average by Factor B HSR_{05} B = 2.7	79.2	76.9	88.5	85.2	84.6	–
$HSR_{05\ \text{private differences}}$	3.8					

The number of buds on the shoot was higher for the Siberia variety, than for the Santander variety by 1.1 pcs. (at HSR = 0.06 pcs.). When processing with MCs, there was a significant increase in the number of buds to 5.0–5.3 pieces (Table 9.5).

The length of the bud in Siberia variety averaged 10.5 cm, while the Santander variety was 1.2 cm lower (Table 9.6). The largest length was found in lilies treated with a 0.01% copper–carbon MC suspension.

TABLE 9.5 Number of Buds, pcs (Average for 2015–2016, the Phase of the Beginning of Budding)

Hybrids of Lilies (Factor A)	Concentration of Mesocomposites (Factor B)					Average by Factor A HSR_{05} A= 0.06
	Without Processing (k)	Water (к)	0.01 (%)	0.02 (%)	0.05 (%)	
Siberia (к)	4.7	4.7	5.8	5.6	5.4	5.2
Santander	3.5	3.5	4.7	4.5	4.6	4.1
Average by Factor B HSR_{05} B = 0.1	4.1	4.1	5.3	5.1	5.0	–
$HSR_{05\ private\ differences}$	0.1					

TABLE 9.6 Length of Bud, cm (Average for 2015–2016, the Phase of the Beginning of Budding)

Hybrids of Lilies (Factor A)	Concentration of Mesocomposites (Factor B)					Average by Factor A HSR_{05} A=0.06
	Without Processing (k)	Water (к)	0.01 (%)	0.02 (%)	0.05 (%)	
Siberia (к)	10.1	10.3	10.9	10.6	10.6	10.5
Santander	9.0	9.1	9.6	9.5	9.3	9.3
Average by Factor B HSR_{05} B = 0.09	9.5	9.7	10.3	10.0	9.9	–
$HSR_{05\ private\ differences}$	0.1					

The length of the bud and the number of buds depends on the photosynthetic activity, which is directly related to the width of the leaves, so the width of the lily leaf was measured during the bud budding phase (Table 9.7).

The variety Santander leaves was wider than that of Siberia, which is one of the varietal traits. The increase in the number and length of buds in both varieties when processing bulbs with copper–carbon MC was associated with an increase in the width of the leaves.

The productivity of the lily (lily cut) is determined by the number of buds in the inflorescence, the diameter of the flower and the height of the plant. In this regard, the plants of lilies count the number of buds when cutting plants (Table 9.8).

TABLE 9.7 The Width of the Leaf, cm (Average for 2015–2016, the Phase of the Beginning of Budding)

Hybrids of Lilies (Factor A)	Concentration of Mesocomposites (Factor B)					Average by Factor A HSR_{05} A = 0.05
	Without Processing (k)	Water (к)	0.01 (%)	0.02 (%)	0.05 (%)	
Siberia (к)	3.0	3.1	3.4	3.3	3.1	3.2
Santander	3.4	3.5	3.7	3.6	3.6	3.6
Average by Factor B HSR_{05} B = 0.08	3.2	3.3	3.6	3.5	3.4	–
$HSR_{05\ private\ differences}$	0.1					

TABLE 9.8 Number of Buds When Cutting Plants, pcs (Average for 2015–2016)

Hybrids of Lilies (Factor A)	Concentration of Mesocomposites (Factor B)					Average by Factor A HSR_{05} A = 0.04
	Without Processing (k)	Water (к)	0.01 (%)	0.02 (%)	0.05 (%)	
Siberia (к)	4.8	4.7	6.0	5.6	5.5	5.3
Santander	3.6	3.5	4.7	4.5	4.6	4.2
Average by Factor B HSR_{05} B = 0.06	4.2	4.1	5.4	5.0	5.0	–
$HSR_{05\ private\ differences}$	0.09					

When cutting lilies, the number of buds on the plant remains the same as in the bud budding phase. There were also a greater number of buds in the Siberia lily variety than in Santander. Depending on the treatment with various concentrations, the maximum number of buds was noted at 0.01% concentration of copper–carbon MC.

The diameter of the open flower in the Santander variety was significantly larger than that of the Siberia variety (21.3 cm) (Table 9.9). Processing of bulbs with MCs promoted a substantial increase in the diameter of the open flower. The largest diameter of the flower was noted when treated with 0.01% MC (22.0 cm).

The height of the stem when cutting lilies increased by treatment with copper–carbon MC to 131–141 cm in comparison with the control (99.0–102.3 cm). The sort of lily Siberia was slightly higher than the Santander variety.

Thus, the treatment of bulb lily copper–carbon MCs contributes to the increase in plant height, the number and diameter of the flowers.

According to the results of scientific and industrial experiments, processing of bulb lily copper–carbon MCs promotes an increase in the height of the flower sprout, the height of the stem during the cutting of plants, the number of buds and the diameter of the open flowers.

TABLE 9.9 Diameter of the Open Flower, cm (Average for 2015–2016, Cutting Plants)

Hybrids of Lilies (Factor A)	Concentration of Mesocomposites (Factor B)					Average by Factor A HSR_{05} A = 0.3
	Without Processing (k)	Water (к)	0.01 (%)	0.02 (%)	0.05 (%)	
Siberia (к)	18.7	18.3	21.3	20.0	20.0	19.7
Santander	20.0	20.0	22.7	22.0	21.7	21.3
Average by Factor B HSR_{05} B = 0.5	19.3	19.2	22.0	21.0	20.9	–
$HSR_{05\ \text{private differences}}$	0.7					

The use of 0.01% copper–carbon MC in the cultivation of lilies led to an increase in the cost of sales and, accordingly, profit and profitability of production. Treatment of lily bulbs with a 0.01% copper–carbon MC is energetically effective, since allowed to get products of high quality.

The mechanism of Me–C MC action on the development of plants can be caused by the following reasons:

- the definite quantity of copper with the introduced copper–carbon mesocomposite;
- the magnetic properties of mesocomposite;
- the active electrons presence on the carbon shell of copper–carbon mesocomposite.

It's possible these factors may be substantial for the development of vital processes in plants and also in other vital organisms.

This trend of investigations needs in the development of serious theoretical elaborations in the direction to new mesoscopic biology and medicine. In these fields the Metal-Carbon mesoscopic composites application for the proceeding of vital processes, compared with the studies of mechanisms of their action, can give new impulse in the phenomena understanding, and also in the expansion of application fields of these exceptional mesoscopic particles in different trends of biology and medicine.

However, at present the widest development takes place concerning to the application of Me–C MC as modifiers of different materials. The MCs application possibilities as modifiers of inorganic and organic materials are considered in succeeding sections.

9.5 MAGNETIC METAL-CARBON MESOSCOPIC COMPOSITES AS MODIFIERS OF CONCRETES, GLASSES, AND ANALOGOUS MATERIALS

The MCs activity in the different media (materials) is changed in the dependence on polarity or polarization of their media. Therefore, the modification conditions for the different materials can be differed from each other.

According to the scheme of possible polarization the increasing of medium (material) density takes place owing to the regular orientation of material fragments with the creation of super molecular and crystalline structures takes place (Figure 9.6).

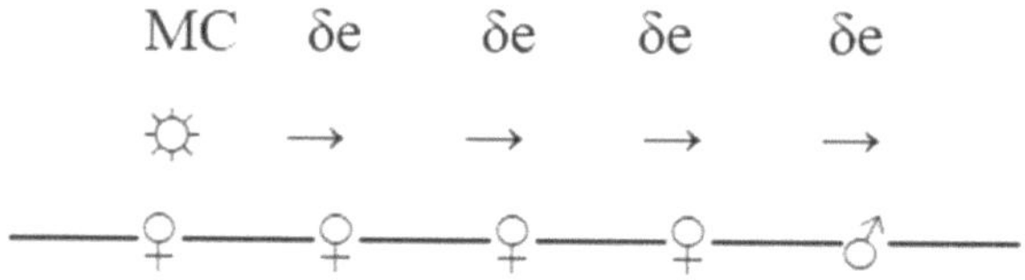

Designations: MC (☼) is the MC, δe is the negative charges quant (electron), → the polarization direction, ♀——♀ is the macromolecule fragment with functional groups.

Polarization growth can be expressed as:

$$P_{com} = \Sigma p_{fg} + p_{NC} \quad (1)$$

where; P_{com} is the common (summary) polarization; Σp_{fg} is the sum of functional groups polarizations; and p_{NC} is the polarization (or dipole moment) of MC (Scheme 9.1).

SCHEME 9.1 Scheme of polarization at charge quantization with expansion of quant influence on materials polar groups.

The polarization extent depends on the quants EM radiation phase velocity. It's necessary to note that this velocity will be decreased in the media with high dielectric constant according to following formula:

$$v = c/\sqrt{\varepsilon} \quad (2)$$

where; 'v' is the phase velocity of electromagnetic radiation; 'c' is the light velocity; and 'ε' is the dielectric constant.

When the dielectric constant is increased, the decrease of MC influence on the media arises and the self-organization process is finished.

Depending on the development of self-organization process (single measured–1D, double measured–2D, third measured–3D) the super molecular structures (mesoparticles) of correspondent forms and sizes are organized. The surface energy of embryos increased influences on the mesoparticles formation. This energy can be expressed as the sum of energetic parts for the realization of different movements:

$$E_S = E_{(tr)} + E_{(r)} + E_{(osc)} + E_{(em)} \qquad (3)$$

where; E_S is the surface energy of macromolecule (mesoparticle); $E_{(tr)}$ is the part of translational motion energy; $E_{(r)}$ is the part of rotary motion energy; $E_{(osc)}$ is the part of oscillatory motion energy; and $E_{(em)}$ is the part of electron motion in surface layer.

In accordance with the formation of mesoparticles which have the identical orientation with each other, the parts of translational motion energy and of rotary motion energy will be near to zero. Therefore, the main contribution to the mesoparticle surface energy will be bring the oscillatory motion and transport of electrons in the surface layer of macromolecules (mesoparticles). Then the change of character of quants radiation wave propagation from 2D (in the surface plane) to 3D (in the space field at surface).

The modification of foam concrete and dense concrete is usually carried out for the increasing of compressive strength. Therefore, the investigations of foam concrete strength changes in depending on the content and nature of Me–C MC are accomplished. The dependence of foam concrete strength on Cu–C MC content (in %) is given in Figure 9.6.

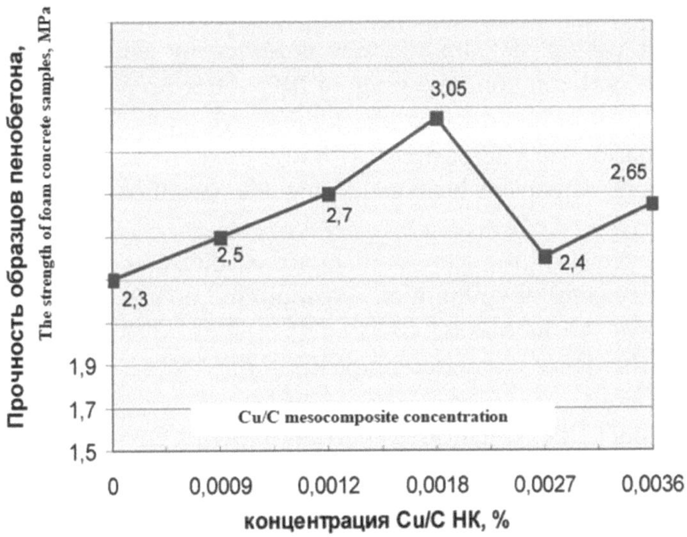

FIGURE 9.6 The dependence of foam concrete strength on copper–carbon mesocomposite content (%).

The introduction of super small quantities of MCs leads to the increasing of foam concrete surface layer density (Figure 9.7).

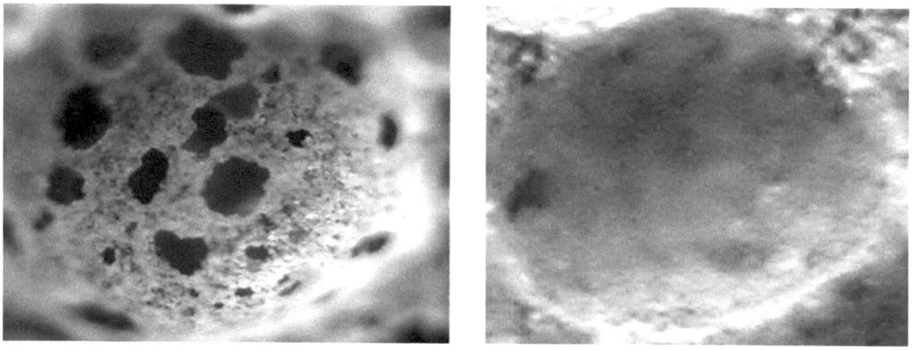

FIGURE 9.7 The foam concrete structure (×90) without mesocomposite and with the addition of mesocomposite (0.002%).

The foam concrete or gas concrete production at their modification by Me–C MC leads to improvement of building constructions durability in 1.5–2 times. The hard concrete production with the Me–C MC using increases the strength characteristics on 30–50%.

At the same time, the kinetics of these materials' strength growth is studied. In this case, the modification of foam concrete is realized with the use of MCs finely dispersed water suspension. According to the results, the strength increasing of foam concrete modified by MC is more on 80% in comparison on foam concrete without MC. After 56 days this increasing is decreased to 30%. It is possible during the period of composition hardening the coordination of MC particle with composition molecules takes place. An analogous effect is observed during the modification of dense (hard) concrete.

After seven days the growth of modified dense concrete strength more than the increasing of strength of non-modified dense concrete on 66% and after 28 days the increasing of strength makes 43%. Owing to the large density of the dense concrete composition the MC re coordination process became difficult.

According to investigation results the following conclusion may be made:

- The production of foam concrete modified by metal-carbon nano-structures leads to the improvement of its characteristics in 1.5–2 times and to the increasing of its durability.

- The strength characteristics of dense concrete modified by Me–C MC increase by 30–50%.

The introduction of a modifier based on Me–C MC into the composition results in medium structuring, decrease in the number of defects, thus improving the material physical and mechanical characteristics.

The definite peculiarities encountered at the modification of silicates and, especially, liquid glass. The interest to liquid glass is conditioned, first of all, by its ecological friendliness and production and application simplicity, inflammability, and, practically, the absence of toxicity, biological stability, and raw material availability. Such nanostructures as Me–C MC can be used to improve such characteristics of paints as elasticity, adhesion to the basic construction, hydrophobic behavior and also helps solving the problems connected with coating flaking-off from the base, discoloration, limited color range, etc. Due to the unique properties of Me–C MC it's possible to apply new properties to silicate paints, for example, to produce the coating protecting from EM action. The application of nanostructures in silicate materials to improve their stability to external action is known [1]. The introduction of nanostructures into the liquid glass allows qualitatively changing the material behavior in a positive way. This possibly occurs in the process of binder super molecular structure change.

Based on the aforesaid, it is important to improve operational properties of compositions on liquid glass basis modifying them with Me–C MC, providing a wider field of their application.

Before the liquid glass modification by Me–C MC the correspondent preparation of reactive composition needs. It' concluded in the ultrasound processing of reactionary masses for the fine dispersion obtaining. In this case it's necessary to define the processing time interval during which the optical density will be at the maximum, i.e., it will correspond to the maximal saturation of MC suspension. To define the time interval of ultrasound processing four suspensions were prepared for each soaking period – 3, 5, 10, 15 min, respectively, and also the reference solution without ultrasound processes – for comparison. The concentration in all solutions was 0.003%.

Thus, the optimal time interval for ultrasound processing is 5 minutes. Further processing of suspensions for investigation will be carried out within this interval. After the preparations of correspondent suspensions or glue samples with Iron or Nickel containing MCs the optical (visible) spectra investigations are carried out. In accordance with the results of investigations of modified films in the glue sample with Fe and glue with Ni, the shift of optical density is observed at some wavelengths, indicating the changes taking place when nanostructures are introduced (Figure 9.8).

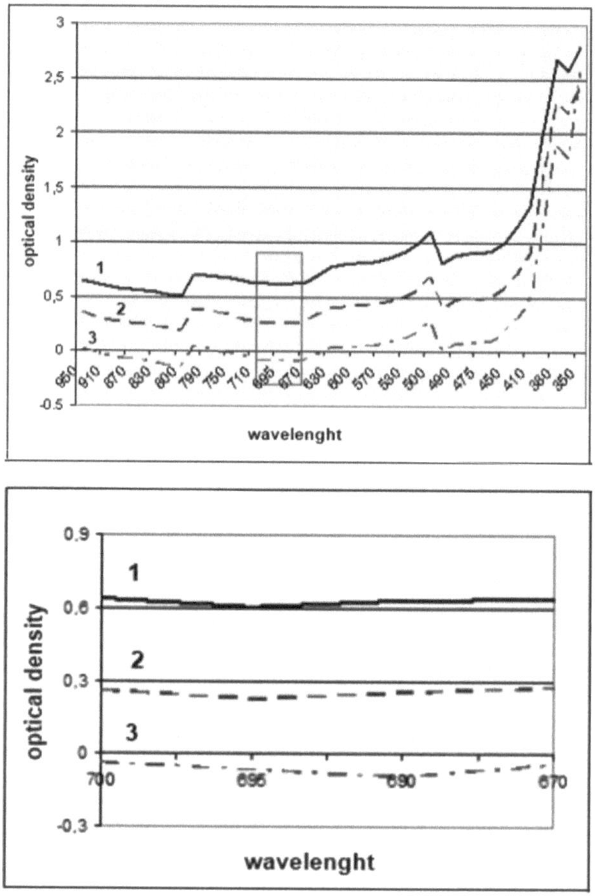

FIGURE 9.8 Changes of liquid glass (3) optical density in comparison with optical densities of liquid glasses, modified by Fe–C (1); or Ni–C (2) mesocomposites.

At other wavelength values, the optical density only increases indicating the possibility to apply nanostructures as coloring pigments in paints. At the same time, the increase in the optical density of films with nanostructures in comparison with liquid glass films possibly indicates the increase in the density of compositions and the formation of new structural elements in them.

Then the heat-physical properties of samples obtained are studied. At the beginning to obtain the details of changes in heat-physical characteristics, the heat capacity and thermal conductivity of the samples based on cardboard and modified liquid glass are investigated. The samples are prepared by gluing several layers of cardboard with liquid glass modified with MCs. The sample dimensions are found by the technique for measuring heat capacity and thermal conductivity. The sample being tested was 15 mm in diameter and 10 mm in height. The sample dimensions are measured with a micrometer with 0.01 mm accuracy. The sample mass was measured with the allowance not exceeding 0.001 g. Heat capacity of the samples is investigated on c-calorimeter IT-c-400. To find thermal conductivity the calorimeter IT-λ-400 is used. Further the specific thermal conductivity of the sample was calculated as follows:

$$\lambda = h/R_s \tag{4}$$

where; 'λ' is the specific thermal conductivity, W/m*K; and 'h' is the sample thickness, m.

The results of thermal physical investigations are given in Table 9.10.

TABLE 9.10 Thermal Physical Characteristics of the Samples

Samples	Cardboard/Glue without Fe and Ni Mesocomposites	Cardboard/ Glue with Fe (Change in %)	Cardboard/ Glue with Ni (Change in %)
Density (kg/m³)	624.5	744 (↑ 19%)	669 (↑ 7%)
Heat capacity C_{spec} (J/kg*K)	1,790	2,156 (↑ 20%)	2,972 (↑ 66%)
Thermal conductivity λ (W/m*K)	0.083	0.061 (↓ 27%)	0.064 (↓ 23%)

Heat capacity of the sample, containing Fe–C MC, increased on 20% on the relation to the standard sample, and sample with Ni–C MC – on 66%. At the same time, thermal conductivity decreased by 27% and 23% for the samples with Fe and Ni, respectively. So, when nanostructures are introduced, in the average, the characteristics change as follows: density increases by 13%, heat capacity by 40%, thermal conductivity decreases by 25%. Further, using the experimental results, the temperature conductivity is calculated by the following formula:

$$a = \lambda/c\rho \tag{5}$$

where; 'a' is the temperature conductivity coefficient; 'λ' is the thermal conductivity coefficient; 'c' is the heat capacity; and 'ρ' is the density.

Inserting the experimental data into the formula, we can define the temperature conductivity values (in %):

$$a = \lambda/c\rho = 0.75\lambda_0/1{,}4c_0{*}1.13\rho_0 = 0.47a_0 \quad (6)$$

or calculate separately:

$$a_1/a_0 = \lambda_1\rho_0C_0/\lambda_0\rho_1C_1 = 0.51 \quad (7)$$

$$a_2/a_0 = \lambda_2\rho_0C_0/\lambda_0\rho_2C_2 = 0.43 \quad (8)$$

where; a_1, λ_1, ρ_1, C1 are the characteristics of the sample with Fe/C NC; a_2, λ_2, ρ_2, C_2 are the characteristics of the sample with Ni/C NC; and a_0, λ_0, ρ_0, C_0 are the characteristics of non-modified sample. Thus, temperature conductivity decreased by nearly 50% in comparison with the initial values (a_0).

When nanostructures are introduced, self-organization takes place. Mesoparticles structure the silicate matrix leading to the formation of new elements in the structure, thus increasing the material density and influencing its heat-physical characteristics. When additional structural elements and new bonds are formed, the system internal energy increases leading to heat capacity elevation and, consequently, temperature conductivity decrease. Thermal conductivity decreases of silicate paints when applied as a coating allows improving heat-physical characteristics of the whole protective structure of a building. In turn, temperature conductivity decreases results in decreasing the amount of heat passing through the coating, thus preserving the adhesive characteristics of the coating for a long time.

9.6 THE APPLICATION OF MAGNETIC METAL-CARBON MESOCOMPOSITES (ME–C MCS) AS MODIFIERS OF POLYMERIC MATERIALS AND POLYMERIC COMPOSITES

When the additive sizes are decreased to nanometer sizes, the phenomena of mesoscopic particles are appeared. According to Ref. [41], the phenomena such as interference, spectrum quantization, charge quantization occur when the mesoparticles have the limitation in motions or in the energetic possibilities realization. In this case the mesoparticle can only vibrate and also the electron transport is possible.

The media (or polymeric compositions) properties changes under the mesoparticles influence can be achieved at equal distribution of these particles in composition volume and at its coagulation absence. Last is possible at the following conditions:

- certain polarity and dielectric constant of medium;
- minute concentration of mesoparticles;

- ultrasound action on the correspondent suspension for the proportional distribution of mesoparticles.

The assignment of active nanostructures (mesoparticles) during the composition's modification is concluded in the activation of matrices self-organization in needful direction. For the realization of this goal, the determination of organized phase part is necessary. The Avrami-Kolmogorov equation is applied for the organized phase determination:

$$W = 1 - \exp(-k\tau^{n}) \quad (9)$$

where; 'W' is the part of organized phase; 'k' is the parameter defined the rate of organized phase growth; 'τ' is the duration of organized phase growth; and 'n' is the fractal dimension.

In some chapters [9–17] the positive results on materials properties improvement are presented when the minute quantities of Me–C MC are introduced in these materials. In this chapter [14], the hypothesis about nanostructures influence transmission on macromolecules of polymeric matrices is proposed. This hypothesis is complied with mesoscopic physics principles which consider quantum effects at the certain conditions of mesoparticle existence.

The composition polarization is possible because of there is the charge quantization with the wave expansion on polar (functional) groups of media (e.g., polymer macromolecule).

The quantum charge wave expansion leads to the functional groups' polarization (dipole moments) change as well as the extinction increasing. Last bring growth of peaks intensities in IR spectra. The individual peaks growth effects in IR spectra are observed at the introduction of MCs minute quantities (*see*, Table 1.3 from Chapter 1). Let us note, that the peaks intensity growth in IR spectra is observed when the quantity of introduced MC is decreased. This fact is complied with fundamental principles of chemical mesoscopic. In an illustrated case, the instance of fine dispersed suspension Cu–C MC (hardener for ERs). According to data of Table 1.3, the decreasing of MC quantity to 0.001% leads to the growth of some peaks intensity in IR spectra.

At the second day of that suspension existence the floccules are formed, and peaks intensity sharply drops. However, the suspension activity can be increased with the use of ultrasound treatment. The treatment optimal duration determined as 7 minutes. In this case, the peak intensity in IR spectra is increased in 2–4 times (*see*, Table 1.4 from Chapter 1). The investigations by X-ray PES show that the films based on polycarbonate have more changes of electron structure at the minute quantities introduction of copper–carbon MC in comparison with other polymeric films because these films are more polarized. The C1s spectra for Cu/C nanocomposite and for nanostructured polycarbonate are presented in Figure 9.9.

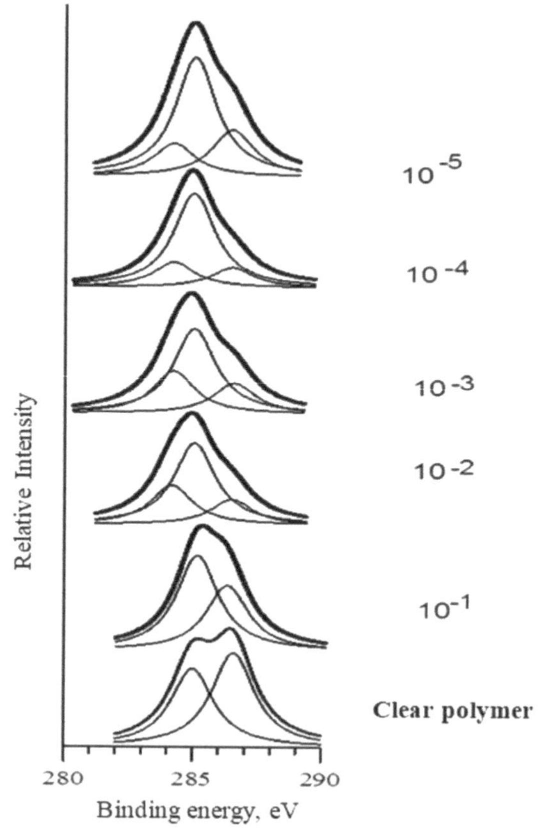

FIGURE 9.9 X-ray photoelectron C1s spectra of nanostructured polycarbonate modified by minute quantities (10^{-1}–10^{-5}%) of Cu–C mesocomposite.

The expansion of C1s spectra for nanostructured polycarbonate is possible owing to the determination of the energetic states for sp, sp^2, and sp^3 satellites. According to the results of C1s spectra for polycarbonate, the different minute quantities of Cu–C MC can contain concentration equaled to 10^{-2}% of Cu/C MC. The peaks correspondent to sp^2 and sp^3 peaks are appeared in these spectra. In other words, the "stamp" of MC which is used during modification is appeared. That "stamp" is observed also at the MC containing, equaled to 10^{-5}%, in polycarbonate film. It's noted that the relation between sp^2 and sp^3 peaks changes. For instance, the intensity of sp^2 hybridization carbon peak is upper the intensity of sp^3 peak in the concentration interval from 0.01 to 0.001% of MC. The change of concentration to 10^{-4}% bring the proximity of intensities sp^2 and sp^3 peaks.

For the decision of question about the nanostructure influence on sub molecular composition structures the AFM method is applied. Some images of polycarbonate nanostructured films surface are presented in Figure 1.1 (*see*, Chapter 1). Polycarbonate is modified by Cu–C MC minute quantities (from 10^{-1} to 10^{-4}%) (*see*, Figure 1.1 from Chapter 1).

It's interesting to observe the direction of carbon fibers in comparison with the direction of sub molecular structures orientation in the nanostructured polycarbonate surface layers.

Thus, it's possible that the wave which initiates the self-organization process in polymeric composition is expensed from these fibers associated with metal cluster. The last leads to the correspondent orientation of sub molecular structures in nanostructured composite surface layers.

The self-organization mechanism for polymeric compositions modified by the Me–C MC minute quantities is concluded in the conditions creation for composition polarization, which bring the great change of electron and sub molecular structures of materials. Certainly, these changes influence on the modified materials properties.

Below the example of results for that modification with using of above considered suspensions will be presented. In this example, the epoxy compositions with different additives including MCs were investigated. As cross-linking agent is used the fine dispersed suspension on based of isomethyl tetra hydro phthalates anhydrate and copper–carbon MC.

In Figure 9.10, the results of modification epoxy compounds (materials 1 and 2) are given. The comparison of adhesion strength for materials 1 (green) and 2 (dark blue) before modification (Figure 9.10(a)) and after modification by copper–carbon MC (Figure 9.10(b)) is shown.

The modification by copper–carbon MC in quantity equaled to 0.005% improves the adhesion characteristics for the material 1 (green) on 59.77% and for the material 2 (dark blue) on 47.17%.

It's possible also other applications that nanostructures owing to its uncial structure and properties.

Below the results of some working-outs [40, 41] show:

1. The introduction of metal-carbon nanostructures (0.005%) in the form of fine suspension into PEPA or the mixture of amines into

epoxy compositions allows increasing the thermal stability of the compositions by 75–100° and consequently increase the application range of the existing products. This modification contributes to the increase in adhesive and cohesive characteristics of glues, lacquers, and binders.

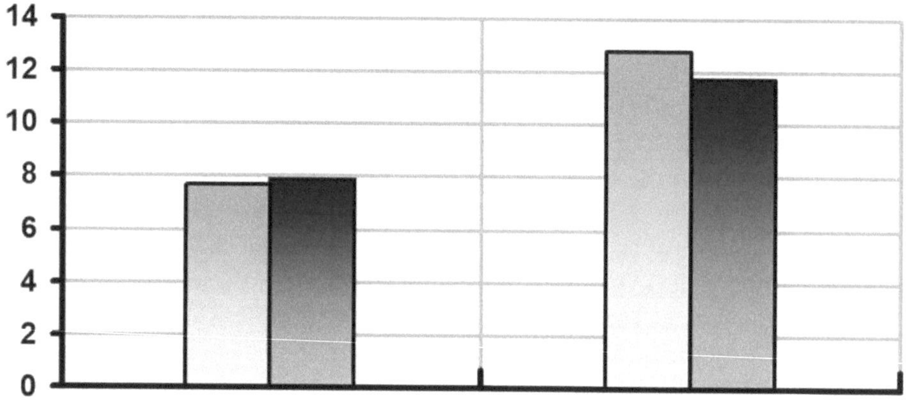

FIGURE 9.10 Comparison of adhesion strength for materials 1 (green) and 2 (dark blue) before (a); and after (b) the modification by copper–carbon mesocomposites.

2. Hot vulcanization glue was modified with copper–carbon and nickel–carbon nanostructures with using Toluene as a base for fine suspensions. On the test results of samples of four different schemes the tear strength $\sigma_{t\ increased}$ at increased up to 50% and shear strength τ_s – up to 80%, concentration of Me–C MC introduced was 0.0001–0.0003%.
3. Fine suspension of nanostructures was produced in dichloromethane and dichloroethane solutions to modify polycarbonate-based compositions. The introduction of 0.01% of copper–carbon MCs leads to the decrease in temperature conductivity of the material (in 1.5 times). The increase in the transmission of visible light in the range 400–500 nm and a decrease in the transmission in the range 560–760 nm was observed.
4. When modifying polyvinyl chloride film by the fine suspension containing Iron Carbon MC, the increase of the crystalline phase in the material was observed. The modified PVC film containing 0.0008% MC does not accumulate the electrostatic charge on its surface at the decreasing of electrostatic quantity more than two times. The material

obtained completely satisfies the requirements applied to PVC films for stretch ceilings.

5. The introduction of nickel–carbon MC (0.01% of the mass of polymer filled on 65% of silver microscopic particles) into the epoxy polymer hardened with PEPA leads to the decrease in electric resistance to 10^{-5} Ohm·cm (10^{-4} Ohm·cm without MC).

9.7 THE APPLICATION POSSIBILITIES OF MAGNETIC METAL-CARBON MESOCOMPOSITES (ME–C MCS) AS INHIBITORS OF CORROSION

From the foregoing follows that the metal-carbon mesoscopic composites radiate the negative charged quants within media and stimulate the media polarization. This MCs energetic action on media leads to media self-organization and the density growth, and also chemical bonds formation. It's possible the high surface energy of Me–C MC influences on the corrosion inhibitor reactivity and the adsorbed layer protective properties.

According to Ref. [38], the addition of copper–carbon MC to such corrosion inhibitor as 1-morpholine methyl cyclohexyl amine is effective at the quantity equaled to 0.001 mg/m^3. In this chapter, it's shown that the joint using of copper–carbon MC with previous corrosion inhibitor improves the protective properties in different corrosive media. The MC content increasing in the mixture with corrosion inhibitor promotes to the growth of protection degree at the corrosion.

The dependence of the inhibitor protection degree from copper–carbon MC concentration logarithm is presented in Figure 9.11.

The experimental results show that the increasing of concentration of mesocomposite (Cu–C MC) leads to the growth of inhibitor protection efficacy.

The MC containing inhibitor efficacy relations in comparison with initial inhibitor are increased. These relations changes are shown as follows:

Inhibitor Content:				
Relations of combined (with Cu–C MC)	25	15	10	5
Inhibitor efficacy in comparison with the initial inhibitor:	1.06	1.33	1.5	5.4
Protection Degree:				
Inhibitor, Z%	80	60	50	5
Inhibitor with 0.001 mg/dm^3 of Cu–C MC, Z%	85	80	75	27

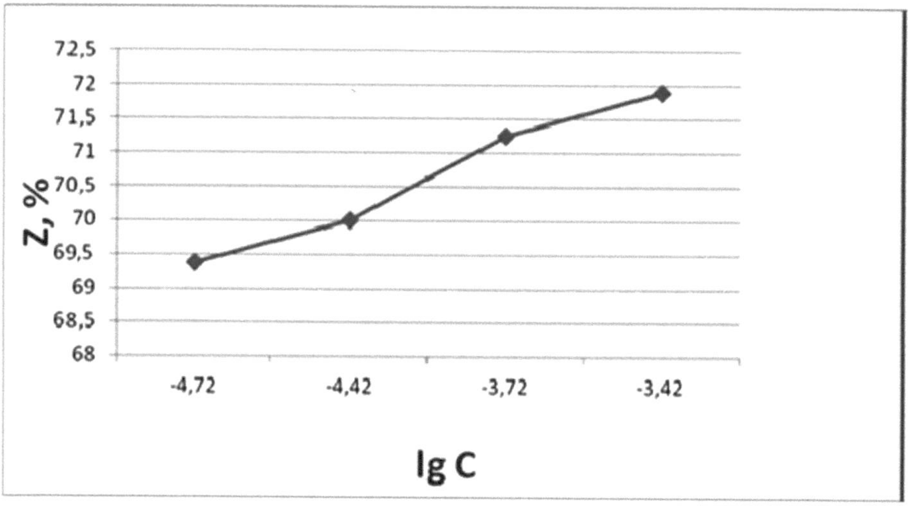

FIGURE 9.11 The dependence of the inhibitor protection degree at steel corrosion from copper–carbon mesocomposite concentration logarithm.

These changes are also shown on the Figure 9.12.

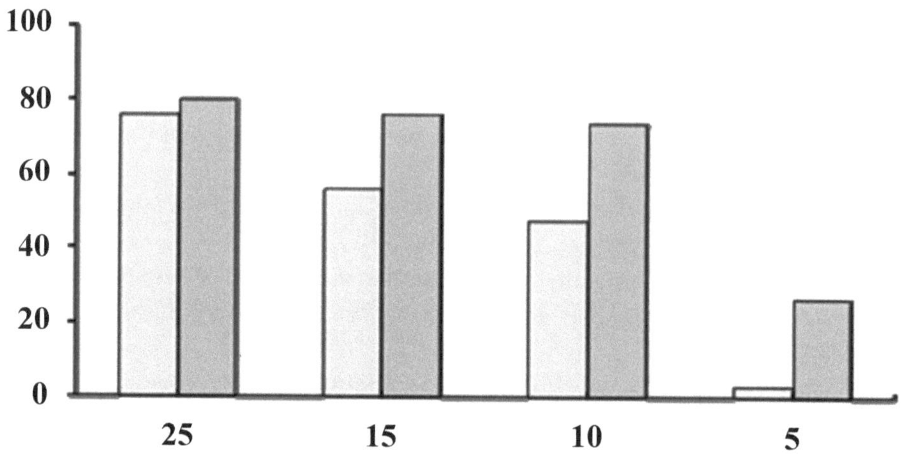

FIGURE 9.12 The dependence of the inhibitor protection degree from concentrations of inhibitor and the mixture of inhibitor and mesocomposite.

Note: The inhibitor concentration (mg/m^3); (1) without mesocomposite; and (2) with mesocomposite (0.001 mg/m^3).

Therefore, the following conclusion can be made Cu–C MC can be an effective inhibitor of corrosion. MC fulfills two functions: (i) for corrosion active agent decrease; and (ii) for the protective film creation on the material surface. These functions are carried owing to the presence on the MC carbon shell of the unpaired electrons and double bonds. Actually, the stable coating with the protection degree approximately 95% can be obtained from the oil fine suspension of the Phosphorus containing MC at the multi graded thermal processing (with the temperature increasing to 500°C).

9.8 MAGNETIC FIREPROOF MATERIALS ON THE BASE OF MAGNETIC MESOPARTICLES

9.8.1 FIRE RESISTANT GLUES BASED ON PHENOL-FORMALDEHYDE RESINS

In this paragraph, the results of modification of fireproof materials, fire resistant intumescent coatings and glues, modified by nanostructures are considered. In the beginning, the example of application of Me–C MC for the obtaining of fire-resistant materials is brought. In this case, the modification of phenol-formaldehyde glues for the obtaining from them intumescent fire-resistant glues are done.

The glues based on phenol-formaldehyde resins (BF-19) are modified by copper–carbon MC and also by Phosphorus containing analog. It's noted that the introduction of these MCs into the glue significantly decreases the material flammability. The samples with Phosphorus containing MCs have better test results. In the case, when Phosphorus containing MC is introduced into the glue, foam coke is formed on the sample surface during the fire exposure. The coating flaking off after flame exposure is not observed at the coating preserved good adhesive properties even after the flammability test.

The Phosphorus presence on the MC surface allows improving the MC structure and increases the activity in different liquid media thus increasing their influence on the material modified. The modification of coatings by MCs obtained finally results in improving their fire-resistance and physical and chemical characteristics.

The modification of BF-19 glues by Me–C MC and Phosphorus containing analogs is considered below.

The glue BF-19 is intended for gluing metals, ceramics, glass, wood, and fabric in hot condition, as well as for assembly gluing of cardboard, plastics, leather, and fabrics in cold condition. The glue composition: organic solvent, synthetic resin (phenol-formaldehyde resins of new lacquer type), synthetic rubber.

When modifying the glue composition, at the first stage the mixture of alcohol suspension (ethyl alcohol + Cu–C MC_{mod}) and APPh is prepared. At the same time, the mixtures containing ethyl alcohol, Cu–C MC and APPh, ethyl alcohol and APPh are prepared. At the second stage the glue composition is modified by the introduction of phosphorus containing compositions prepared into the glue BF-19.

For flammability determination the following method of preparation samples is used.

The samples to be tested are the plates with the dimensions 150×15×3 mm. The plates consist of foam polyethylene and paper glued together with phosphorus containing glue modified with Me–C MC with and without phosphorus. At the same time, check samples are prepared. These are the plates of foam polyethylene and paper glued together with the glue BF-19 filled with APPh with phosphorus content in the glue 3, 4, 5% from its mass.

For the testing of sample flammability, the estimation method on the determination of the lengths of carbonized parts with 1.5 minute flame exposure is used. To compare the results of coating flammability, three samples of glue are selected and tested.

The test results revealed that the length of carbonized part of the samples containing APPh and exposed to burner flame for 1.5 minutes can be about 8.5 cm with 3% phosphorus content in the sample (Table 9.11).

TABLE 9.11 Results of Testing Samples Containing Ammonium Polyphosphate

Sample No.	Sample Composition	Phosphorus Content (%)	Flame Exposure Time (min)	Length of Carbonized Part of Samples (mm)
1.	APPh	5	1.5	65.33
2.	APPh	4	1.5	82
3.	APPh	3	1.5	84.67
	Average			**77.33**

The tests of check samples confirmed that with the phosphorus content increase in composition, the length of carbonized parts of samples goes down.

The composition of sample coating: *glue BF-19 + APPh + Cu–C* MC_{pure}

The next step was to test samples containing MCs with the decreasing of phosphorus content. The average value of the carbonized part of the samples was 21.81 mm. The test results (Table 9.12) allow making the conclusion that nanocomposite inclusion significantly decreases the material flammability (in 3.5 times).

TABLE 9.12 Results of Testing Samples of Modified Compositions with Mesocomposite

Sample No.	Sample Composition	Phosphorus Content (%)	MC Content (%)	Flame Exposure Time (min)	Length of Carbonized Part of Samples (mm)
4.	Cu–C MC_{pure} + APPh	5	0.00025	1.5	15.33
5.	Cu–C MC_{pure} + APPh	4	0.0002	1.5	23.43
6.	Cu–C MC_{pure} + APPh	3	0.00015	1.5	26.66
	Average				**21.81**

The composition of sample coating: glue BF-19 + APPh + Cu–C MC_{mod}.

Phosphorus containing samples of Cu–C MC had better flammability test results than samples with APPh and samples containing APPh and Cu–C MC_{pure}. The length of the carbonized part of the samples was less by 3 mm in the average [39]. The average value of the carbonized part of the samples was 18.89 mm (Table 9.13). Thus, it can be concluded that the inclusion of Phosphorus in MC decreases the material flammability to a greater extent than MC which do not contain phosphorus.

TABLE 9.13 Results of Testing Samples of Compositions Modified by Phosphorus Containing Mesocomposites

Sample No.	Sample Composition	Phosphorus Content (%)	NC Content (%)	Flame Exposure Time (min)	Length of Carbonized Part of Samples (mm)
7.	Cu–C MC_{mod} + APPh	5	0.00025	1.5	14.67
8.	Cu–C MC_{mod} + APPh	4	0.0002	1.5	16.67
9.	Cu–C MC_{mod} + APPh	3	0.00015	1.5	25.33
	Average				**18.89**

From the data demonstrated based on the test results it can be concluded that MC inclusion into the glue composition significantly decreases the material flammability. The length of the carbonized part of the samples modified with MCs was in 4.1 times in the average less in comparison with similar parameters of the samples not containing MCs. The samples with phosphoric groups in MCs have better test results (Figure 9.13).

The coating flaking off after flame exposure is not observed, i.e., the coating preserves good adhesive properties even after the flammability test.

When the intumescent glue composition is modified with nanostructures, the material is structured with the formation of crystalline regions. In turn, such structuring under the influence of mesoscopic systems results in the increased physical and mechanical characteristics, including their stability against high and low temperatures.

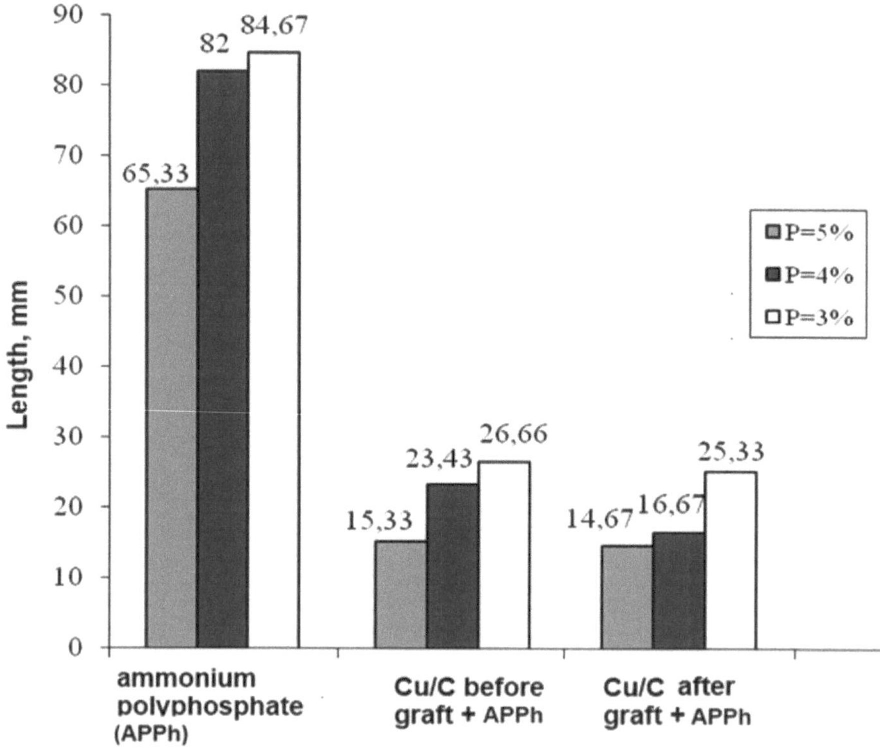

FIGURE 9.13 Diagram of the lengths of carbonized parts of the samples depending on phosphorus content in the composition.

9.8.2 PROPERTIES STUDIES OF FIREPROOF COATING BASED ON EPOXY AND MELAMINE FORMALDEHYDE RESINS

The obtaining and investigations of properties for the intumescent fireproof materials based on epoxy and melamine formaldehyde resins are considered. In both cases the Phosphorus containing copper–carbon MC together with APPh (P →Cu–C MC) is used. At the production of fireproof materials based on the ERs, the MC P →Cu–C MC is introduced into PEPA for the fine dispersion preparation. In the case when the fireproof coating is based on the melamine formaldehyde resin the P →Cu–C MC is introduced into melamine formaldehyde lacquer. For the testing of correspondent coatings such metal substances as steel and aluminum are applied. It's necessary to note that the last coating will be simultaneously fireproof and anticorrosive coating. This is especially significant for the protection of aluminum. Both coating is studied on the adhesive durability and fire proof efficiency. These tests are necessary because it's known that the coating adhesive durability is decreased at the fire action on coating. However, when the heat capacity of coating is increased at the influence of MC the coating adhesion strength is conserved. At the investigations of correspondent characteristics of coatings, the following methods are applied: for the tests on the fire proof efficiency–ASTM E-14-10 method, and for the tests on the adhesive durability–GOST 14753–73 method (Russian).

The studies on the ASTM E-14-10 method are carried out on samples which have plate forms with the following sizes: length–50–55 mm, width–10–15 mm, thickness–1–1.3 mm. The fire-resistant composition covers the correspondent plate. The coating layer thickness on plate compiles 1 mm. The samples are cross linked at normal conditions during 24 h. In the case when fire resistant composition is presented as ERs modified by APPh with MCs or without them the flame of gas burner is used at these tests. The characteristics of fire-resistant protection can be the height of foam coke formed from above fire resistant compositions. The height of foam coke is measured from the surface of the steel plate to upper level of foam coke layer. The results of the studies of fireproof efficiency for the coatings containing APPh in the comparison with the coatings modified by the copper–carbon MC (Cu–C MC) or by the Phosphorus containing Cu–C MC (P→Cu–C MC) are presented in Figure 9.14.

In correspondence with results obtained, the height of foam cokes samples which are modified by means of phosphorus containing copper–carbon MC is approximately 47% upper than initial sample.

For the determination of copper–carbon MCs introduction efficiency the following compounds are prepared:

1. ED-20 – APPh (20%) with polyethylene polyamine initial samples.
2. ED-20 – APPh (20%) with fine dispersed suspension of Cu–C MC.
3. ED-20 – APPh (20%) with fine dispersed suspension of P containing Cu–C MC.

The quantity of MCs which are introduced in compositions is equaled to 0.1%. This quantity of MCs is considered as maximum for the getting of positive effect without negative consequences and forms is presented in Figure 9.15.

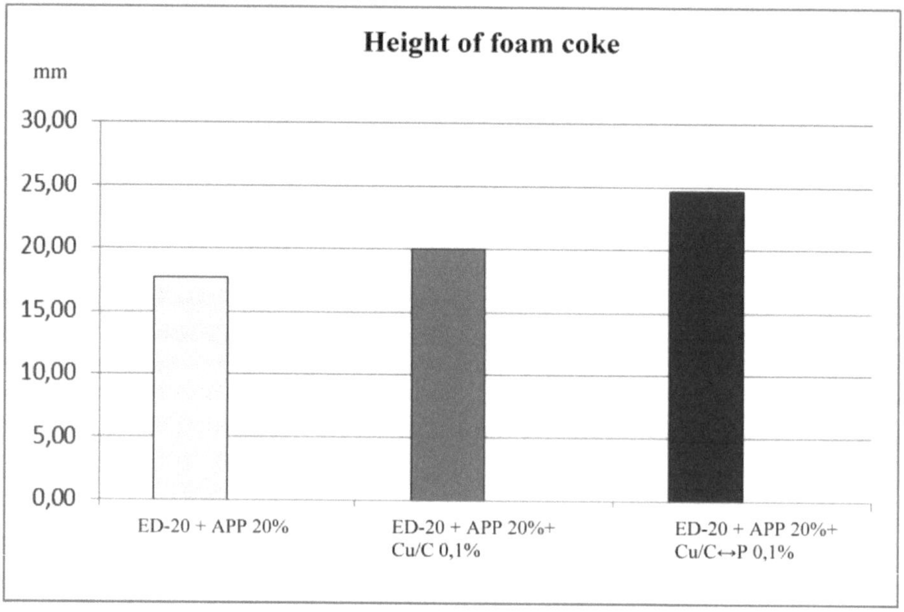

FIGURE 9.14 The dependence of the samples foam cokes height from the fireproof coating content.

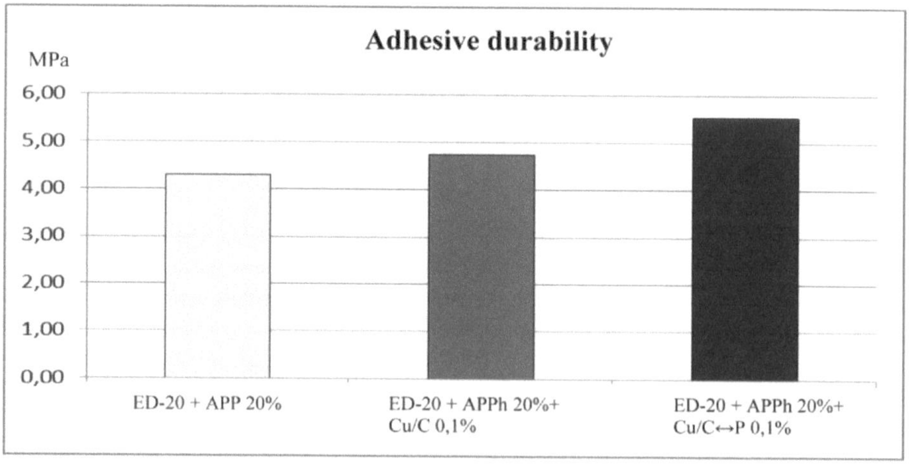

FIGURE 9.15 The dependence of samples adhesive durability from contents of them.

According to Figure 9.15, the introduction phosphorus containing copper–carbon MC in coating composition (sample c, red) leads to the increase of adhesive durability on 30% in the comparison with the fireproof coatings without the presence of MC (sample a, yellow).

Thus, on the basis of the above experiments, more effective fireproof coating is intumescent coating based on ERs with APPh modified by phosphorus containing copper–carbon MC that testify about the increasing of copper–carbon MC activity after modification of initial MC by APPh. The application of these active nanosystems for the obtaining of fireproof intumescent coatings containing ERs and APPh is perspective.

At the investigations of coatings based on melamine formaldehyde lacquer modified by APPh with copper–carbon mesocomposite (Cu–C MC) the following optimal quantities of modifiers were established: APPh – 15%; Cu–C MC – 0.008%. In this case, the dense protective layer formation on steel is determined by means of AFM method.

The coating fireproof efficiency properties improvement is observed:

- The mass losses decreasing on 43% at the burning;
- The burnt part length abridgement on 66% in the burning conditions;
- The burnt velocity decreasing on 45% in the same conditions.

Thus, the coating based on melamine formaldehyde lacquer modified by 15% of APPh with 0.008% of copper–carbon mesoscopic composite can be applied as intumescent fireproof and anticorrosive coating for metals including steel and aluminum.

9.9 CONCLUSION

The presented review is dedicated to the consideration of properties and the possible applications of a new class of such mesoscopic particles as Metal-Carbon mesoscopic composites and their modified analogs. This review consists of row paragraphs in which the consecutive description of structures and properties for Me–C MC and their modified analogs is done. The paragraphs, beginning with the second, are accompanied by the correspondent examples of the described MCs applications in the different spheres. In the first paragraph the structures and properties for initial Me–C MC are considered. The unique structures of Me–C MC are explained by the reactivity of Carbon shells which contain the poly acetylene and carbine fragments with unpaired electrons on the joints of fragments. The stability of shells is possible because of the interactions of double bonds, which take place in fragments, with metal clusters within MCs. The increasing of

Carbon shell activity is obtained after the reduction oxidation reactions at the modification of initial Metal-Carbon mesoscopic composites by reagents containing the positive charged atoms (or oxidizers). These processes as its established in experiments lead to the shift of electrons on the higher energetic levels and the formation of unpaired electrons that is accompanied by the growth of atomic magnetic moments of MC cluster metals. These phenomena of structures and energetic characteristics of MCs obtained cause their possibilities for applications in the different fields.

The examples of following applications in radical, red-ox and addition processes as catalysts, reagents, and also inhibitors as well as additives and modifiers improving properties of materials (inorganic and organic polymeric materials), adhesives and glues, fireproof systems, corrosion inhibitors, medicine magnetic transport remedies, stimulators of plant growth are presented. The metal-carbon mesoscopic composites owing to their magnetic characteristics can be used in the EM radiation focal systems. This unique scientific trend discovers a new era in the development of novel nanostructures application widening.

KEYWORDS

- **adhesives**
- **annihilation**
- **charge quantization**
- **corrosion inhibitors**
- **fire retardants**
- **inorganic and organic polymeric materials**
- **interference**
- **magnetic transport**
- **mesocomposites**
- **modifiers**
- **phase coherency**
- **red-ox synthesis**
- **self-organization**
- **stimulators**

REFERENCES

1. Kodolov, V. I., & Khokhryakov, N. V., (2009). *Chemical Physics of the Processes of Formation and Transformation of Nanostrustures and Nanosystems* (Vol. 1, 2, pp. 360, 415). Izhevsk: Izhevsk State Agricultural Academy.
2. Shabanova, I. N., Kodolov, V. I., Terebova, N. S., & Trineeva, V. V., (2012). *X Ray Electro Spectroscopy in Investigation of Metal/Carbon Nanosystems and Nanostructured Materials* (p. 252). Izhevsk–Moscow: Publ. "Udmurt University."
3. Kodolov, V. I., & Trineeva, V. V., (2013). Fundamental definitions for domain of nanostrustures and metal/carbon nanocomposites. In: *Book: Nanostructure, Nanosystems and Nanostructured Materials: Theory, Production and Development* (pp. 1–42, 559). Toronto, Canada – New Jersey, USA: Apple Academic Press.
4. Khokhryakov, N. V., Kodolov, V. I., Korablev, G. A., et al., (2013). *Prognostic Investigations of Metal or Carbon Nanocomposites and Nanostructures Synthesis Processes Characterization* (pp. 43–99). Ibid.
5. Kodolov, V. I., Trineeva, V. V., Blagodatskhikh, I. I., et al., (2013). *The Nanostructures Obtaining and the Synthesis of Metal or Carbon Nanocomposites in Nanoreactors* (pp. 101–146). Ibid.
6. Kodolov, V. I., Akhmetshina, L. F., Chashkin, M. A., et al., (2013). *The Functionalization of Metal or Carbon Nanocomposites or the Introduction of Functional Groups in Metal/ Carbon Nanocomposites* (pp. 147–175). Ibid.
7. Shabanova, I. N., Terebova, N. S., Kodolov, V. I., et al., (2013). *The Investigation of Metal or Carbon Nanocomposites Electron Structure by X Ray Photoelectron Spectroscopy* (pp. 177–230). Ibid.
8. Kodolov, V. I., Khokhriakov, N. V., Trineeva, V. V., et al., (2013). *Computation Modeling of Nanocomposites Action on the Different Media and on the Composition Modification Processes by Metal/Carbon Nanocomposites* (pp. 231–286). Ibid.
9. Kodolov, V. I., Lipanov, A. M., Trineeva, V. V., et al., (2013). *The Changes of Properties of Materials Modified by Metal/Carbon Nanocomposites* (pp. 327–373). Ibid.
10. Kodolov, V. I., & Trineeva, V. V., (2013). Theory of modification of polymeric materials by super small quantities of metal/carbon nanocomposites. *Chemical Physics & Mesoscopy, 15*(3), 351–363.
11. Kodolov, V. I., & Trineeva, V. V., (2012). Perspectives of idea development about nanosystems self-organization in polymeric matrixes. In: *"The Problems of Nanochemistry for the Creation of New Materials* (pp. 75–100). Torun, Poland: IEPMD.
12. Akhmetshina, L. F., Lebedeva, G. A., & Kodolov, V. I., (2012). Phosphorus containing metal/carbon nanocomposites and their application for the modification of intumescent fireproof coatings. *Journal of Characterization and Development of Novel Materials, 4*(4), 451–468.
13. Kodolov, V. I., Kovyazina, O. A., Trineeva, V. V., Vasilchenko, Yu. M., Vakhrushina, M. A., & Chmutin, I. A., (2010). On the production of metal/carbon nanocomposites, water and organic suspensions on their basis. *VII International Scientific-Technical Conference "Nanotechnologies to the Production – 2010": Proceedings* (pp. 52, 53). Fryazino.
14. Chashkin, M. A., (2012). *Peculiarities of Modification by Metal/Carbon Nanocomposites for Cold Hardened Epoxy Compositions and the Investigation of Properties of Polymeric Compositions Obtained* (p. 17). Thesis of cand. diss.–Perm: PNSPU.

15. Kodolov, V. I., & Kodolova, V. V., (Trineeva), Semakina, N. V., Yakovlev, G. I., Volkova, E. G., et al., (2008). Patent 2337062 *Russia Technique of Obtaining Carbon Nanostructures from Organic Compounds and Metal Containing Substances*.
16. Kodolov, V. I., Trineeva, V. V., Kovyazina, O. A., & Vasilchenko, Yu. M., (2012). Production and application of metal/carbon nanocomposites. In: *"The Problems of Nanochemistry for the Creation of New Materials* (pp. 23–36). Torun, Poland: IEPMD.
17. Akhmetshina, L. F., Kodolov, V. I., Tereshkin, I. P., & Korotin, A. I., (2010). The influence of carbon metal containing nanostructures on strength properties of concrete composites. *Internet Journal "Nanotechnologies in Construction," 6*, 35–46.
18. Kodolov, V. I., Khokhriakov, N. V., Trineeva, V. V., & Blagodatskikh, I. I., (2008). Activity of nanostructures and its display in nanoreactors of polymeric matrixes and in active media. *Chem. Phys.& Mesoscopy, 10*(4), 448–460.
19. Kodolov, V. I., Khokhriakov, N. V., Trineeva, V. V., & Blagodatskikh, I. I., (2010). Problems of nanostructures activity estimation, nanostructures directed production and application. In: *Nanomterials Yearbook – 2009: From Nanostructures, Nanomaterials and Nanotechnologies to Nanoindustry* (pp. 1–18). N.Y.: Nova Science Publ., Inc.
20. Kodolov, V. I., Blagodatskikh, I. I., Lyakhovitch, A. M., et al., (2007). Investigation of the formation processes of metal containing carbon nanostructures in nanoreactors of polyvinyl alcohol at early stages *Chem. Phys.& Mesoscopy, V., 9*(4), 422–429.
21. Trineeva, V. V., Lyakhovitch, A. M., & Kodolov, V. I., (2009). Forecasting of the formation processes of carbon metal containing nanostructures using the method of atomic force microscopy. *Nanotechnics, 4*(20), 87–90.
22. Khokhriakov, N. V., & Kodolov, V. I., (2005). Quantum-chemical modeling of nanostructure formation *Nanotechnics, 2*, 108–112.
23. Kodolov, V. I., Khokhriakov, N. V., Nikolaeva, O. A., & Volkov, V. L., (2001). Quantum-chemical investigation of alcohols dehydration and dehydrogenization possibility in interface layers of vanadium oxide systems *Chemical Physics and Mesoscopy, 3*(1), 53–65.
24. Kodolov, V. I., Didik, A. A., Volkov, A. Yu., & Volkova, E. G., (2004). Low-temperature synthesis of copper nanoparticles in carbon shell *Bulletin of HEIs. Chemistry and Chemical Engineering, 47*(1), 27–30.
25. Lipanov, A. M., Kodolov, V. I., Khokhriakov, N. V., et al., (2005). Challenges in the production of nanoreactors for the synthesis of metallic nanoparticles in carbon shells *Alternative Energetic and Ecology (ISJAEE), 2*(22), 58–63.
26. Nikolaeva, O. A., Kodolov, V. I., Zakharova, G. S., et al., (2004). *Method of Obtaining Carbon-Metal-Containing Nanostructures*. Patent of the RF 2225835.
27. Didik, A. A., Kodolov, V. I., Volkov, A. Yu., et al., (2003). *Inorganic Materials, 39*(6), 693–697.
28. Kodolov, V. I., & Trineeva, V. V., (2015). The metal/carbon nanocomposites influence mechanisms on media and compositions. In: *Nanostructures, Nanomaterials and Nanotechnologies to Nanoindustry* (pp. 171–185). Toronto, New Jersey: Apple Academic Press.
29. Shabanova, I. N., Kodolov, V. I., Terebova, N. S., et al., (2016). x-Ray photoelectron study of the influence of the amount of carbon nickel containing nanostructures on the degree of the poly methyl metacrylate modification. In: *Book Multifunctional Materials and Modeling* (pp. 211–218). Canada-USA: AAP.
30. Kodolov, V. I., & Trineeva, V. V., (2016). Self-organization in processes under action of super small quantities of metal/carbon nanocomposites: Review on investigation results. In: *Book Multifunctional Materials and Modeling* (pp. 263–331, 343). Canada-USA: AAP.

31. Kodolov, V. I., Trineeva, V. V., Pershin, Yu. V., et al., (2020). *Method of Metal Carbon Nanocomposites Obtaining from Metal Oxides and Polyvinyl Alcohol.* Pat. RU 2018122 001.
32. Mustakimov, R. V., Kodolov, V. I., Shabanova, I. N., & Terebova, N. S., (2017). Modification of copper carbon nanocomposites with the use of ammonium polyphosphate for the application as modifiers of epoxy resins. *Chemical Physics & Mesoscopics, 19*(1), 50–57.
33. Kodolov, V. I., Trineeva, V. V., Kopylova, A. A., et al., (2017). Mechanochemical modification of metal carbon nanocomposites. *Chemical Physics & Mesoscopics, 19*(4), 569–580.
34. Kodolov, V. I., Trineeva, V. V., Terebova, N. S., et al., (2018). The change of electron structure and magnetic characteristics of modified copper carbon nanocomposites. *Chemical Physics & Mesoscopics, 20*(1), 72–79.
35. Merzlyakova, V. M., Lapin, A. A., & Kodolov, V. I., (2019). Efficiency of metal carbon nanocomposite application in lily growing under protected soil conditions. In: *Nanoscience and Nanoengineering: Novel Applications* (pp. 177–188, 355). Toronto-New Jersey: Apple Academic Press.
36. Karavaeva, N. M., Pershin, Yu. V., Kodolov, V. I., et al., (2019). Change of morphology and swelling of cured epoxy compositions upon their modification with minute quantities. *Pol. Sci., Ser. D., 12*(2), 179–181.
37. Kodolov, V. I., et al., (2019). Modified magnetic metal carbon mesoscopic composites with bio active substances. *Chemical Physics & Mesoscopics, 21*(3), 446–454.
38. Pletnev, M. A., Ovechkina, O. A., Buldakova, N. S., et al., (2014). The influence of metal carbon nanocomposites on protective action of corrosion inhibitors. *Intellectual Systems in Engineering, 1*(23), 150–152.
39. Mustakimov, R. V., (2019). *The Method of Modification of Metal Carbon Nanostructures by Ammonium Polyphosphate.* Pat. N 2694092.
40. Kodolov, V. I., Semakina, N. V., & Trineeva, V. V., (2018). Introduction in science about nanomaterials. *Monograph* (p. 476). Izhevsk: Publisher – M.T. Kalashnikov Izhevsk State Technical University.
41. Kodolov, V. I., & Kodolova–Chukhontzeva, V. V., (2019). Fundamentals of chemical mesoscopics. *Monograph* (p. 218). Izhevsk: Publisher – M.T. Kalashnikov Izhevsk State Technical University.
42. Pat. RUN 2490027, 2017. Volume 9, Issue 4, August 2021, pp. 97–104.
43. Wang, J. Q., Wu, W. M., & Feng, D. M., (1992). *The Introduction of Electron Spectroscopy* (p. 640). Beijing: National Definite Industry Press.

CHAPTER 10

Epoxy Composites with 5 wt.% Nanodispersed Magnetites and Ferroxides: Strength, Heat Resistance, and Morphology

D. STAROKADOMSKY,[1,2] M. RESHETNYK,[3,1] N. BODUL,[4] and L. KOKHTYCH[5,6]

[1]*M. P. Semenenko Institute of Geochemistry, Mineralogy and Ore Formations, National Academy of Sciences (NAS) of Ukraine, Ukraine*

[2]*Chuyko Institute of Surface Chemistry, NAS, Ukraine*

[3]*National Natural History Museum NAS, Ukraine*

[4]*Ukrayinsky Phyzico-Mathematic Litseum, T. Schevchenko Kyiv National University, Ukraine*

[5]*School of Engineering and Architecture, Lucerne University of Applied Sciences and Arts, Horw, Switzerland*

[6]*Institute of Physics, NAS, National Academy of Sciences (NAS) of Ukraine, Ukraine*

ABSTRACT

The morphology and physico-mechanical properties of magnetically sensitive ferro-containing epoxycomposites, were studied. Optical microscopy indicates the ability of magnetites and ferroxides to form carcass-like structures that can enhance strength. AFM microscopy of the composites indicates a change in the surface morphology after filling and integration of structure

Modern Magnetic Materials: Properties and Applications. Iuliana Stoica, Ann Rose Abraham, & A. K. Haghi (Eds.)

heterogeneities. SEM and optical microscopy show significant morphological changes after filling. A larger number of pores and large agglomerates, as well as air bubbles stabilized by nanoparticles (NPs), appear in the structure of composites.

The data obtained indicate a negligible effect of 5 wt.% – filling on compressive strength, both during the normal treatment of cured composites (50–60°C) and after exposure under aggressive conditions (250°C or water-endurance). After heat treatment, the compressive strength of filled composites decreases as a rule, while the elastic modulus can increase. But the effect of thermo-strengthening is observed for ferric red pigment (compressive strength), and ferrite (modulus). After water-endurance, compressive strength falls for unfilled, and can be stable for same ferro-filled composites.

Approximately such effect of these fillers is seen in the assessment of flexural strength. Therewith, the modulus of elasticity (both in compression and flexural) can be increased by 10–15% after filling. Therewith, the filling can significantly (2–2.5 times) increase the fire resistance of polyepoxide. The filling increases the resistance of composites to aggressive organic media (acetone, ethylacetate), but weakens resistance in oxidizers (60% H_2O_2). These results open the way to create magnetically sensitive polymer nanocomposites with susceptible strengths.

10.1 INTRODUCTION

Epoxy resins (ERs) are used in almost all spheres of human activity, in the manufacture and repair of devices, as adhesives and compounds in construction, as osteoprostheses in medicine, hulls for yachts, aircraft, and machines in the aerospace shipbuilding and engineering industries, in demand in the manufacture of toys (handmade) [1–4]. Therefore, on the basis of ERs, new compositions are created that are resistant to various conditions of use and for solving various problems. In our work, the task was set to create a composite more resistant to aggressive environments and high loads, which included magnetite. Epoxy composites with mineral fillers—is a popular area of research today [1–24].

Iron-containing epoxy composites have found application in industries requiring special adhesives and polymer composites – magnetically sensitive, heat-conducting, or similar in properties to iron and its alloys [8–13]. To a large extent, they are needed when repairing chips and equipment defects, when welding methods are expensive or unacceptable [2–4]. In particular, epoxy-magnetite composites have attracted ongoing scientific attention. Magnetite (α-Fe_2O_3) is iron oxide, an ore mineral common in eastern Europe among other iron oxide minerals [6–8].

Articles about the effect of ferrites, iron particles and magnetic powders in organic coatings and compositions constantly appear in scientific periodicals [9–24].

Wu [11] placed carbon nanotube fibers with Fe_3O_4 (own production) in ERs. In this way, they obtained magnetic and electrically conductive composites. The authors consider such composites to be useful for industry. At the same time, the authors state that, as a rule, there is a decrease in strength and other properties, or rather, their discrepancy with the declared theoretically high estimates (Introduction [11]). They explain this by imperfect distribution, poor interphase interactions, and poor structuring of the filler.

In Ref. [13], magnetite powder with different weight percent (4, 8, and 12 wt.%) was dispersed into epoxy composition (with ratio 2:1 epoxy and hardener) matrix mixture and poured into samples (22.86 width × 10.16 height ×2 thickness) mm. The 12 wt.% composite had the highest value of real part of permittivity due to greater reflection coefficient and also highest dielectric loss factor. Each of the composite had very low magnetic loss mechanism, and the value of μ_r was nearly unity. The σ of the composites increased with frequency where the 8 and 12 wt.% contents showed the highest value of conductivity.

The particle size of nanomagnetites does not allow to introduce a lot of them into the ERs, and already at 5–10 wt.% the composition is significantly thickened. In this case, there are many structural changes that can significantly change the strength of the final cured product. It is known [3] that the ERs are well combined with iron powders of micro- and nanoscale, and even capable of better curing in their presence.

Our laboratory has experience in the development of magnetites (as elements of targeted drug transport), and some experience in studying them in polymers [15, 25]. When planning this work, it was decided to compare the properties of composites with several types of magnetites, as well as with yttrium ferrite and ordinary minimum (α-Fe_2O_3). Their comparison can provide very complete and thereby valuable information about what to expect from the introduction of magnetites and ferroxides into ERs (and this is often practiced on an industrial and repair-service scope). At the same time, the obtained data is logically correlated with visual information from various microscopy methods.

10.2 METHODS AND REAGENTS

For testing, Italian resin EPOSiR-7120 was used, which is characterized by frost resistance (does not freeze at 0–10°C) and heat resistance of the polymer. The reason for this heat resistance is evident in the characteristics of the resin modifiers. The manufacturer claims the presence of a certain "epoxyalkyl diluent," which gave it such properties. The same chemical modification obviously led to a decrease in compressive strength by almost a factor of 2 – in our case, from 400 + –50 kgf (for the Czech and Soviet/Russian trademark ED-20) to 270 + –20 kgf. And this same additive interferes with determining the adhesion for EPOSiR to steel, since it does not allow the surfaces to be glued (possibly due to the release of alkyl on the interphase surface). As can be seen from Table 10.1, after 250°C the properties of the polymer from EPOSiR, unlike the ED-20 and its analogs, remain the same; even the plasticity of the material is not violated (as can be seen from the diagrams).

The following fillers were used for the study:

1. (M) Magnetite M–nanoscale, with an average size of primary particles 30 nm;
2. (A) Magnetite A–nanoscale, with an average size of primary particles 50 nm;
3. (T) Magnetic toner for Hewlett Packard printers;
4. (Y) Nano-microdispersed Yttrium Ferrite of the general formula $Y_4Fe_5O_{12}$;
5. (C) Ferroxide Fe_3O_4 micron-sized industrial "Meerkat red pigment" (production of Ukraine).

Strength tests were carried out in accordance with or taking into account standard methods (GOST or ASTM).

Tensile strength (GOST 56810-2015, ASTM D790). For bending tests, plates 6 × 1 × 0.2 cm in size were made. Their fracture during bending was carried out on the basis of L = 3 cm of the DI-1 bending testing machine. Based on the test results, the strength i was calculated (i = 3PL/2h2b, P is the resulting load in kgf on a scale of 1 cm = 1 mm, L is the length of the fracture base equal to 30 mm, h is the thickness of 2 mm, b is the width of the sample equal to 10 mm) and modulus elasticity in bending I (I = PL3/4bh3W, where W is displacement on a scale of 1 cm = 20 μm).

Compression tests (ISO 604: 2002) were subjected to cylinder-shaped samples with a diameter of 6.5 mm and a height of 10–12 mm (using a Louis Shopper press machine) manufactured at 25°C and heat-treated. The adhesive peel tests (GOST 14760-69) were subjected to gluing of metal cylinders with a diameter of 2.2 cm on a test binder, on a UMM-10 Armavir installation. All rounding, including averaging, is done towards a larger value, and the smallest 1–2 values are not taken into account.

10.3 THE RESULTS OF THE EXPERIMENT

10.3.1 MICROSCOPIC MORPHOLOGY

10.3.1.1 AFM MICROSCOPY OF THE SURFACE OF COMPOSITES

The AFM does not always provide complete information about the structure but is very useful for assessing surface changes—roughnesses, pores, smooth zones. It can be seen from them (Figure 10.1) that composites with magnetite tend to enlarge and group "micro-islands" of the surface, to form zones of a smooth surface – up to their dominance (Figure 10.1-2).

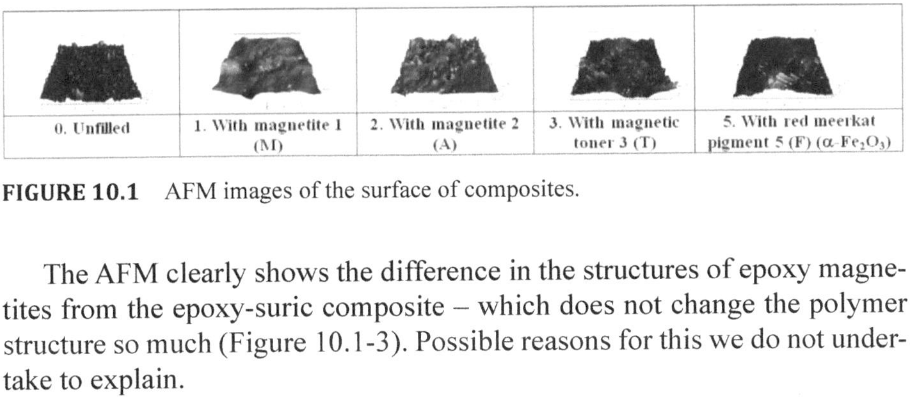

FIGURE 10.1 AFM images of the surface of composites.

The AFM clearly shows the difference in the structures of epoxy magnetites from the epoxy-suric composite – which does not change the polymer structure so much (Figure 10.1-3). Possible reasons for this we do not undertake to explain.

10.3.1.2 OPTICAL MICROSCOPY

It can be seen from Figure 10.2 that magnetite powders have a granular aggregate structure. Initial magnetite powders are formed into aggregates and agglomerates, reaching 500 microns (Figure 10.2) and often forming chains and dendritic structures. This is especially noticeable in the microstructure of the toner powder (No. 3), where, in addition to magnetic nanoparticles (MNPs), there are half-dimensional additives. The iron oxide pigment forms magnetite-like structures (Figure 10.2, No. 5). –Fe_2O_3 (meerkat).αSometimes dendritic structures are manifested in magnetite agglomerates, which are even

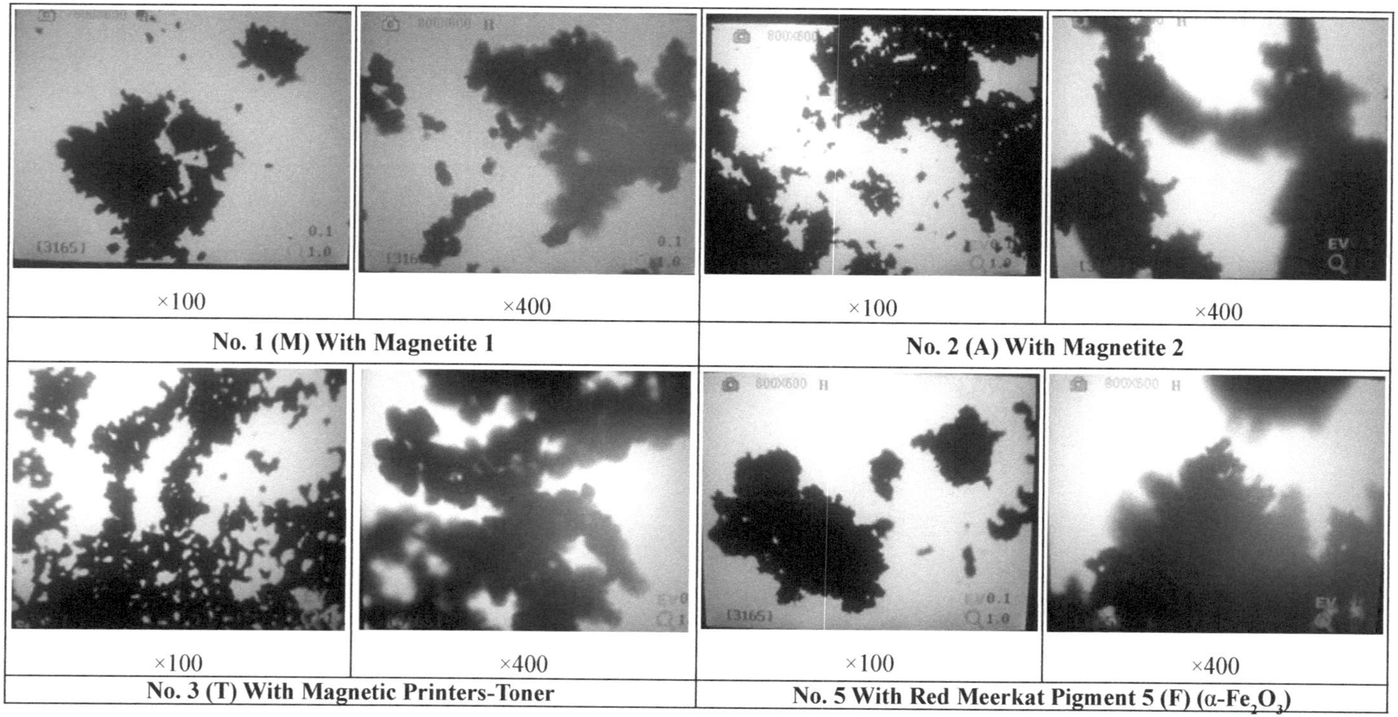

FIGURE 10.2 Optical microphoto of magnetite powders, with an increase of 100 or 400 times (with a screen length of 7 cm).

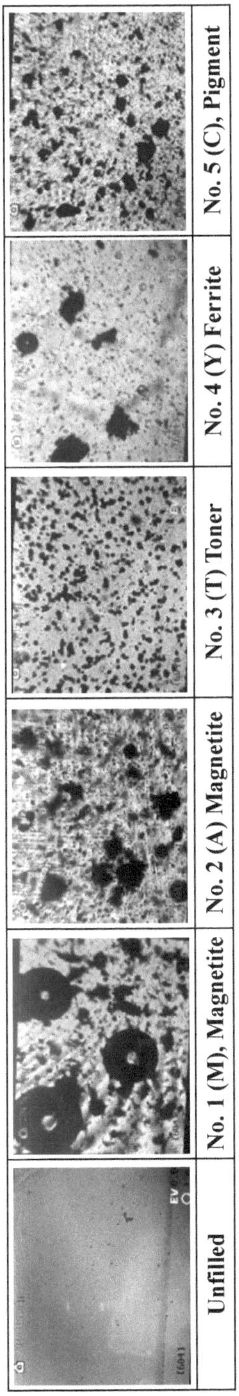

FIGURE 10.3 Optical microphotographs of epoxy compositions with magnetites, with an increase of 100 times (with a screen base length of 7 cm).

more common in This gives hope for the formation of soft semi-organized matrices in the ERs and later in the polymer. This makes it possible to predict a very good compatibility and distribution of these powders in the epoxide, which may improve the strength and resistance of composites.

From the optical photos (Figure 10.4) of epoxy compositions, a fairly uniform distribution of almost all magnetites (except for coarsely dispersed No. 4) and meerkat is noticeable. And sometimes magnetite can "tighten" in the system and stabilize very large air bubbles by the surface layer of NPs (sm. No. 1 and No. 4). As we see, the unfilled composite has practically no serious defects (bubbles, inhomogeneities). The uniform distribution of magnetite can be useful for enhancing hardened composites in a number of ways.

10.3.1.3 SEM MICROSCOPY

The initial unfilled composite is characterized by a fairly uniform distribution of pores and irregularities. That generally corresponds to modern ideas about the fibrillar-pack structure of three-dimensional thermosetting plastics.

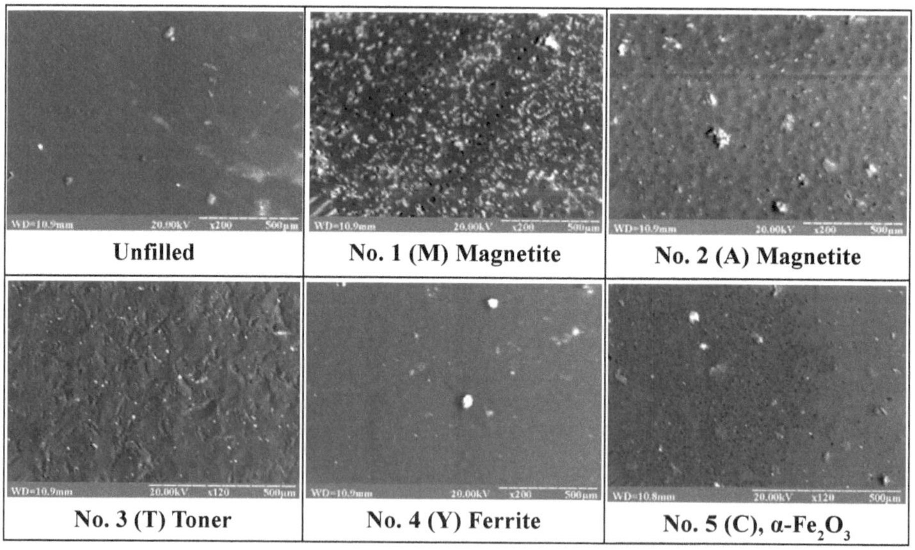

FIGURE 10.4 SEM photo of composites (× 200).

Although the initial resin does not contain obvious inhomogeneities in the photo of non-hardened compositions, in the hardened epoxy polymer, they are still visible as separate shapeless inclusions up to 50 microns (Figure 10.4).

10.3.2 STRENGTH

It can be seen from the experiments that, according to the compressive strength, the unfilled polymer based on EPOSiR-7120 resin (heat-resistant) is insensitive to either hard heating or 7-day exposure in water (7 days). The strength practically does not change (by 2–3% – see Table 10.1, samples "H"), and if the module E decreases slightly then by 2–3% (to 10.8 instead of 11.1. Table 10.1 see samples "H"). This is uncharacteristic for standard epoxides, since from our recent work (on resin ED20 and Epoxy520 [1–4]), it is clear that heat treatment and holding in water led to a noticeable drop in the indicators of unfilled epoxy polymers. Recall that an ordinary resin like ED20 after such heating loses strength by a third or more (see our early works Refs. [1–4]).

TABLE 10.1 Strength Parameters of Samples of Composites with 5 wt.% of Fillers#

	H	1 (M)	2 (A)	3 (T)	4 (Y)	5 (F)
Soft Thermo-Treatment (55°C, 5 h)						
Compression's load C^d (kgf)	285^{450}	270^{330}	280^{430}	290^{410}	300^{420}	280^{480}
Max. compression's load C (kgf)	290	280	290	300	310	290
Modulus E*1000 (kg/cm^2)	11.1	12.8	–	11.2	12.0	12.0
Fire-resistance (seconds)	1	2	2	2	2	2.5*
Hard Thermo-Treatment (55°C, 5 h, after that 250°C, 1 h)						
C (kgf) after 250°C	300^{450}	270	280	160*	160*	310^{360}
Modulus E*1000 (kg.cm^2)	10.9	10.6	12.4	10.3	12.8	11.7
Aqua-Treatment (55°C, 5 h, after that – 7 days in H_2O)						
C (kgf)*	270	270	260	270	250	290
Modulus E*1000 (kg.cm^2)	10.8	11.2	10.6	–	–	–

#*Designations of the samples correspond to the numbering of magnetites in the section "methods and reagents." For limit of plasticity C at compression's load C^d index ^{d} shows load of filly destruction.*

*– *estimation.*

It can be seen that the presence of magnetite particles (as well as iron oxide) is not very significant on the compressive strength (*see* Table 10.1, samples 1 (M) and 2 (A). True, for a mixture of magnetite with a thermoplastic (toner 3 (T), Table 10.1), the load of yield stress C increases by 5–6% (for compression tests this also matters), and for ferrite 4 (Y) it grows even by almost 10% (Table 10.1). It does not change, only in some cases changing by 5–10% (Table 10.1). On the contrary, the modulus of compression elasticity can noticeably change. So, it grows by 5–8% for ferrite 4 (Y) and ferroxide pigment 5 (C), and for magnetite 2 (M) – even by 15%. That is, iron oxides, even at 5 wt.%, can give the polymer much higher elasticity, and this, by the way, can also be seen from the diagrams (Figures 10.2–10.4). This is also true for samples aged in water. That can be considered a very acceptable result for the tasks of creating magnetic, or iron-containing epoxides. After harsh heat treatment, the compressive strength of the filled polymers drops, sometimes substantially (Table 10.1). The elastic modulus E, on the contrary, can noticeably increase with filling (Figure 10.4). For several templates (N 2 (A) and 4 (Y)) E increases after hard heat-treatment (Table 10.1). Thus, for epoxy-ferroxides we can observe the effects of "thermo-hardening of composites," which we described earlier for filled epoxides after destructive heating 250–300°C [1, 3, 4, 9].

A diagrams "load-compressive deformation" for composites shows a certain increase in elasticity after filling. Indeed, an unfilled polymer after a load of the plastic limit (letter 'P' in Figure 10.5) already weakly resists further loading (Figure 10.5). This can be seen by the small angle of the slope before the final destruction (letter 'D' in Figure 10.5). But almost all filled composites have a steeper slope angle before D (Figure 10.6).

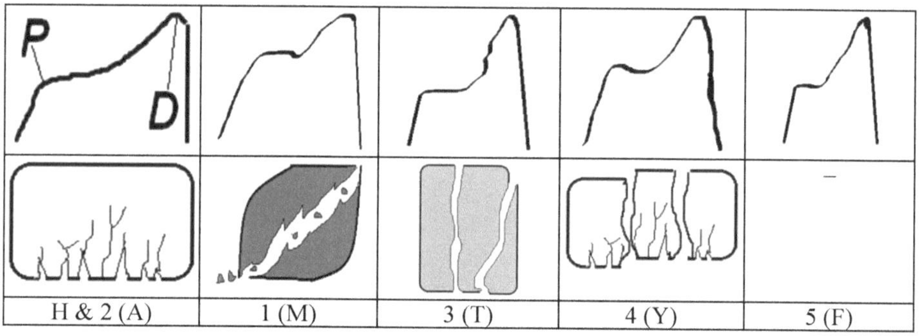

FIGURE 10.5 The type of compression diagram "load-deformation" for different composites, after conventional heat treatment (50–60°C).

From Table 10.2, it is seen that the tensile strength deteriorates after filling (which is typical after filling of polyepoxides). At the same time, the elastic modulus can appreciably increase.

10.3.3 SWELLING AND RESISTANCE IN AGGRESSIVE ENVIRONMENTS

Resistance in acetone and ethyl acetate. Acetone is a very aggressive environment for polyepoxides (especially unheated ones). It follows from our early work that exposure to acetone leads to rapid swelling on the 1st day, and then (if there is no reinforcing filler or 100° heat treatment) the sample is destroyed [1, 3, 4, 9, 16]. Ethyl acetate also gives a similar effect [3, 4, 9].

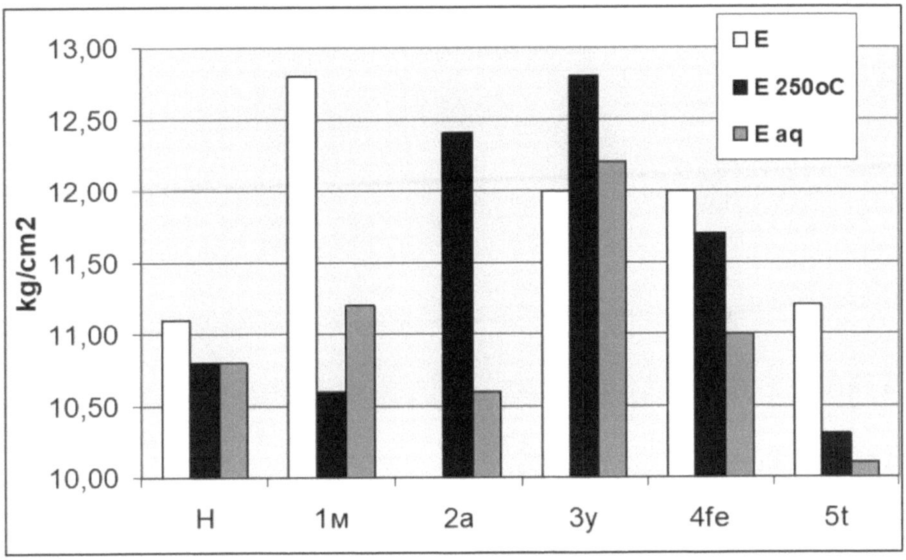

FIGURE 10.6 A comparative histogram of the values of the modulus of elasticity of compression of the samples at different exposure modes–the usual 60°C; hard 250°C and exposure in water for seven days (Eaq).

TABLE 10.2 Strength and Modulus of Elasticity in Bending Plates (1.5 mm Thick, 1 cm Wide)

	H	1 (M)	3 (T)
Bending Strength (kgf/mm^2)	3.8	3.1	2.8
Bending Modulus (1,000 kgf/cm^2)	19	16	23

Heat treatment of the epoxy polymer leads to a marked improvement in resistance to these solvents, including a decrease in swelling in the later stages of curing (1–2 days, Table 10.3). A similar effect can be achieved by simple filling with magnetite (Table 10.3), without heat treatment. This is important in cases where heat treatment is not possible (for example, when coating walls). Table 10.3 also shows (on the example of sample 3 (T)), that not every ferro-powder is able to increase resistance.

Swelling and destruction in peroxide. In a strong oxidizing agent, the unfilled polymer is slightly more stable than the magneto-filled ones (Table 10.4). The composites collapsed after 3–6 days of exposure, showing a double increase in the degree of swelling.

TABLE 10.3 Swelling (%) and Destruction of Composite Tablets in Acetone-Ethylacetate Mix

	H	H_2 (thermo)	3 (T)	1(M)
0	0	0	0	0
0.1	6.0	6.2	8.0	9.5
0.2	9.5	8.7	14.3	9.5
1	31.0	11.8	28.6	14.4
2	Destruct	18.0	Destruct	11.9
3	–	24.2	–	19.9
6	–	–	–	19.9

TABLE 10.4 Swelling (%) and Destruction of Composite Tablets in 60% H_2O_2

	H	H_2 (thermo)	T	M
0	0.0	0.0	0.0	0.0
0.08	31.3	7.6	6.3	4.7
0.21	41.6	12.9	14.5	7.7
1	68.7	35.9	44.9	27.7
2	94.0	75.3	78.3	48.9
3	111.4	101.8	108.7	100.9
6	100.0	105	112.6	Destruct
8	Destruct	–	Destruct	–

10.4 CONCLUSIONS

1. Epoxy composites with 5 wt.% of iron-oxide nanodispersed fillers are characterized by a high modulus of elasticity and acceptable (100–115% compared to unfilled polymer) compressive strength. Bending strength is reduced, while the modulus of elasticity in bending can increase markedly.

   The introduction of fillers increases the resistance of composites to open fire by 2–2.5 times.
2. The hard heat-treatment (250+–10°C), leads to increase of compression strength for unfilled polymer (load increases from 285 to 300

kgf), but not for most of filled composites. An interesting exception is only a composite with the cheapest filler - iron oxide (growth from 280 to 310 kgf). This indicates the manifestation of the effect of thermo-strengthening for the unfilled polymer and the composite with Fe_2O_3. Also, we see the effect of thermo-strengthening for compression-modulus in composite with ferrite (4(Y)). Filling can enhance resistance of the compressive strength to water-endurance (what can be seen from composites with magnetite and iron oxide).

3. Analysis of AFM, SEM and optical microimages shows a noticeable effect of fillers on the morphology of the composites. According AFM, magnetites group aggregates and "micro-islands" of the composite surface, and form zones of a smooth surface. And epoxy-Fe_2O_3 system is able to form a large crystallites in a native surface of composite. According SEM, the initial resin does not contain obvious inhomogeneities in the photo of non-hardened compositions, in the hardened epoxy polymer they are still visible as separate shapeless inclusions up to 50 microns. The optical microscopy visualizes air bubbles stabilized by nanoparticles (NPs), that appear in the composites after filling.
4. The work shows the possibilities and limitations (strength, durability, resistance, morphology) in the preparation of magnetically sensitive epoxy-composites. An interesting combination of conclusions can be considered the high prospects for the production of composites with the cheapest filler-pigment iron oxide.

KEYWORDS

- **AFM microscopy**
- **epoxy resins**
- **microscopic morphology**
- **microscopy**
- **nanoparticles**
- **polymer composites**
- **SEM microscopy**
- **thermo-strengthening**

REFERENCES

1. Starokadomsky, D., & Reshetnyk, M., (2019). Microfilled epoxy-composites, capable of thermo-hardening and thermo-plasticization after hard heating (200–300°C)–for "in-field/offroad" use in bio-, agro-, medservice. *Biomedical Journal of Scientific & Technical Research, 19*(1), 14118–14123. https: doi.org/10.26717/BJSTR.2019.19.003257.
2. Matykiewicz, D., Barczewski, M., & Mousa, M., (2021). Impact strength of hybrid epoxy–basalt composites modified with mineral and natural fillers. *Chem Engineering, 5*(3), 56–65. https: doi.org/10.3390/chemengineering5030056.
3. Starokadomsky, D. L., (2019). Possibilities of creating fire-resistant, heat-hardening and thermoplastic at 250°C epoxy-composite plastics with micro dispersions of SiC, TiN and cement. *Plasticheskie Massy,* 5–6, 40–43. https: doi.org/10.35164/0554- 2901-2019-5-6-40-43.
4. Starokadomsky D., Halysh V., Starokadomska A., Nikolaychuk A., Shulga S., Sigareva N., & Reshetnyk M. (2020). Restorative biocompatible polymer composites on the base of epoxy-resin filled by surface-modified disperse utilizated solid waste of industrial paper products. *Journal of Applied Surfaces and Interfaces. 8*(1–3), 1–6. http://revues.imist.ma/?journal=jasi.
5. Mostovoy, A. S., Nurtazina, A. S., Burmistrova, I. N., & Kadykova, Yu. A., (2018). Effect of finely dispersed chromite on the physicochemical and mechanical properties of modified epoxy composites. *Russian Journal of Applied Chemistry, 91*(11), 1758−1766.
6. Reshetnyk, M., Starokadomsky, D., & Popov, C. (2012). The complex magnetic scanning as an effective method to investigate the exposures of Precambrian basement: example of Ukrainian Shield. *Scientific Journal of Pure and Applied Sciences, 1*(1), 22–29.
7. Reshetnyk, M., (2020). Rock magnetic properties of enderbites in Gaisin block of Ukrainian shield. *International Journal of Geography, Geology and Environment, 2*(2), 23–28.
8. Reshetnyk, M., & Starokadomsky, D., (2021). Rock magnetic properties of enderbite-charnokites in Gaisin block of Ukrainian shield *The Scientific Heritage, 62*, 20–27. doi: 10.24412/9215–0365-2021-62-1-20-27. https://ru.calameo.com/read/005059769b0b 36329d 938 (accessed on 16 January 2022).
9. Starokadomskii, D. L., (2017). Epoxy composites with 10 and 50 wt% micronanoiron: Strength, microstructure, and chemical and thermal resistance. *Russian Journal of Applied Chemistry, 90*(8), 1337–1345.
10. Ulbrich, M., & Kalendova, A., (2013). Properties of organic coatings with nonisometric ferrite particles. *Physics Procedia, 44*, 247–255.
11. Wu, S., Ladani, R., Zhang, J., & Wang, C., (2015). Epoxy nanocomposites containing magnetite-carbon nanofibers, aligned using a weak magnetic field. *Polymer, 68*, 25–34.
12. Atta, A., El-Saeed, A., El-Mahdy, G., & Al-Lohedan, H., (2015). Application of magnetite nano-hybrid epoxy as protective marine coatings for steel. *RSC Advances, 2015*, 123.
13. Chan, Y. L., Esa, F., You, K., & Jusoh, A., (2017). Electromagnetic properties of magnetite/epoxy resin composites at x-band frequency. *Conference: 2017 Progress in Electromagnetics Research Symposium.*–Fall (Singapore). doi: 10.1109/PIERS-FALL.2017.8293649. https:www.researchgate.net/publication/276271709_ (accessed on 16 January 2022).
14. Chigna, T., Zannoni, M., Jones, M., & Simonetto, A. (2020). *Permittivity and Permeability of Epoxy-Magnetite Powder Composites at Microwave Frequencies.* https:arxiv.org/pdf/2001.02336.pdf (accessed on 16 January 2022).

15. Starokadomsky, D., Reshetnyk, M., & Rassokhin, D., (2020). Epoxy composites with 5 wt.% of nanodispersed magnetites and ferroxides. Strength, heat resistance, morphology. *J. Material & Environmental Sciences, 11*(8), 1241–1249. https: www.jmaterenvironsci.com/Document/vol11/vol11_N8/JMES-2020-11109-Staromsky.pdf (accessed on 16 January 2022).
16. Starokadomsky, D., Reshetnyk, M., & Terebilenko, A., (2021). Restorative and industrial reinforced epoxy composites with micro-nano-dispersed (Si, Ti, Zr, Cr, Mo, and Nb)-carbide fillers. *The Scientific Heritage, 2*(64). doi: 10.24412/9215-0365-2021-64-2-10-19.
17. *Kakhramanov, N. T., Azizov, A. G., Osipchik, V. S., Mamedly, U. M., & Arzumanova*, N. B., (2016). Nanostructured composites and polymeric materials technology. *Plasticheskie Massy, 1, 2,* 49–57. https://doi.org/10.35164/0554-2901-2016-1-2-49-57; https:www.plastics-news.ru/jour/article/view/18 (accessed on 16 January 2022).
18. Kakhramanov, N. T., Allahverdieva, K. V., & Koseva, N. S., (2022). Adhesive features of functionalized metal–polymer systems based on polyolefins. *Polym. Sci. Ser. D, 15,* 19–24. https: doi.org/10.1134/S1995421222010105.
19. Danchenko, Y., Andronov, V., Barabash, E., Rybka, E., & Khmyrova, A., (2019). Acid-basic surface properties of dispersed fillers based on metal oxides TiO_2, Al_2O_3, CaO, and Fe_2O_3. *IOP Conference Series: Materials Science and Engineering, 708*, 012083. doi: 10.1088/1757-899X/708/1/012083.
20. Danchenko, Andronov, V., Barabash, E., Obigenko, T., Rybka, E., Meleshchenko, R., & Romin, A., (2017). Research of the intramolecular interactions and structure in epoxyamine composites with dispersed oxides. *Eastern-European Journal of Enterprise Technologies, 6*(12). 4–12. https: doi.org/10.15587/1729-4061.2017.118565.
21. Ramajo, L. A., Cristóbal, A. A., Botta, P. M., PortoLópez, J. M., Reboredo, M. M., & Castro, M. S., (2009). Dielectric and magnetic response of Fe_3O_4/epoxy composites. *Composites Part A: Applied Science and Manufacturing, 40*(4), 388–393 https: doi.org/10.1016/j.compositesa.2008.12.017.
22. Darwish, M., Trukhanov, A., Senatov, O., Morchenko, A., Saafan, S., & Singh, C., (2020). Investigation of AC-measurements of epoxy/ferrite composites. *Nanomaterials (Basel), 10*(3), 492. doi: 10.3390/nano10030492.
23. Zaidi, M. G. H., Sah, P. L., Alam, S., & Rai, A. K., (2009). Synthesis of epoxy ferrite nanocomposites in supercritical carbon dioxide *Journal of Experimental Nanoscience, 4*(1), 55–66. doi: 10.1080/17458080802656515.
24. Kanapitsas, A., Tsonos, C., Psarras, G., & Kripotou, S., (2016). Barium ferrite/epoxy resin nanocomposite system: Fabrication, dielectric, magnetic and hydration studies. *eXPRESS Polymer Letters, 10*(3), 227–236. doi: 10.3144/expresspolymlett.2016.21; https:www.researchgate.net/publication/289378337_Barium_ferriteepoxy_resin_nanocomposite_system_Fabrication_dielectric_magnetic_and_hydration_studies (accessed on 16 January 2022).
25. Chang, C., Su, S., Chang, T., & Chang, C., (2021). Frequency-induced negative magnetic susceptibility in epoxy/magnetite nanocomposites. *Sci. Reports, 11*, 3288. https: doi.org/10.1038/s41598-021-82590-w.
26. Starokadomsky, D., (2021). New effects of thermo-hardening and thermo-plasticization after hard heating in epoxy-composites with optimal micro-nano-fillers. *Globus an International Journal of Medical Science, Engineering and Technology, 10*(2), 55–62. doi: 10.46360/globus.met.320212010.

CHAPTER 11

Magnetic Particles and Their Role in Polymer Composites: From Molecular Modeling to Applications

RALUCA MARINICA ALBU

"Petru Poni" Institute of Macromolecular Chemistry, Laboratory of Physical Chemistry of Polymers, Iasi, Romania

ABSTRACT

The domain of electromagnetism is very vast and led to noteworthy advances in the field of material science. The chapter presents important aspects related to magnetic particles and their contribution to the performance of magnetic composites with polymer matrix. Principal classes of single/multi-component magnetic materials are described, together with the known types of magnetic behavior. Molecular modeling studies of polymer matrix and filled polymer systems are presented to understand the advantages introduced by the selected matrix on the final composite features. Certain recent applications of polymer magnetic composites in several technical domains and biomedical areas are reviewed.

11.1 INTRODUCTION

The design of materials for an engineering product or a targeted use requires deep knowledge on the chemical and physical properties. Aside from the thermal resistance, mechanical strength and dimensional stability, the electrical or magnetic characteristics are of paramount importance in many practical cases. Among these, the magnetic properties of materials display

Modern Magnetic Materials: Properties and Applications. Iuliana Stoica, Ann Rose Abraham, & A. K. Haghi (Eds.)

a particular role in several applications, such as absorbers based on magnetorheological (MR) fluids [1], drug release carriers [2], cancer curing [3], bio-separation [4], hyperthermia treatment [5], sensors [6, 7] and electrical machines [8]. In the past years, the advances in the area of magnetic materials have brought outstanding breakthroughs, which have contributed to the progress in the multidimensional fields [6]. There is a broad spectrum of magnetic compounds, which have facilitated the disclosure of some spectacular new phenomena. Furthermore, besides the influence on magnetic performance induced by the chemical composition, a remarkable importance is observed as a function of the morphology and the dimensions of the magnetic particles [10].

In order to enhance the functionality and applicability of a product that concerns the modern technologies, the principles of nanotechnology, electromagnetism, material science, plastics technology, and chemical engineering have been interfering and inspiring scientists to fabricate multifunctional composites with appropriate matrices [11]. Magnetic composites generally contain magnetic inclusions (of micro- or nano-dimensions) inserted in non-magnetic or magnetic matrix. In this context, polymers have gained huge interest owing to their benefits, like lightweight, reduced expenses, flexibility, wear resistance, and ease of processing in precise conditions [12]. Different types of synthetic or natural polymers have been employed over the time to produce novel engineering materials. The continuous phase represented by the polymer has the role to bring the filler together and ensure an adequate transfer of load between them. So, the matrix acts as a platform to distribute the micro- or nanoparticles (NPs) uniformly throughout the structure. As a consequence, the mechanical and physicochemical properties of such polymer composites are affected by the content of the matrix, reinforcing agent and the interphase interactions/compatibility [13–15]. Another relevant factor resides from the thermal behavior of the polymer during composite processing. Derived from this aspect, literature indicates that there are two main classes, namely thermoset and thermoplastic matrix composites. The first class is more common, while the second one appears to experience a faster development. The benefits introduced by the composites with thermoplastic matrix in regard to the thermoset ones arise from diminished preparation costs since it is no need for curing, the materials display prolonged shelf life, possibility of re-processing, good weldability, combined with reduced moisture content [16, 17]. Relatively recent trends are devoted to the combination of magnetism and additional functions. Thus, such approaches have attracted tremendous attentiveness since they helped to make progress in the domain of smart magnetic materials [18–20]. One problem could reside in the fact that magnetic reinforcements present a powerful tendency to make agglomerates for lowering the energy ascribed to the high surface area-to-volume ratio of the micro/nano-sized fillers. To resolve the issue of aggregation of magnetic reinforcements, certain strategies have been proposed and they are relying on surface protection. These approaches are meant to chemically stabilize the pristine magnetic filler by grafting of or covering with organic compounds (i.e., surfactants or macromolecules) or overlay with an inorganic layer (i.e., silica or carbon). The embedding of these functionalized magnetic fillers in polymers or other matrices demonstrated a higher efficiency of the composite [21]. Further refinement of the magnetic polymer composites requires not only complex synthesis routes, but also deep elucidation of the interactions among the phases of the system. The latter can be predicted prior preparation step by performing molecular modeling experiments, which allow extracting data on the quantitative structure-activity relationships (QSAR) properties.

In this context, this work presents a current state-of-art in the area of magnetic materials and polymer composites containing magnetic inclusions. The chapter starts by describing some introductive aspects concerning the main types of magnetic materials that are utilized as fillers. Additionally, the types of magnetic behavior are depicted. Molecular modeling of several important commercial polymers is performed to depict the implications of the conformational properties on the composite performance. The practical uses of the magnetic composites in many technical domains and biomedical areas are reviewed.

11.2 GENERALITIES ON MAGNETIC MATERIALS

11.2.1 PRINCIPAL CLASSES OF SINGLE/MULTI-COMPONENT MAGNETIC MATERIALS

Current advances in the field of magnetism are indicating that there are many basic sorts of materials that display magnetic behavior, namely [22]:

1. **Metals:** Iron, nickel, cobalt, steel, gold, rare earth metals represent the most relevant metals known for their magnetic properties;
2. **Metal Oxides:** Iron oxides (e.g., magnetite, hematite, and maghemite) and not only are included here. Dilute magnetic semiconductors (DMS) denote a category of substances that exhibit semiconducting abilities, as well as magnetic features. In DMS, a part of the cations

in the network are replaced with magnetic ions, so that their atomic spin gives the possibility of interaction with the carriers in the material structure to render global ferromagnetic order. Therefore, these materials depict unique magnetic features owing to the existence of separated magnetic ions in semiconducting network. Considerable research has been done to develop new DMS products but also for clarifying the origin of magnetism in such compounds since the 1980s. Among the DMS materials can be considered simple oxides like tin (IV) oxide, zinc oxide, titanium oxide, or mixed oxides reinforced with particular transition metals (Fe, Co, Ni, Mn) [23] or rare earths metals (Eu, Dy, Er) [24];

3. **Ceramics:** These are oxide materials which have a particular sort of permanent magnetization, such as barium titanate, strontium titanate, etc.;
4. **Polymer Magnets:** It is a nonmetallic magnet constructed based on a polymer, like DNA and other proteins. This can be regarded as a novel class of magnetic compounds that have drawn the interest of researchers. Torrance & collaborators [25] have prepared poly(1,3,5-triaminobenzene) that was oxidized with iodine to render a ferromagnetic property up to 400°C. Subsequently, Rajca et al. [26] have fabricated an organic polymer with π-conjugated structure that enabled a tremendous magnetic moment and magnetic order at cold temperatures under – 263°C. Another relevant investigation was performed by Zaidi et al. [27] which made a new magnetic polymer based on polyaniline (PANi) combined with an acceptor component (i.e., tetracyanoquinodimethane). This was the first polymer system with magnetism emphasized at room temperature. Their new material mingles the advantages of the conjugated nitrogen from the main chain with molecular charge transfer side groups. Such approach generates a stable polymer with a significant density of localized spins that could contribute to the rise to coupling. Magnetic assessments reveal that the polymeric system is ferri- or ferromagnetic with a curie temperature of beyond 77°C and highest saturation magnetization of 0.1 JT^{-1} kg^{-1}. Crayston & co-workers [28] have also studied the synthesis of polymer-based magnets;
5. **Other Materials:** Water, hydrogen, crown glass, alcohols, solutions of salts of iron and oxygen.

Regarding the architecture of composites with magnetic properties, literature [22] emphasizes four types of categories, namely:

- Core-shell inorganic composites;
- Self-assembled composites;
- Silica-based magnetic composites;
- Organic-inorganic composites.

Among these, organic-inorganic magnetic filled systems have raised an immense interest because of the combination of peculiar features of organic and inorganic counterparts from the material. Hybrid organic-inorganic magnetic composites could be attained by in situ, ex situ, co-precipitation, microwave reflux, melt mixing, ceramic-glass synthesis, and plasma polymerization approaches. The magnetic behavior is depending on the used materials for the preparation of the new magnets, temperature of processing, or sometimes of the system composition.

11.2.2 TYPES OF MAGNETIC BEHAVIOR

Literature survey [29–31] on this topic describes several major classes of magnetic properties of the materials, namely:

- Diamagnetism;
- Paramagnetism;
- Ferromagnetism;
- Ferrimagnetism;
- Antiferromagnetism;
- Metamagnetism.

The acknowledged distinct forms of magnetic phenomena appear to be the consequence of the modalities in which moments of the negative charges in molecular and supramolecular arrays might be found. In the next paragraphs, a brief presentation concerning the peculiarities of each sort of bulk magnetic behavior is made:

1. **Diamagnetic Compounds:** These are those that have occupied orbital shells and are lacking a magnetic moment in the absence of an external field. When the latter is present, the spinning electrons process and such movement generates a magnetization along the reverse direction. When such a magnetic substance is under the influence of a non-uniform magnetic field, it is repulsed from the zone of bigger field strength to the area of smaller field strength. The diamagnetic materials are also known for their reduced and negative susceptibility (independent of temperature), but also for their subunitary values

of the relative permeability. A diamagnetic solid substance is able to levitate under powerful magnetic fields, as shown in transducer applications. Many materials display diamagnetic properties, though it is frequently screened by the stronger paramagnetic or ferromagnetic term [29–31].

2. **Paramagnetic Materials:** These are those that present partially filled orbitals and the unpaired spins make electrons to act as magnetic dipoles. In the presence of a magnetic field, these dipoles arrange themselves parallel to the direction of imposed field, producing a magnetization in the same direction. Under the action of non-uniform magnetic fields, the paramagnetic substance is drawn by the zones of bigger field strength from smaller field strength areas. Such compounds present a low positive susceptibility (decreases with increasing temperature) and relative permeability slightly larger than unity [29–31].
3. **Ferromagnetism:** This happens in materials with unpaired electrons and powerful exchange interaction, all these resulting in orientation of magnetic dipoles over extensive zones. This is observed in certain ordered lattices, where atomic magnetic moments are interacting to line up parallel to each other. Two distinctive features of ferromagnetic substances are their spontaneous magnetization and the occurrence of magnetic ordering temperature. Hence, such substances have areas with magnetization even when the magnetic field is missing. Inside these regions, the magnetic field is powerful, but in overall, the substance is not magnetized since most domains are randomly disposed in regard to one another. This sort of magnetic behavior manifests itself in the fact that a weak external field is able to line up magnetic regions so that the material gains magnetization. The driving magnetic field will be enhanced by a huge factor which is often reflected in the relative permeability for the substance. Ferromagnets can remain magnetized to an extent beyond being under the action of the external magnetic field and the property to memorize magnetic history is generally named hysteresis. The amount of the saturation magnetization which is retained when the driving field is stopped is referred to as the remanence of the material and is a paramount key in the operability of permanent magnets. This can be ascribed to crystalline anisotropy or shape anisotropy. In order to fully demagnetize such a material, it is mandatory to utilize a magnetic field of strength that acts in the reverse direction (named

coercive field). The dissimilarity among the spontaneous magnetization (net magnetization noticed in the absence of a field) and the saturation magnetization is arising from the magnetic domains. The saturation magnetization represents an intrinsic feature, not relying on particle size, but sensitive to the temperature. The electronic exchange forces within the ferromagnets can be strong; however, thermal energy overcomes this and induces a randomizing effect. The temperature at which ferromagnetic property vanishes is named the Curie temperature. Ferromagnetic magnetic compounds present relative permeability values of the order of 1,000 or larger [29–31].

4. **Antiferromagnetism:** It is a property highlighted at room temperature only by chromium. The main distinction in regard to ferromagnets resides in the fact that the exchange interaction among the adjacent atoms determines the anti-parallel orientation of the atomic magnetic moments. Consequently, the magnetic field annuls, and the substance seems to behave analogously to the paramagnetic compounds [29–31].
5. **Ferrimagnetism:** It is mainly noticed in substances having a highly complex crystal structure in comparison to the pure elements. Inside such materials, the exchange interactions produce parallel disposition of atoms in certain part of the crystal and anti-parallel one in others. The substance breaks down into magnetic regions similarly to ferromagnetic compound, and the magnetic character is analogous, but ferrimagnetic materials often have smaller saturation magnetizations. Another feature is that the susceptibility is large under the Curie temperature. Ferrimagnetic, ferromagnetic, and anti-ferromagnetic substances display hysteresis, which denotes an irreversibility of magnetic property as the applied magnetic field is modified [29–31].
6. **Metamagnetism:** It is an abrupt and sometimes dramatic augmentation of the magnetization of a substance with a low variation in an externally applied magnetic field. The sudden variation might be caused by the transient spontaneous magnetization generated as a result of high magnetic field breaking the antiferromagnetic spin orientation. So, metamagnetism can be regarded as the transition under magnetic field to a saturated ferromagnetic phase via a first-order transformation. Oppositely to normal ferromagnetism, no hysteresis is remarked. The metamagnetic property might display very distinct physical causes for many sorts of metamagnets [29–31].

In order to make a conclusive picture of the categories of compounds with magnetic properties and the aforementioned types of magnetic behavior, Figure 11.1 illustrates a general scheme on these aspects.

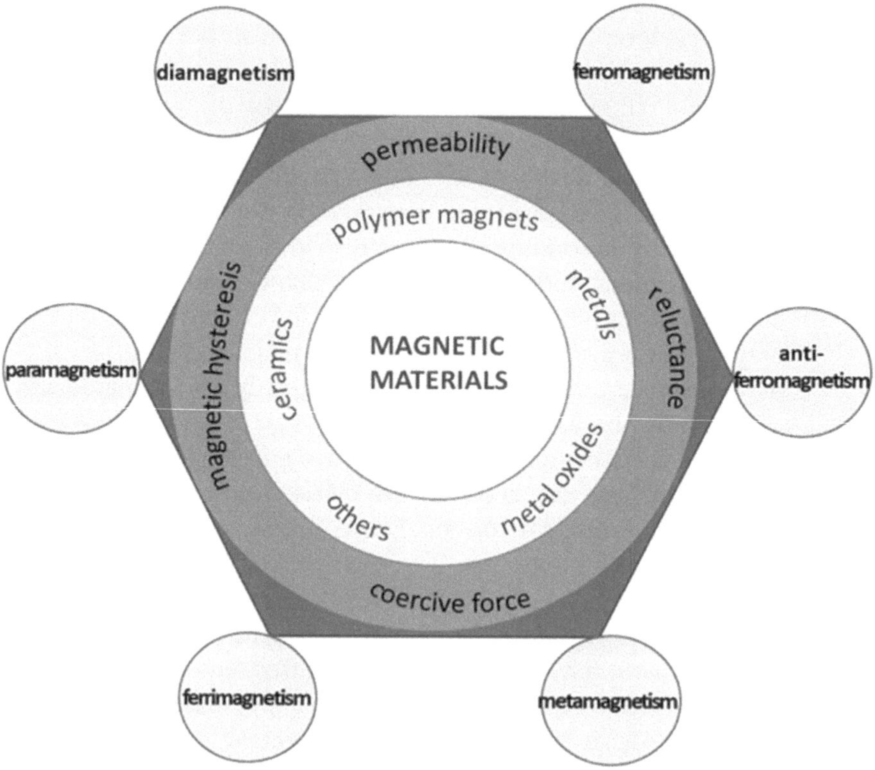

FIGURE 11.1 General scheme of the categories of substances with magnetic properties and the known types of magnetic behavior.

Also, Figure 11.1 depicts the most relevant magnetic properties of magnetic substances. These features influence the ability of the compounds to be appropriate for a targeted magnetic application. Among the typical magnetic features of engineering materials, the following one must be mentioned [32–34]:

1. **Permeability:** The feature of a magnetic substance that reveals the facility level to build up a magnetic flux inside the material. It is established to be the ratio of magnetic flux density to magnetizing force generating the magnetic flux density;

2. **Retentivity or Magnetic Hysteresis:** It refers to materials whose magnetic domains get arranged along the magnetic field and after its action stops, still a magnetization is observed, denoting the magnetic retentively of the compound;
3. **Coercive Force:** It is related to the fact that certain compound display retentivity even when the external field is turned off and in order to cancel the residual magnetization another field of reverse direction is applied, and the new forces (coercive forces) are overcoming the residual magnetism;
4. **Reluctance:** The feature of a magnetic substance to resists to occurrence of magnetic flux in the bulk.

The substance depicting a big value of residual magnetization and coercive force are named magnetically hard materials, while those displaying the opposite properties are known as magnetically soft materials. In Table 11.1, some important magnetic materials and their basic properties are listed [35–43].

11.3 MOLECULAR MODELING OF POLYMERS AND THEIR MAGNETIC COMPOSITES

The design of a magnetic composite requires profound knowledge on the properties of each constituting phase. Conformational characteristics of the matrix dictate its physical parameters and thereby the performance of the composite. Table 11.2 presents some physical properties of common polymers used for composite fabrication [44–49].

The presence of the magnetic particles in a polymer is expected to change its properties as a function of the filler size, morphology, and amount. In certain situation, it is preferable to perform molecular modeling to understand the conformational modifications induced by magnetic fillers and how these could affect the performance of the loaded polymer. In the following paragraphs, some case studies of molecular modeling are presented. Two polymer matrices are selected for computations, namely poly(m-toluidine) (PmT) (*see* Figures 11.2 and 11.3) and Kapton polyimide (*see* Figures 11.4 and 11.5). These polymers are simulated to interact with several amounts of magnetic fillers (i.e., Fe_3O_4 and Co_3O_4).

As seen in these figures, the presence of the fillers slightly affects the conformational properties of the polymer matrix. Further, from these computations, the quantitative structure-activity relationships (QSAR) properties

TABLE 11.1 Common Properties of the Most Important Magnetic Materials

Formula	Density (g/cm3)	Size (mm)	Magnetic Behavior	Magnetic Susceptibility	Band Gap (eV)	References
Strontium ferrite ($SrFe_{12}O_{19}$)	5.3	131 ± 3	Ferrimagnetic	0.0004 emu/mol	~ 2	[35]
Barium ferrite ($BaFe_{12}O_{19}$)	5.28	55	Ferrimagnetic	0.0002 emu/g	1.86	[36]
Neodymium-iron-boron (NdFeB)	7.51	98 ± 2	Ferrimagnetic	0.05–10 ($\chi_{\perp}$)	1.56	[37]
Cobalt ferrite ($CoFe_2O_4$)	5.23	8.5–9.6	Ferromagnetic	15,000 emu/gG	2.27	[38]
Iron nitrate ($Fe\text{-}(NO_3)_3{\cdot}9H_2O$)	1.68	–	Paramagnetic	$15{,}200\times10^{-6}$ cm^3/mol	2.53	[39]
Cobalt nitrate ($Co\text{-}(NO_3)_2{\cdot}6H_2O$)	2.49	–	Ferromagnetic	10,000 cm^3/g	–	[40]
Fe	7.87	1–30	Ferrimagnetic	$15{,}900\times10^{-8}$ m^3/kg	–	[41]
Ferrous oxide (FeO)	5.74	14.6	Ferrimagnetic	$7{,}200\times10^{-6}$ c.g.s. units	2.57–2.75	[41]
Ferric oxide (Fe_2O_3)	5.242	85	Ferrimagnetic	0.0055 emu/cm^3	2.2	[41]
Iron oxide (magnetite) Fe_3O_4	5.18	78 ± 10	Ferrimagnetic	$50{,}000\times10^{-8}$ m^3/kg	2.6	[41]
Pyrite (FeS_2)	5	1.9 µm	Paramagnetic	30×10^{-8} m^3/kg	0.7–2.6	[42]
Ferrihydrite ($5Fe_2O_3{\cdot}9H_2O$)	3.77 gm/cc	9–16	Paramagnetic	40×10^{-8} m^3/kg	2.47	[42]
Ilmenite ($FeTiO_3$)	4.3–4.6	48	Superparamagnetic	200×10^{-8} m^3/kg	2.5	[42]
Copper nickel alloy	8.5–8.95	34	Ferromagnetic	1×10^6, 30% Ni	2.41	[43]
Cobalt oxide (Co_3O_4)	6.11	<50	Antiferromagnetic	10×10^{-3} emu/(mol Co^{2+})	1.66–2.12	[43]
Carbonyl iron (CI)	7.86	6 µm	Ferromagnetic	4.5–4.8 (χ_s)	–	[43]

for all fillers, polymers, and polymer/filler systems are obtained and listed in Tables 11.3 and 11.4. It can be seen that the addition of the magnetic fillers increases the surface area, van der Waals volume (vwv), refractivity, and polarizability of all studied systems. In case of Kapton-based composites, the computed parameters present larger values, probably to its higher aromatic content in the structural unit in regard to PmT.

TABLE 11.2 Main Magnetic Properties of Common Polymers Used for Composite Fabrication

Polymer	Magnetic Susceptibility	Magnetic Behavior	References
Poly(m-toluidine)	97 (emu/mol)	Diamagnetic	[44]
Polyaniline	24 (emu/mol)	Diamagnetic	[44]
Poly(o-ethylaniline)	115 (emu/mol)	Diamagnetic	[44]
Poly(o-propylaniline)	158 (emu/mol)	Diamagnetic	[44]
Polystyrene	<1	Diamagnetic	[45]
Polypyrrole	1.52×10^{-7} (emu/g)	Diamagnetic	[46]
Polythiophene	1.33×10^{-7} (emu/g)	Diamagnetic	[46]
Kapton	8.02×10^{-7} (emu/g)	Diamagnetic	[47]
Polyethyleneimine-based structures	$11\div41\times10^{-6}$ (emu/g)	Paramagnetic	[48]
Vanadoarenophane-derived structures	–	Paramagnetic	[49]

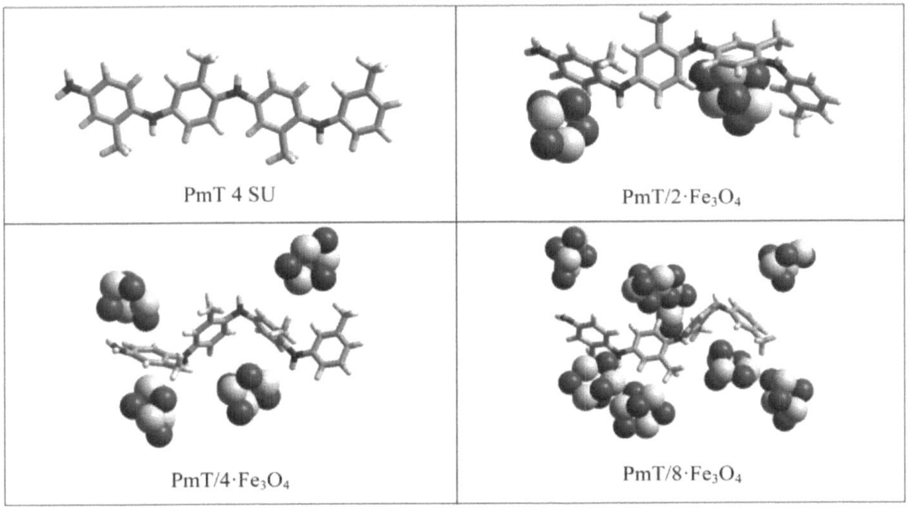

FIGURE 11.2 The conformational optimized structures of poly(m-toluidine) (PmT) having 4 structural units (SUs) with 2, 4, and 8 Fe_3O_4 fillers (overlapping spheres–fillers, tubes – polymer).

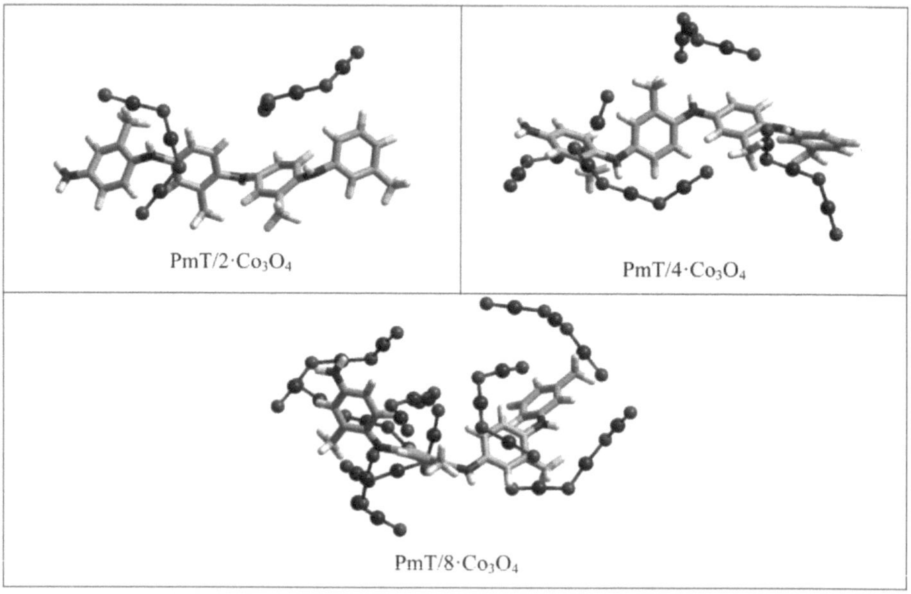

FIGURE 11.3 The conformational optimized structures of PmT (4 SUs) with 2, 4, and 8 Co_3O_4 fillers (balls and cylinders – fillers, tubes – polymer).

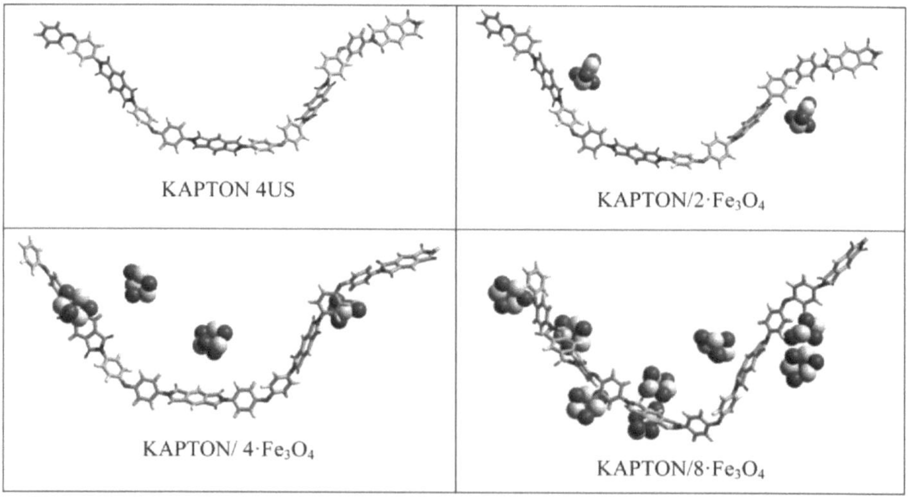

FIGURE 11.4 The conformational optimized structures of KAPTON (4 SUs) with 2, 4, and 8 Co_3O_4 fillers (overlapping spheres – fillers, tubes – polymer).

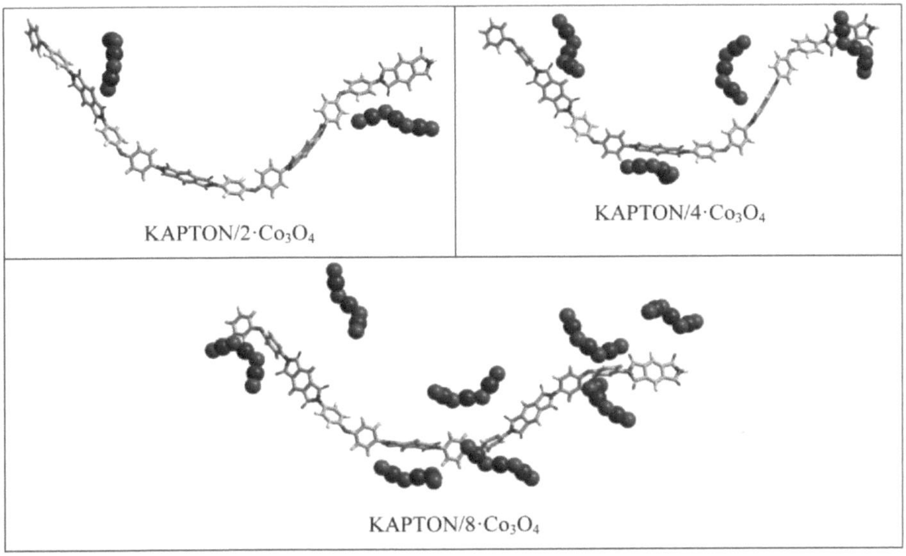

FIGURE 11.5 The conformational optimized structures of KAPTON (4 SUs) with 2, 4, and 8 Co_3O_4 fillers (overlapping fillers – solvents, tubes – polymer).

TABLE 11.3 QSAR Data of the Fillers, PmT, and PmT/Filler Systems

Compound/System	Surface Area (Grid) (Å²)	Volume (vwv) (Å³)	Refractivity (Å²)	Polarizability (Å³)
Fe_3O_4	208.76	47.97	6.48	2.60
Co_3O_4	290.86	49.38	5.59	2.51
PmT	285.28	114.18	35.80	13.62
PmT 4SU	753.24	422.70	136.38	52.16
PmT/2·Fe_3O_4	918.57	520.12	147.18	57.35
PmT/4·Fe_3O_4	934.24	616.76	157.99	62.54
PmT/8·Fe_3O_4	1264.74	808.08	179.59	72.93
PmT/2·Co_3O_4	1093.53	524.04	147.55	57.17
PmT/4·Co_3O_4	1244.94	593.95	158.73	62.18
PmT/8·Co_3O_4	1767.60	819.26	181.07	72.20

The molecular modeling experiments also enabled to determine the log P parameter (partition coefficient), which is related to hydrophobicity degree of the system. The obtained log P values against the amount of the considered magnetic fillers from the aforementioned composites are presented in Figures 11.6 and 11.7. The positive values reflect a hydrophobic character, and it ranges as a function of the used magnetic particles. The presence of Fe_3O_4 increase the hydrophobicity, while Co_3O_4 leads to its decreasing.

TABLE 11.4 QSAR Data of KAPTON and KAPTON/Filler Systems

Compound/System	Surface Area (Grid) (Å²)	Volume (vwv) (Å³)	Refractivity (Å²)	Polarizability (Å³)
KAPTON	603.93	321.98	101.50	39.23
KAPTON 4SU	2110.03	1254.64	401.83	154.59
KAPTON/2·Fe_3O_4	2437.94	1350.58	412.63	159.79
KAPTON/4·Fe_3O_4	2828.75	1446.98	423.43	164.98
KAPTON/8·Fe_3O_4	3304.65	1638.17	445.04	175.36
KAPTON/2·Co_3O_4	2691.33	1353.42	413.00	159.60
KAPTON/4·Co_3O_4	3011.79	1452.08	424.18	164.61
KAPTON/8·Co_3O_4	3901.94	1648.43	446.52	174.63

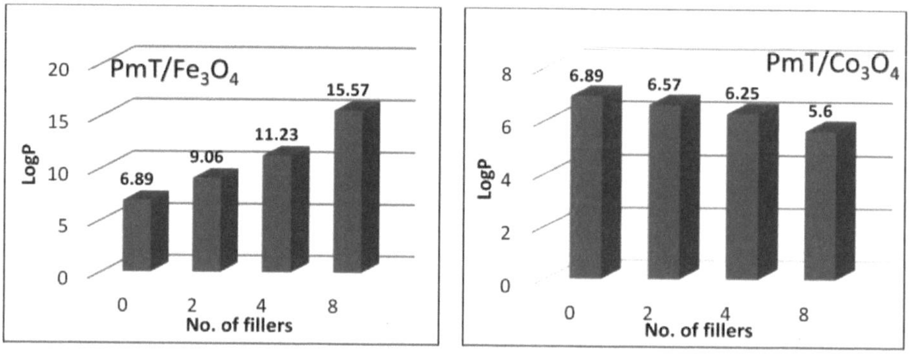

FIGURE 11.6 Log P data for PmT/filler systems.

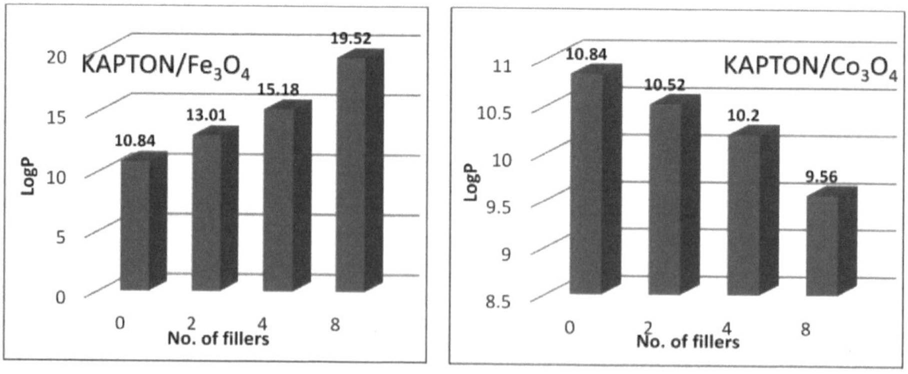

FIGURE 11.7 Log P data for KAPTON/filler systems.

11.4 APPLICATIONS

Practical use of the magnetic composites with polymer matrix is very vast. This part of the chapter is describing several technical applications of magnetic filled polymers by analyzing the most representative achievements in this research area.

Several years ago, Tawansi et al. [50] studied the interfacial effects on the magnetic properties of polymethylmethacrylate (PMMA) loaded with several sorts of magnetic particles attained by casting approach. The PMMA systems filled with Fe present no significant variations of the magnetic susceptibility with temperature (0–100 K), while for the composites with $Y_3Fe_5O_{12}$ a slight decrease is noted after ~230 K and those with $Er_3Fe_5O_{12}$ am increase in susceptibility is noticed up to 180 K and after that it remains almost invariable. At room temperature this magnetic feature at many filler amounts could not fit the magnetic dilution model. The differences between measured values and the estimated ones could be ascribed to the long-distance ferromagnetic interactions occurring among the moment of the filler phase and the spin moment of the macromolecular phase at the interfacial zones. Such investigations are essential for understanding how to design composites with polymer matrix in order to control their magnetic performance.

A highly relevant property for magnetic registering applications is the coercivity. This feature can be enhanced by the shape anisotropy via the following procedures [51]:

- Acicular filler forms growing;
- Magneto-crystalline anisotropy via the non-null angular momentum ion incorporation;
- Surface anisotropy via the absorption of specific ions at the outer layers of the particles.

Dai et al. [52] have made self-assembled polymer composites containing cobalt ferrite fillers for magnetic recording uses. The magnetic inclusions were introduced in a solution of an amphiphilic diblock copolymer, whose acid groups allow linking to the surface of the filler, whereas other segments of the copolymer facilitate the filler-polymer complex occurrence. In this way, the colloidal stability is enhanced by lowering the forces among the magnetic phase responsible for their aggregation. The system was spin-coated onto a silicon layer to facilitate self-organized films, where the particle inter-distance was adjusted by the copolymer molecular mass. The solid composites were exposed to a magnetic field and concomitantly heated beyond the glass transition temperature of one block of the matrix to enable filler rotation and orientation of easy axes parallel to the direction of the external field. The attained self-assembled magnetic filled polymer was adequate as a magnetic recording media and read/write cycles were proved.

Another applicative direction of magnetic composites is represented by energy storing/retrieving. This was studied by Sanida et al. [53], which prepared four series of epoxy nanocomposites loaded with several sorts of magnetic particles. Experimental analyzes prove that the inserting of hard magnetic hexaferrite fillers (such as $BaFe_{12}O_{19}$ and $SrFe_{12}O_{19}$) determines an optimum thermomechanical performance of the materials, whereas the systems loaded with soft magnetic fillers (such as Fe_3O_4 and $ZnFe_2O_4$) having spinel structure, reveal large values of dielectric permittivity and AC conductivity, as well as energy storage features. The reported systems demonstrated versatile magnetic characteristics that were shown to be advantageous in energy-related applications.

Another important use of magnetic polymer composites is for sensing applications. Fullerene/polymer composites were reported by Konchits & collaborators [54] to have paramagnetic behavior. They evidenced that there is a sudden variation in the electron paramagnetic resonance (EPR) data when the temperature ranged between 20°C and 300°C. This seems to be the result of the interactions among the carbon defects and adsorbed gas molecules in the samples. It is proved that the paramagnetic centers in the incorporated fillers in the composite are heterogeneous. The analysis of EPR in regard to absorbed structures is useful for construction of gas sensors. Qin et al. [55] inserted ferromagnetic microwires in a polymer for sensing uses. The materials display a huge magnetoimpedance effect of up to 320% within the frequency domain of 1–100 MHz. This is caused by stress augmented transverse magneto-anisotropy. When addition a larger percent of magnetic phase, the magnetoimpedance is further influenced showing that the prepared polymer system is good for sensing applications. The microwire-doped system also presents remarkable stress-sensing resolution being around 134 kHz/microstrain, which is extremely higher in regard to another short-range radar sensor. As proved by the structural evaluations and tensile data, the low volume fraction of filler (0.01 vol.%) enable the wire-composites to gain outstanding stress sensing features. The ferromagnetic/dielectric composites display remarkable microwave tunable features and electromagnetic (EM) interference shielding abilities. So, these multi-functional materials present great potential for a range of engineering applications. Alfadhel et al. [56] have made nanocomposites based on wires introduced in polydimethylsiloxane (PDMS) or UV sensitive SU-8. Their materials employ the permanent magnetic behavior derived from the filler enabling remote operation in the absence of a supplementary magnetic field to magnetize the reinforcement agent. This is good for constructing miniaturized devices. The composite has also the advantage of easy patterning rendering corrosion resistant, very elastic, combined with permanent magnetic product suitable for sensitive applications. Nanocomposite pillars are made and incorporated in the magnetic sensor parts to accomplish very sensitive, flow, and tactile sensors. The reported sensor could register air or water flow at considerable small power consumption. A tactile sensor unit prototype is fabricated based on the same concept, where a pressure of about 0–169 kPa is ascertained with good resolution.

An interesting applicative direction of magnetic composites is that of smart materials [57]. Thévenot et al. [20] have reviewed the responsive behavior of some magnetic composites with polymer matrix. Magnetic actuation offers special capabilities since it is possible to spatially and temporally control it, and to be triggered outside the system, making possible a non-invasive remote control. According to their work [57], there are several categories of magnetic responsive reinforced materials which are divided as a function to their activation mode and possible use:

- Composites with elastomeric matrix that depict the ability to be deformed (stretching, bending, rotation) after subjecting to magnetic field which is highly desirable for many technical applications.
- Composites with biocompatible matrix have an ability to remotely drag to a desired area, named magnetic guidance, which is especially useful for biomedical uses.
- Composites with the capacity to employ magnetic induction for thermoresponsive actuation sustained by the matrix, which is suitable for controlled substance release and shape memory devices.

Regarding the first category, Zrinyi et al. [58] have prepared ferrogels by loading magnetite in glutaraldehyde cross-linked poly(vinyl alcohol) (PVA). In the presence of inhomogeneous static magnetic field such materials perform great stretching, contraction or bending. It was shown that at low strains, the elongation is ranging with the square of steady current intensity, while the elastic modulus of composites is not affected by the magnetic field. Zamora-Mora et al. [59] reported the preparation and characterization of chitosan/agarose (CS/AG) ferrogels. The incorporation of oleic-acid-coated iron oxide filler led to larger degradation temperatures in regard to neat CS/

AG. Mechanical experiments proved that the elastic moduli of the reinforced ferrogels indicated that the insertion of magnetic filler avoid the AG gelation observed during cooling the CS/AG matrix solution. Also, the composites tend to heat as a consequence of exposure to an alternating magnetic field so that they can be regarded as good candidates for magnetic hyperthermia treatments (MHTs). Magnetic polymer actuators can be attained by introducing CI in silicone-based elastomers. Such material present small Young's modulus (~10^6 Pa) combined with a large magnetization (>10^4 A·m^{-1}), but also wide deformation, in both field gradients and homogeneous magnetic fields [60, 61].

The second category involves a controlled displacement of a material, made of magnetic filler combined with a polymeric matrix. In a variable field, the polymer has the role of multifunctional structure that is capable of adsorbing distinct species, permitting their ulterior separation or just displacement. Levy et al. [62] have shown the magnetic guidance of endothelial cells bearing magnetic fillers. The magnetic platform relied on iron oxide particles and albumin stabilized poly(lactide). Full and uniform bovine aortic endothelial overlay was attained in the presence of the uniform magnetic field. Oppositely, when no magnetic exposure is used, the magnetic filler loaded cells were not detected.

The third category of composites requires the use of inductive heating under the action of the alternative magnetic field. In such situation, magnetic filler produce heat via hysteresis losses (ferromagnetic substances) or via oscillation of the magnetic moment because of the Néel and Brownian relaxations (superparamagnetic fillers). Shape memory materials might perform significant shape modification, upon application of a programming process after which they regain the initial shape. Lendlein et al. [53] have obtained triple-shape polymer composites based on containing poly(ε-caprolactone) (PCL) and poly(cyclohexyl methacrylate) (PCHMA) useful for non-contact actuation. The prepared systems display two thermal transitions. The insertion of magnetic filler did not affect much the thermal performance of the matrix but influenced the mechanical performance by reducing the elongation at break. Therefore, the amount of filler must be selected to have a good balance among the heating characteristics (bigger reachable temperature) and the mechanical ones. An essential factor that must be considered in order to attain the desired control on shape recuperation is the surface to volume ratio (S/V). As a function of the S/V parameter, the heat loss (caused by differences with environment) might range, involving considerable differences regarding the magnetic field strength demanded to acquire the transition temperature. The remarkable triple shape features with full recuperation of the initial shape were attained for samples with 40 wt.% of PCL. A stepwise enhancement of the magnetic field strength facilitated the recuperation of the intermediate and initial shapes in a gradual manner. Such studies open novel perspectives on actuation devices with futuristic design.

11.5 CONCLUSIONS

The magnetic composites are important materials in many technical applications. This work presents the main classes of magnetic materials and the sorts of magnetic behaviors. Some molecular modeling data are described for some magnetic fillers and their polymer composites by emphasizing the influence of the filler on the QSAR properties and log P parameter. It can be noted that these are changed as a function of both polymer type and filler nature. Finally, some application of magnetic polymer composites in sensors, magnetic recording, actuators, and energy storing/retrieving are briefly reviewed.

KEYWORDS

- **dilute magnetic semiconductors**
- **magnetic filler**
- **magnetorheological**
- **molecular modeling**
- **poly(m-toluidine)**
- **polymer**
- **structural units**

REFERENCES

1. Kumbhar, B. K., Patil, S. R., & Sawant, S. M., (2015). Synthesis and characterization of magneto-rheological (MR) fluids for MR brake application. *Eng. Sci. Technol. an Int., 18*, 432–438.

2. Adams, S. A., Hauser, J. L., Allen, A. C., Lindquist, K. P., Ramirez, A. P., Oliver, S., & Zhang, J. Z., (2018). Fe_3O_4@SiO_2 Nanoparticles functionalized with gold and poly (vinylpyrrolidone) for bio-separation and sensing applications. *ACS Appl. Nano Mater., 1*, 1406–1412.
3. Baghban, R., Soleymani, M. A. J., & Rahimi, M., (2021). Were magnetic materials useful in cancer therapy? *Biomed. Pharmacother, 144*, 112321.
4. Sun, C., Lee, J. S. H., & Zhang, M., (2008). Magnetic nanoparticles in MR imaging and drug delivery. *Adv. Drug Delivery Rev., 60*, 1252–1265.
5. Gorodov, V. V., Kostrov, S. A., Kamyshinskii, R. A., Kramarenko, E. Y., & Muzafarov, A. M., (2018). Modification of carbonyl iron particles by carboxyl-containing polydimethylsiloxanes. *Russian Chem. Bull., 67*, 1639–1647.
6. Ding, L., Xuan, S., Feng, J., & Gong, X., (2017). Magnetic/conductive composite fiber: A multifunctional strain sensor with magnetically driven property. *Compos. Part A Appl. Sci. Manuf., 100*, 97–105.
7. Khan, M. A., Alfadhel, A., Kosel, J., & Bakolka, M., (2016). Fabrication and characterization of magnetic composite membrane pressure sensor. *IEEE Sensors Applications Symposium (SAS).* doi: 10.1109/sas.2016.7479871.
8. Guo, Y. G., & Zhu, J. G., (2006). Applications of soft magnetic composite materials in electrical machines. *Aust. J. Electr. Electron., 3*, 37–46.
9. Kumar, U. N., Kratz, K., Wagermaier, W., Behl, M., & Lendlein, A., (2010). Non-contact actuation of triple-shape effect in multiphase polymer network nanocomposites in alternating magnetic field. *J. Mater. Chem., 20*, 3404–3415.
10. Hassnain, J. G., Ekiert, T., Unruh, K. M., & Ismat, S. S., (2012). Effect of particle size distribution on the magnetic properties γ-Fe_2O_3 nanoparticles. *Mater. Sci. Eng. B, 177*, 935–941.
11. Wu, S., Hu, W., Ze, Q., Sitti, M., & Zhao, R., (2020). Multifunctional magnetic soft composites: A review. *Multifunct. Mater., 3*, 042003.
12. Liew, K. B., Goh, C. F., Sajid, A. H., & Syed, K., (2021). Overview of mechanical and physicochemical properties of polymer matrix composites. In: Brabazon, D., (ed.), *Encyclopedia of Materials: Composites* (Vol. 1, pp. 565–576). Elsevier: USA.
13. Filimon, A., Albu, R. M., Stoica, I., & Avram, E., (2016). Blends based on ionic polysulfones with improved conformational and microstructural characteristics: Perspectives for biomedical applications. *Compos. B. Eng., 93*, 1–11.
14. Barzic, A. I., Albu, R. M., Stoica, I., & Hulubei, C., (2022). New shielding covers based on transparent polyimide/ferrous sulfide composites that reduce optical losses in solar cells. *Compos. Sci. Technol., 218*, 109140.
15. Albu, R. M., Avram, E., Stoica, I., & Ioan, S., (2012). Polysulfones with chelating groups for heavy metals retention. *Polym. Compos., 33*, 573–581.
16. Freile-Pelegrín, Y., & Madera-Santana, T. J., (2017). Characterization techniques for algae-based materials. In: Zia, K. M., Zuber, M., & Ali, M., (eds.), *Algae Based Polymers, Blends, and Composites: Chemistry, Biotechnology and Materials Science* (pp. 649–670). Elsevier: USA.
17. Mahesh, V., Joladarashi, S., & Kulkarni, S. M., (2021). A comprehensive review on material selection for polymer matrix composites subjected to impact load. *Def. Technol., 17*, 257–277.
18. Liao, X., Liao, Q., Zhang, Z., Yan, X., Liang, Q., Wang, Q., Li, M., & Zhang, Y., (2016). A highly stretchable ZnO@ fiber-based multifunctional nanosensor for strain/temperature/UV detection. *Adv. Funct. Mater., 26*, 3074–3081.

19. Huang, G., Liu, L., Wang, R., Zhang, J., Sun, X., & Peng, H., (2016). Smart color-changing textile with high contrast based on a single-sided conductive fabric. *J. Mater. Chem. C, 4*, 7589–7594.
20. Thévenot, J., Oliveira, H., Sandre, O., & Lecommandoux, S., (2013). Magnetic responsive polymer composite materials. *Chem. Soc. Rev., 42*, 7099.
21. Kango, S., Kalia, S., Celli, A., Njuguna, J., Habibi, Y., & Kumar, R., (2013). Surface modification of inorganic nanoparticles for development of organic–inorganic nanocomposites: A review. *Prog. Polym. Sci., 38*, 1232–1261.
22. Kalia, S., Kango, S., Kumar, A., Haldorai, Y., Kumari, B., & Kumar, R., (2014). Magnetic polymer nanocomposites for environmental and biomedical applications. *Colloid. Polym. Sci., 292*, 2025–2052.
23. Gopinadhan, K., Kashyap, S. C., Pandya, D. K., & Chaudhary, S., (2007). High temperature ferromagnetism in Mn-doped SnO_2 nanocrystalline thin films. *J. Appl. Phys., 102*, 113513–113518.
24. Kant, K. M., Sethupathi, K., & Rao, M. S. R., (2004). Magnetic properties of 4f element doped SnO_2. *International Symposium of Research Students on Materials Science and Engineering, ISRS-2004*. Chennai, India. Technical Proceedings.
25. Torrance, J. B., Bagus, P. S., Johhannsen, I., Nazzal, A. I., Parkin, S. S. P., & Batail, P., (1998). Ferromagnetic interactions in organic solids: An overview of theory and experiment. *J. Appl. Phys., 63*, 2962–2965.
26. Rajca, A., Wongsriratanakul, J., & Rajca, S., (2001). Magnetic ordering in an organic polymer. *Science, 294*, 1503–1505.
27. Zaidi, N. A., Giblin, S. R., Terry, I., & Monkman, A. P., (2004). Room temperature magnetic order in an organic magnet derived from polyaniline. *Polymer, 45*, 5683–5689.
28. Crayston, J. A., Devine, J. N., & Walton, J. C., (2000). Conceptual and synthetic strategies for the preparation of organic magnets. *Tetrahedron, 56*, 7829–7857.
29. Scepka, T. (2016). *Noninvasive Control of Magnetic State in Ferromagnetic Nanodots by Hall Probe Magnetometry*. https:www.researchgate.net/publication/301549615 (accessed on 16 January 2022).
30. *Magnetic Materials: Types of Magnetism, Applications, and Origin of Magnetism*. https:msestudent.com/magnetic-materials-types-of-magnetism-applications-and-origin-of-magnetism/ (accessed on 16 January 2022).
31. Khan, M. A., (2019). *Magnetic Polymer Composite Transducers for Integrated Systems*. Dissertation thesis, King Abdullah University of Science and Technology Thuwal, Kingdom of Saudi Arabia.
32. Cullity, B. D., & Graham, C. D., (2008). *Introduction to Magnetic Materials* (2nd edn.). Wiley: USA.
33. Coey, J. M. D., (2010). *Magnetism and Magnetic Materials*. Cambridge University Press: Dublin.
34. Shokrollahi, H., (2017). A review of the magnetic properties, synthesis methods and applications of maghemite. *J. Magn. Magn. Mater., 426*, 74–81.
35. Urogiova, E., Hudec, I., Bellusova, D., & Suri, P., (2007). Structure and properties of rubber blends with ferromagnetic fillers. *Polym. Adv. Technol., 18*, 128–134.
36. Pullar, R. C., (2012). Hexagonal ferrites: A review of the synthesis, properties and applications of hexaferrite ceramics. *Prog. in Mater. Sci., 57*, 1191–1334.
37. Önal, M. A. R., Dewilde, S., Degri, M., Pickering, L., Saje, B., Riaño, S., Walton, A., & Binnemans, K., (2020). Recycling of bonded NdFeB permanent magnets using ionic liquids. *Green Chem., 22*, 2821.

38. Loan, N. T. T., Lan, N. T. T., Hang, N. T. T., Hai, N. Q., Anh, D. T. T., Hau, V. T., Tan, L. V., & Tran, T. V., (2019). $CoFe_2O_4$ nanomaterials: Effect of annealing temperature on characterization, magnetic, photocatalytic, and photo-Fenton properties. *Processes, 7*, 885.
39. Căpraru, A., Moacă, E. A., Păcurariu, C., Ianoş, R., Lazău, R., & Barbu-Tudoran, L., (2021). Development and characterization of magnetic iron oxide nanoparticles using microwave for the combustion reaction ignition, as possible candidates for biomedical applications. *Powder Technol., 394*, 1026–1038.
40. Rahmana, A., Yusufa, A., Putria, R. A., Anggrainia, Y., Pujaningsihb, F. B., Kurniaa, D., Wonorahardjoc, S., & Sutjahja, I. M., (2021). Effect of static magnetic field on nucleation of cobalt nitrate hexahydrate. *Mater. Res., 24*, e20210088.
41. Ganapathe, L. S., Mohamed, M. A., Yunus, R. M., & Berhanuddin, D. D., (2020). Magnetite (Fe_3O_4) nanoparticles in biomedical application: From synthesis to surface functionalization. *Magnetochemistry, 6*, 68.
42. Van, D. R. L., Hendrickx, J. M. H., Harrison, B., Borchers, B., Norman, D. I., Ndur, S., Jasper, C., et al., (2004). Spatial variability of magnetic soil properties. In: Harmon, R. S., Broach, J. T., & Holloway, Jr. J. H., (eds.), *Detection and Remediation Technologies for Mines and Minelike Targets XI* (Vol. 5415, pp. 665–676). Proceedings of SPIE: Bellingham, WA.
43. Lu, Q., Choi, K., Nam, J. D., & Choi, H. J., (2021). Magnetic polymer composite particles: Design and magnetorheology. *Polymers, 13*, 512.
44. Kahol, P. K., Spencer, W. R., Pinto, N. J., & McCormick, B. J., (1994). Magnetic-susceptibility analysis of polyaniline and its derivatives in terms of triplet bipolarons. *Phys. Rev. B, 50*, 18647–18650.
45. https:www.rgpballs.com/en/polystyrene-ps-balls/ (accessed on 16 January 2022).
46. Nalwa, H. S., (1989). Phase transitions in polypyrrole and polythiophene conducting polymers demonstrated by magnetic susceptibility measurements. *Phys. Rev. B, 39*, 5964–5974.
47. Jiang, J., Han, H., Xia, W., Guo, Y., Zhang, Y., Fang, H., & Yan, L., (2018). *Observation of Meisser Effect in Kapton Tapes*. https:arxiv.org/ftp/arxiv/papers/1805/1805.12458.pdf (accessed on 16 January 2022).
48. Yu, X., Yuan, X., Zhao, Y., & Ren, L., (2015). Synthesis of paramagnetic polymers based on polyethyleneimine (PEI). *RSC Adv., 5*, 92207–92211.
49. Braunschweig, H., Adams, C. J., Kupfer, T., Manners, I., Richardson, R. M., & Whittell, G. R., (2008). Synthesis of a paramagnetic polymer by ring-opening polymerization of a strained. *Vanadoarenophane, 47*, 3826–3829.
50. Tawansi, A., & Zidan, H. M., (1991). Interfacial effects on the magnetic susceptibility of PMMA composites. *Internat. J. Polym. Mater., 15*, 45–56.
51. Kronmüller, H., Parkin, S., Waser, R., Böttger, U., & Tiedke, S., (2007). *Handbook of Magnetism and Advanced Magnetic Materials*. Wiley: USA.
52. Dai, Q., Berman, D., Virwani, K., Frommer, J., Jubert, P. O., Lam, M., Topuria, T., Imaino, W., & Nelson, A., (2010). Self-assembled ferrimagnet−polymer composites for magnetic recording media. *Nano Lett., 10*, 3216–3221.
53. Sanida, A., Stavropoulos, S. G., Speliotis, Th., & Psarras, G. C., (2021). Evaluating the multifunctional performance of polymer matrix nanodielectrics incorporating magnetic nanoparticles: A comparative study. *Polymer, 236*, 124311.
54. Konchits, A. A., Shanina, B. D., Krasnovyd, S. V., Burya, A. I., & Kuznetsova, O. Y., (2017). Paramagnetic properties of fullerene-derived nanomaterials and their polymer composites: Drastic pumping out effect. *Nanoscale Res. Lett., 12*, 475.

55. Qin, F., Peng, H. X.,Tang, J., & Qin, L. C., (2010). Ferromagnetic microwires enabled polymer composites for sensing applications. *Compos.-A: Appl. Sci. Manuf., 41*, 1823–1828.
56. Alfadhel, A., Li, B., & Kosel, J., (2014). Magnetic polymer nanocomposites for sensing applications. *IEEE Sensors Proceedings*, 2066–2069.
57. Filipcsei, G., Csetneki, I., Szilágyi, A., & Zrínyi, M., (2007). Magnetic field-responsive smart polymer composites. In: *Oligomers-Polymer Composites-Molecular Imprinting: Advances in Polymer Science* (Vol. 206, pp. 137–189). Springer: Berlin, Heidelberg.
58. Zrinyi, M., Barsi, L., & Buki, A., (1996). Deformation of ferrogels induced by nonuniform magnetic fields. *J. Chem. Phys., 104*, 8750–8756.
59. Zamora-Mora, V., Soares, P., Echeverria, C., Hernández, R., & Mijangos, C., (2015). Composite chitosan/agarose ferrogels for potential applications in magnetic hyperthermia. *Gels, 1*, 69–80.
60. Bellan, C., & Bossis, G., (2002). Field dependence of viscoelastic properties of MR elastomers. *Int. J. Mod. Phys. B, 16*, 2447–2453.
61. Nguyen, V. Q., Ahmed, A. S., & Ramanujan, R. V., (2012). Morphing soft magnetic composites. *Adv. Mater., 24*, 4041–4054.
62. Polyak, B., Fishbein, I., Chorny, M., Alferiev, I., Williams, D., Yellen, B., Friedman, G., & Levy, R. J., (2008). High field gradient targeting of magnetic nanoparticle-loaded endothelial cells to the surfaces of steel stents. *Proc. Natl. Acad. Sci. USA, 105*, 698–703.

CHAPTER 12

Magnetism Towards Smart Materials

V. N. ARCHANA,[1] N. G. DIVYA,[2] and REYHA BENEDICT[3]

[1]*Mar Athanasius College, Kothamangalam, Ernakulam, Kerala, India*

[2]*Bharat Mata College, Thrikkakkara, Ernakulam, Kerala, India*

[3]*St. Teresa's College, Ernakulam, Kerala, India*

ABSTRACT

'Smart materials' have awakened an increased interest in recent years since they can be exploited to achieve the 'respond to stimuli' nature. The stimuli can be a magnetic field, electric field, or temperature. The idea of changing the mechanical properties of materials by applying a magnetic field is not at all a new concept. This chapter gives a brief introduction to magnetism and nanomagnetism. Also, the magnetic properties of smart nanomaterials are discussed in detail.

12.1 MAGNETISM AND NANOSCALE MAGNETISM

Magnetism is a fundamental phenomenon of nature, and the history of magnetism dates back to ancient civilizations. It is a property of a material to react to a useful magnetic field that causes the material to be either attracted or repelled. Some practical materials that show simply detectable magnetic properties (called magnets) are:

- nickel;
- iron;
- cobalt; and
- gadolinium and their alloys.

Modern Magnetic Materials: Properties and Applications. Iuliana Stoica, Ann Rose Abraham, & A. K. Haghi (Eds.)

Meanwhile, all materials are formed to much greater or somehow, lesser degree by the presence of a magnetic field. However, the potentiality of magnets to act on objects at a distance makes them valuable tools of science.

There was an earlier saying that history cannot be changed. But the technology especially nanotechnology make revolutions in history. Thus, miniaturization became the technological trend to produce smaller, lighter, and cheaper devices having improved functionality using less energy. Nanotechnology is defined by the understanding and manipulation of the material world at the nanoscale or near atomic scale. Nano which means dwarf is the 1 billionth of a meter (10^{-9} m). Nanoscience and nanotechnology are concerned with the science and engineering of creating materials, functional structures, and devices with at least one of the dimensions in the size range of 1–100 nm. The main idea of nanotechnology and nanoscience is first introduced by the physicist Richard P. Feynman in his famous lecture titled "There's plenty of room at the bottom" in 1959. He thus first paved the idea of manipulation of matter at a very small scale. Eric Drexler, an American Engineer introduced and popularized the word 'nanotechnology' through his book *Engines of Creation: The Coming Era of Nanotechnology* in 1986.

Nanomagnetism is now at the frontiers of nanoscience and technology, and nanosized magnetic materials exhibit a rich variety of magnetic phenomena in comparison with the bulk counterparts, giving rise to a novel range of applications. It basically involves the study of ferromagnetic materials behavior when they are analytically controlled in at least one dimension. The physical properties observed at such reduced dimensions are strongly sensitive to slight variations of size, shape, composition, structure, and morphology as well as myriad surface effects. Compared to the bulk counterparts, magnetic nanoparticles (MNPs) exhibit a dramatic change in their magnetic properties.

According to Weiss's theory, a ferromagnet consists of small regions called domains, where dipoles are aligned parallel to one another. All the domains align in the same direction and there exists a net non-zero magnetization. Magnetization process in a ferromagnetic material can be explained using domain theory. Initially the domains are aligned in such a way that net magnetization is "zero." In the existance of "Magnetic Field," domains grow at the expense of other domains by domain wall motion. When the field is sufficient to eliminate all domain walls, a single domain exists [1, 2]. Superparamagnetism and single domain particles are the two major aspects related to nanomagnetism. Single domain is a state of ferromagnet in which the magnetization direction is the same throughout the material. When the particle size is reduced and when it is closer to a critical size, the formation of domain walls will not be energetically favorable, and the particles become single domain. For a single domain particle magnetization will change by the rotation of spins. Superparamagnetism is seen in ferromagnetic or ferrimagnetic nanoparticles composed of a single domain. Zero remanence and zero coercivity are the two characteristic features of superparamagnetism.

12.2 MAGNETIC PROPERTIES OF SMART NANOMATERIALS

The term "Smart Materials" is frequently used for special materials that have the potential to sense changes in their environment and can particularly respond in a very controlled and practical manner. Incorporating such materials in structures makes the structures smart, in the sense that they have a capability to respond to external stimuli to compensate for undesired effects, or to enhance their performance. The terms "active," "adaptive," and "intelligent" materials are also used to describe this kind of material [3, 4]. Examples of smart materials that are commercially available are piezoelectric materials (which change length with the application of an electric field), magnetostrictive materials (which change length with the application of a magnetic field), shape memory alloys (which regain their original shape when heated, if they have been deformed at ambient temperature), biomimetic materials (which imitate biological tissues and structures by changing their dimensions or chemical diffusivity as a response to a chemical or an electrical input), electrochromic coatings (which change color when an electric potential is applied), and electrorheological (ER) and magnetorheological (MR) materials [4]. Therefore, these materials should be some types of:

- sensor;
- processor; and
- actuator.

These characteristics must be looped, and the effect of changing the "materials properties" must be done in "real time."

The magnetic structures at the nanoscale size range also possess great potential for the development of smart nanomaterials with huge functional diversity. The optical, chemical, physical, and electrochemical properties of the smart magnetic nanomaterials can be tuned by modifying the externally applied magnetic fields. Thus, by considering the benefits of these smart functional magnetic nanomaterials, several research attempts are recently being made to develop more efficient nanoscale magnetic assembly for technological advancement in various fields.

Magnetic properties of nanomaterials arise from the spin and orbital motions of electrons. Depending on the magnetic properties, the nanomaterials are broadly classified into different categories such as:

- ferromagnetic;
- paramagnetic;
- diamagnetic;
- ferrimagnetic; and
- antiferromagnetic materials.

Among these different kinds of magnetic materials, ferromagnetic and ferrimagnetic materials are preferred for hyperthermia therapy applications. Ferromagnetism is an inherent property of certain materials such as Fe, Ni, Co, and their alloys. In ferromagnetic materials with a large and positive magnetic susceptibility, the magnetic moments existing in the same ferromagnetic domains are aligned parallel to each other at room temperature which gives rise to net magnetic moments in each domain, at the same time the net magnetic moments of various domains are arbitrary oriented at room temperature, hence the net magnetization of the ferromagnetic material will be zero. However, in the presence of an applied magnetic field, these moments get aligned along the direction of the applied field.

The plot of magnetization versus applied magnetic field produces hysteresis loops in ferromagnetic materials. The ferromagnetic materials with thin hysteresis loops are grouped into soft ferromagnets and those with wider hysteresis loops are grouped into hard ferromagnets. Above a critical temperature known as transition temperature T_c, the ferromagnetic materials lose their ferromagnetic properties and become paramagnetic. These functional traits can also be utilized for the magnetic applications of smart functional nanomaterials in industries.

The ferrimagnetism arises in oxides of Fe combined with one or more transition metals such as Zn, Ni, Mn, etc., in which unequal and adjacent magnetic spins are aligned antiparallel to each other which produces a net nonzero magnetic moment even in the absence of an applied magnetic field. In the presence of an externally applied magnetic field, these moments get aligned along the direction of the applied field.

Superparamagnetism with a non-hysteresis curve having zero Hc (coercivity) and Mr (remnant magnetization) occurs in both ferromagnetic and ferrimagnetic materials with particle sizes around a few nanometers. When the size of these materials is of the order of the diameter of a domain, they enter into a superparamagnetic state, which is a promising thermal agent for magnetic fluid hyperthermia applications. These materials are aligned in a preferred arrangement, thus getting magnetized quickly compared to other states, whereas each of the atomic magnetic moments maintains its ordered state. Thus, biocompatible superparamagnetic nanoparticles (SPMNPs) are promising candidates to be utilized in magnetic hyperthermia applications as compared with other magnetic nanomaterials.

12.2.1 SMART MAGNETIC NANOMATERIALS FOR HYPERTHERMIA

Hyperthermia is a therapeutic method for cancer treatment in which cancerous cells or tumors are selectively destroyed by exposure to higher temperatures usually in the range of 41–46°C without destroying the healthy cells. The term hyperthermia is derived from the Greek words *hyper* (rise) and *therme* (heat). This treatment procedure is possible due to the lack of heat resistance capacity of tumor cells compared to normal cells [5]. During this procedure, MNPs are injected near the site of the tumor cells within the vicinity of the external alternating magnetic field which helps to focus the heat energy only on the affected cells. This is known as magnetic hyperthermia or magnetic nanofluid hyperthermia if the colloidal suspension of MNP in a liquid medium is used for hyperthermia applications [6]. This treatment procedure has a long-lasting history in the annals of cancer treatment, either alone or in combination with other therapeutic techniques. The schematic illustration of magnetic hyperthermia application is given in Figure 12.1.

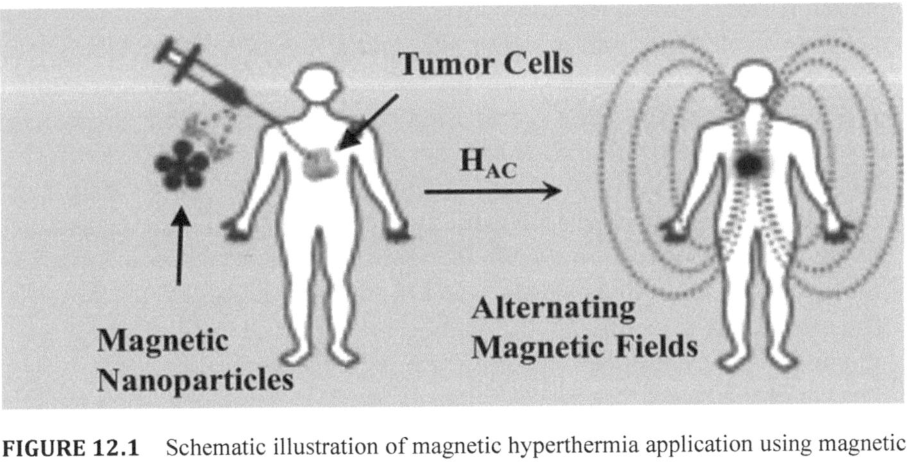

FIGURE 12.1 Schematic illustration of magnetic hyperthermia application using magnetic nanoparticles.

Source: Reproduced with permission from Ref. [6]; Copyright ©2022, Elsevier.

The use of magnetic materials for magnetic hyperthermia applications to produce heat when exposed to an alternating magnetic field was first proposed by Gilchrist et al. in the 1950s [7]. Then, the idea of intracellular magnetic hyperthermia treatment (MHT) was coined by Gordon et al. by employing dextran-coated magnetite materials to increase the temperature of tumor cells in an alternating magnetic field [8]. Later in 2004, the first clinical treatment system was developed in Berlin, and a few years later, magnetic hyperthermia has been clinically approved for brain tumor treatment in Europe. However, this treatment procedure is still not commonly utilized in hospitals, especially in the USA and some other parts of the world, which can be attributed to gaps in the development of optimized MNP hyperthermia agents. Recently, with the advent of nanotechnology, the ongoing scientific progress in cancer treatment research has taken a new dimension and more innovative magnetic nanomaterials have been proposed by several researchers to kill cancer cells [9].

MNPs are generally zero-dimensional, inorganic, magnetic particles in nanosize that can be manipulated in an external alternating magnetic field and consequently employed *in vitro* (outside the body) and *in vivo* (inside the body) bio-medical applications like the MRI or magnetic resonance imaging (MRI), and magnetic hyperthermia, respectively. These materials usually exhibit unique properties, such as superparamagnetism, high saturation magnetization (Ms), biocompatibility, and low toxicity. The biomedical applications of MNPs depend on a number of factors such as physicochemical properties, particle size distribution, synthesis methods, interaction among particles, etc. [6].

Upon excitation in an external alternating magnetic field, MNPs transform the electromagnetic (EM) energy into heat energy, which can be utilized to destroy cancer cells in hyperthermia treatment procedures. In magnetic hyperthermia, the heating of MNPs occurs due to the eddy current heating due to induction in an alternating magnetic field, frictional heating due to mutual interaction between NPs and the surrounding medium, or hysteretic losses and spin relaxation of MNPs. However, the structural changes and inter-particle connections surely affect the heating capability of these NPs. Thus, in-depth scientific investigations are required in this direction for efficient practical applications of these NPs in biomedical fields [9].

One of the major theoretical advantages of MHT includes the possibility of local, regional, or generalized heating while keeping the surrounding healthy cells safe [10]. Research works are still in progress in this area to use MNPs in the medical field more effectively as an efficient medium to produce heat. However, several drawbacks are yet to be addressed, such as the difficulty of temperature monitoring and *in vivo* temperature control during the treatment procedure for more efficient clinical applications [5].

The clinical applications of magnetic hyperthermia also require proper optimization of the magneto-structural properties, the effect of magnetic dipolar interactions, and the reliability of methods used in hyperthermia measurements that need to be addressed on priority. This method is not very effective in fully curing cancer cells due to some serious drawbacks of hyperthermia such as heterogeneous distribution of heat and incapability to prevent overheating on the affected site.

12.2.1.1 COMMON MAGNETIC NANOMATERIALS FOR HYPERTHERMIA APPLICATIONS

MNPs with inherent properties such as magnetic metal oxide NPs and alloys have widely been investigated for the past several years for hyperthermia applications. Upon excitation with an externally applied alternating magnetic field, these nanomaterials produce heat energy that can be utilized to destroy cancer cells in hyperthermia treatment.

Magnetic metal oxide NPs include materials such as iron oxide, ferrites, nickel oxide, etc. Among the different magnetic metal oxide NPs, iron oxide is explored extensively for hyperthermia applications due to its natural abundance, low cost, tunable magnetic properties, biocompatibility, and potential as a diagnostic and therapeutic agent. Iron oxide exists in 16 pure phases comprised of oxides such as α-Fe_2O_3, β-Fe_2O_3, Fe_3O_4, γ-Fe_2O_3, etc., and oxyhydroxides such as akaganeite, goethite, feroxyhyte, etc., and hydroxides. Ferrites are having a general representation of $M(Fe_xO_y)$, where M represents metal cations and x and y are variables. Still, several kinds of research are going on to develop more efficient magnetic nanomaterials with minimum drawbacks for practical applications.

12.2.1.2 EFFECTS OF SIZE AND SHAPE OF MAGNETIC NANOPARTICLES (MNPS) ON HYPERTHERMIA

Magnetic properties of nanomaterials depend on the size, shape, shape anisotropy, and uniformity of the NPs. Normally, small magnetic particles with a size in the range of 5–20 nm, leading to a high magnetization and high induction heating efficiency. However, NPs with a size below 5 nm suffer surface spin disorders, decreasing nanomaterials' magnetization. The wide size distribution adversely affects the magnetic properties such as saturation magnetism, coercivity, etc. Magnetic materials produce heat via eddy current loss, hysteresis losses in multi-domain magnetic particles (size > 1 μm), relaxation losses in single-domain SPMNPs, and frictional losses especially in viscous suspension in an alternating magnetic field.

The majority of magnetic particles have higher electrical resistivity, which causes the induction of very low eddy current loss. The eddy current loss is more prominent in multi-domain magnetic particles which is negligible in MNPs and thus the prominent heat loss mechanism in multidomain magnetic materials with ferromagnetic properties is the hysteresis loss (size > 20 nm). When the size of the particle is small (< 20 nm), the thermal energy barrier for the reflux of magnetization will be decreased, causing a multidomain to single-domain magnetic transition that leads to a relaxation loss mechanism. Once the alternating magnetic field is removed, the magnetic moments undergo Néel relaxation (relaxation of internal magnetic spins) or Brownian relaxation (rotation of individual NPs). Usually, superparamagnetic Néel and Brownian heat loss mechanisms are preferred for magnetic hyperthermia applications which exhibit good magnetic properties with a lack of magnetization in the absence of an applied magnetic field.

12.2.2 MAGNETICALLY RESPONSIVE PLASMONIC NANORODS

Surface plasmon resonance (SPR) refers to the collective motion of conduction electrons oscillating on the surface of metallic nanostructures when irradiated with light, which is highly sensitive to size, shape, and the surrounding medium. One-dimensional anisotropic plasmonic nanostructures produce novel optical band structures due to their anisotropic nature, compared to isotropic spherical nanostructures that can be effectively controlled by employing an externally applied driving force such as the magnetic field. The externally applied magnetic fields have been usually preferred due to positional adjustment, synergistic effects, magnetic separation, etc.

Surface plasmons are sensitive to the refractive index of the dielectric medium that can be utilized for the development of plasmonic biosensors for the detection of analyte molecules. Most of the commercially available SPR biosensors use surface plasmon polaritons (SPPs), which can provide detection up to a limit exceeding 10^{-5} refractive-index units (RIU) [11]. However, the SPP-based method needs further improvement in sensitivity and selectivity, especially size-based selectivity, and selective nano-architectures. Considering all these drawbacks of sensing applications of plasmonics structures, new plasmonic metamaterials, and composite structures that are capable of providing more sensitive detection than SPPs are still under investigation because when the distance between the nanorods is smaller or comparable with the wavelength, the metamaterial supports a guided mode with the field distribution inside the layer depend on the plasmon-mediated interaction between the nanorods. This has resonant excitation conditions similar to SPP of a metal film, with high sensitivity to refractive-index change [11].

SPR of noble metal nanomaterials have been studied for several applications like optoelectronics, photothermal therapy, chemical sensing, etc. Since the oscillation of electrons depends on the size, shape, and neighboring particles, several research investigations have been devoted to controlling the plasmonic property by tuning these parameters during the synthesis of these nanostructures. However, reversible tuning of plasmonic property of colloidal metal nanostructures is possible by changing their orientation using an external magnetic field. The magnetic orientations of these nanostructures can be controlled by binding them to superparamagnetic iron oxide nanorods so that the resulting hybrid nanostructures tend to align in the direction of the externally applied magnetic field to minimize their potential energy, thus enabling the excitation of the plasmon modes of nanostructures through the alteration of the field direction compared to the directions of incidence and polarization of light.

Nanostructures like gold NPs usually exhibit transverse and longitudinal modes of resonance, which produces distinct bands with different wavelengths in the extinction spectrum. The shorter wavelength bands correspond to the excitation of transverse plasmon due to the oscillation of electrons along short axes, and the longer wavelength bands are produced by the excitation of longitudinal plasmon due to the oscillation of electrons along the longer axis. The excitation of these different modes is determined by the orientation of the nanorods concerning the direction of polarization of incident light. When the direction of polarization of incident light is parallel to the longer axes of nanorods, longitudinal plasmonic excitation occurs. And the transverse mode will be excited when the polarization of light is parallel to the short axes of nanorod structures [12].

The presence of an external magnetic field allows tuning the optical properties of nanorod structures even under normal light illumination. The magnetic field enables the instant and reversible alignment of nanorod structures in three dimensions. Moreover, the anisotropic nature of magnetic interactions allows the proper alignment of magnetic dipoles along the direction of the external magnetic field. When the nanorod structure is nonmagnetic, magnetically active nanomaterials need to be incorporated into the nanorods to enable magnetic control. As the magnetic nanorods orient themselves in the direction of the externally applied magnetic field to reduce their potential energy and dipole-dipole interactions, thus the nanostructures also will be aligned in the same direction enabling magnetic control of these nanostructures plasmonic excitation [12]. However, the net magnetization should be small enough to minimize the magnetically induced agglomeration of magnetic nanostructures, at the same time the magnetic response of these nanostructures should be strong enough to ensure the reorientation under the influence of normal magnetic fields. The diameter of nanorods and the size of the nanostructures should be close so that a parallel attachment of these nanostructures is preferred [12, 13]. The colloidal dispersion of such nanostructures shows instant optical switching concerning a change in the orientation or strength of externally applied magnetic fields. The optical switching is highly sensitive, which is possible even under a weak applied magnetic field. These tunable plasmonic hybrid nanostructures provide an efficient platform for developing novel optical components, display devices, and highly efficient sensors [11, 12].

12.2.3 MAGNETORHEOLOGICAL (MR) FLUIDS

Among smart materials, ER and MR materials are an important group. They are a class of materials where rheological properties are rapidly varied by the application of an external electric or magnetic field. The change in their properties is in proportion to the magnitude of the field applied and is immediately reversible.

'Deformation' is the motion of the body relative to other parts so that the body changes its size or shape, i.e., a displacement occurs between points in the body, and a 'flow' is a continuous change of deformation with time. The section of mechanics that deals with such motion is called continuum mechanics or mechanics of deformable media. Many examples of deformation are observed in everyday experiences, the motion of the liquid being poured, of water in a flowing river, of flag fluttering in the wind, of water boiling, of a rubber band being stretched and of a violin string when it is plucked. When an external force is applied to a material, it assumes a deformed shape in equilibrium and returned to their original shape after the forces are removed. Such reversible deformations are known as 'elastic deformation' or 'elasticity' and such materials are known as 'elastic solids.' Other materials can maintain an equilibrium shape when subjected to hydrostatic pressure (the force perpendicular to the surface). These materials, both liquids and gases are known as 'fluids.' Under other types of forces, they deform indefinitely as long as the external force is applied and do not return to their original form even if these forces are removed. Such an irreversible deformation is called 'flow.'

Rheology, the branch of condensed matter physics and takes its name from the Greek word "rei" (to flow) and is defined (by the Society of Rheology) as the "science of deformation and flow"[14]. It is concerned with the mechanical response of the material with the applied force. Most of the scientists and engineers who work in this field, called rheologists are concerned with the mechanics of complex, fluid-like substance that exhibits widely different flow behavior depending on the deformation to which they are subjected.

'Rheology' is a part of continuum mechanism which deals with the formation of constitutive equations, i.e., the relation between force and deformation that describes the mechanical behavior of a given material. Such relation describes an appropriate description of the behavior over a certain range of circumstances. For example, a metal under a light load might be considered as rigid, under heavier load or with more accurate measurement of length-linearly elastic, under a very large load–a plastic solid and under a small oscillatory motion–a linearly viscoelastic solid. Two of the basic concepts of continuum mechanisms are those of strain, the measure of amount of deformation and of stress (in units of force per unit area), as a measure of contact force, i.e., the force exerted by one part of a body on neighboring part [14]. Thus, rheological materials demonstrate dramatic changes in their rheological properties including yield stress, loss, and storage moduli, etc., in response to an externally applied electric field or magnetic field (electrorheology or magnetorheology). Electrorheology is a phenomenon whereby a suspension of fine polarizable particles (0.1–100 μm in diameter) in a dielectric liquid undergoes orders of magnitude increase in viscosity upon application of an electric field. This effect was first reported by Winslow in 1949 and is sometimes termed the "Winslow effect" [15]. As Winslow reported, apparent viscosities can increase by several orders of magnitude when electric fields of the order of 1 kV/mm are applied. Whereas magnetorheology refers to the phenomenon whereby magnetizable particles are dispersed in a non-magnetic matrix in which rheological properties can be changed continuously, rapidly, and reversibly by an applied magnetic field.

The MR fluids are field responsive rheology where their fluid properties can be controlled by varying the external magnetic field. The discovery of MR fluids was credited to Rabinow [16] at the US National Bureau of Standard [16]. MR fluid is the first developed magneto-sensitive smart material, which is a suspension by mixing micro-meter-sized ferromagnetic fillers, non-magnetic fluid, plus some additives together. On applying an external magnetic field, the MR fluid will switch from Newtonian-like fluid to semi-solid material quickly. The magnetic fillers dispersed randomly will be re-arranged to form chain-like ordered micro-structure throughout the magnetic interaction.

Typical MR fluids are the suspensions of micron sized, magnetizable particles (mainly iron) suspended in an appropriate carrier liquid such as mineral oil, synthetic oil, water or ethylene glycol, etc. The carrier fluid serves as a dispersed medium and ensures the homogeneity of the particles in the fluid. A variety of additives (stabilizers and surfactants) are also used in the MR fluid to prevent gravitational settling and to promote a stable particles suspension, which enhances the lubricity and change the initial viscosity of the MR fluids [16]. The stabilizers serve the purpose to keep the particles suspended in the fluid, whilst the surfactants are absorbed on the surface of the magnetic particles in order to enhance the polarization induced in the suspended particles on application of the magnetic field. When there is no applied magnetic field, the ferrous particles in the MR fluid will be randomly dispersed in the medium. In the presence of an externally applied magnetic field, the particles start to align themselves along the lines of magnetic flux [17]. As this change occurs almost instantly, the MR fluids are attractive solutions for real-time control applications, such as shock absorbers, brakes, clutches, engine mounts, valves, etc. The changes of liquid-solid-liquid state or the consistency or yield strength of the MR fluid can be controlled precisely and proportionally by altering the strength of the applied magnetic field [18–20].

The changes in the microstructure while applying an external field, is responsible for the changes in the rheology of the fluid. That is, a stronger external magnetic field will induce a more ordered chain-like micro-structure parallel to the direction of the magnetic field.

MR fluids are being used for various operations in which active vibration control or transfer of torque or force is required. Therefore, considerable efforts have been paid in the development of dampers using MR fluids. The other applications of MR fluids are related to energy management in the automotive industry. Similarly, MR fluid devices for aerospace applications as well as optical polishing are also promising and emerging.

12.3 CONCLUSION

Smart nanomaterials, or stimuli-responsive nanomaterials, have been investigated widely due to their potential applications in diverse fields. Among the different development processes of smart functional nanomaterials, the most widely accepted is the reliable bottom-up approach in which the collective physical and chemical properties of the resulting smart nanomaterials can be tailored by the spatial arrangements of their building blocks and their chemical nature. A smart material can sense changes in their environment and respond in a controlled and useful manner. Smart magnetic materials are of great interest nowadays to enable high end applications since their physical properties can easily be controlled by a suitable magnetic field. The properties of smart materials can be easily and reversibly tuned by changing the environment, i.e., temperature, EM fields, stresses, etc. The engineering applications of smart magnetic materials have to be concerned for the better performance of the device fabricated. Development of new smart materials as well as new ways of their effective control is thus a challenging task.

KEYWORDS

- **magnetic hyperthermia**
- **magneto rheological fluids**
- **nanomagnetism**
- **plasmonic nanorods**
- **smart magnetic nanoparticles**
- **smart nanomaterial**
- **superparamagnetic nanoparticles**
- **surface plasmon resonance**

REFERENCES

1. Vatta, L. L., Sanderson, R. D., & Koch, K. R., (2006). Magnetic nanoparticles: Properties and potential applications. *Pure and Applied Chemistry, 78*, 1793–1801.
2. Owens, F. J., (2015). *Physics of Magnetic Nanostructures*. John Wiley & Sons.
3. Aguirre, M. A., & NeroneandACalvo, N., (2001). *Granular Matter, 2*, 75.
4. Garg, D. P., & Anderson, G. L., (2000). Research in active composite materials and structures: An overview. *Proc. SPIE, 3992*, 2–12.
5. Zhang, W., Wu, C. W., & Ravi, S. R. P., (2018). Proposed use of self-regulating temperature nanoparticles for cancer therapy. *Expert Rev. Anticancer Ther., 18*(8), 723–725.

6. Hedayatnasab, Z., Abnisa, F., & Daud, W. M. A. W., (2017). Review on magnetic nanoparticles for magnetic nanofluid hyperthermia application. *Mater. Des., 123*(5), 174–196.
7. Gilchrist, R. K., Medal, R., Shorey, W. D., Hanselman, R. C., Parrott, J. C., & Taylor, C. B., (1957). Selective inductive heating of lymph nodes. *Ann. Surg., 146*, 596–606.
8. Gordon, R. T., Hines, J. R., & Gordon, D., (1979). Intracellular hyperthermia: A biophysical approach to cancer treatment via intracellular temperature and biophysical alterations. *Med. Hypotheses, 5,* 83–102.
9. Abenojara, E. C., Wickramasinghe, S., Bas-Concepcion, J., & Samia, A. C., (2016). Structural effects on the magnetic hyperthermia properties of iron oxide nanoparticles. *Prog. Nat. Sci., 26*(5), 440–448.
10. Vilas-Boas, V., Carvalho, F., & Espiña, B., (2020). Magnetic hyperthermia for cancer treatment: Main parameters affecting the outcome of *in vitro* and *in vivo* studies. *Molecules, 25*, 2874.
11. Kabashin, A. V., Evans, P., Pastkovsky, S., Hendren, W., Wurtz, G. A., Atkinson, R., Pollard, R., et al., (2009). Plasmonic nanorod metamaterials for biosensing. *Nat. Mater., 8*, 867–871.
12. Wang, M., Gao, C., He, L., Lu, Q., Zhang, J., Tang, C., Serkan, Z., & Yin, Y., (2013). Magnetic tuning of plasmonic excitation of gold nanorods. *J. Am. Chem. Soc., 135*, 15302–15305.
13. Jung, I., Jang, H., Han, S., Acapulco, J. A. I., & Park, S., (2015). Magnetic modulation of surface plasmon resonance by tailoring magnetically responsive metallic block in multisegment nanorods. *Chem. Mater., 27*(24), 8433–8441.
14. Herman, F. M., Norman, G. G., & Norbert, M. B., (1971). *Encyclopedia of Polymer Science and Engineering, 14.*
15. Aguirre, M. A., Grande, J. G., Calvo, A., Pugnaloni, L. A., & Géminard, J.-C. (2010). Pressure independence of granular flow through an aperture. *Phys. Rev. Lett., 104*, 238002.
16. Ashwani, K., & Mangal, S. K., (2013). Modeling, testing and evaluation of magneto-rheological shock absorber. *Int. J. Mech. Eng. & Rob. Res., 2.*
17. Ashwani, K., & Mangal, S. K., (2010). Comparative study of vibration-control systems. *International Conference on Engineering Innovations.*
18. Carlson, J. D., & Jolly, M. R., (2000). MR fluid, foam and elastomer devices. *Mechatronics, 10*, 555.
19. Jolly, M. R., Carlson, J. D., & Munoz, (1996). A model of the behavior of magnetorheological *materials. Smart Mater. Struct., 5*, 607.
20. Bossis, G., Lacis, S., Meunier, A., & Volkova, (2002). Magnetorheological fluids. *J. Magn. Magn. Mater., 252*, 224.

CHAPTER 13

Magnetic Nanoparticles in Biomedical Applications

NAMITHA BINU,[1] RUBY VARGHESE,[2] and YOGESH B. DALVI[1]

[1]*Pushpagiri Institute of Medical Sciences and Research Center, Tiruvalla, Pathanamthitta, Kerala, India*

[2]*Department of Chemistry, School of Sciences, Jain Deemed to be University, Bangalore, Karnataka, India*

ABSTRACT

Nanomedicine introduced various novel materials for the delivery of drugs as well as applications in CT scans and MRI, PET, and CT, where its legitimacy is directly proportional to its molecular signatures. Magnetic nanoparticles (MNPs) have now been believed to have magical properties, which have gained a wide interest present time and have impacted nanomedicine biosensing and analytical chemistry fields. The modification and functionalization of MNPs with various kinds of ligands of biomolecules have already been exploited. The application of magnetic particles can be easily observed in the microscopic manipulation of Nanosized and micro-objects. Manipulating these particles needs immense care, especially on their immunogenicity and biocompatibility, as these are applied in procedures covering from catalysis to delivery of drugs and healing. Appropriate layering and also particle coating with suitable agents can alleviate this hurdle. This chapter describes the general characteristics of MNPs, information regarding their physical and chemical designs, and various factors that are to be taken care of when dealing with the preparation of "MNPs" for biomedical applications. MNPs Find their profound use in theranostics platforms for diagnosis and as an

Modern Magnetic Materials: Properties and Applications. Iuliana Stoica, Ann Rose Abraham, & A. K. Haghi (Eds.)

interventional modality in disease management. Nanocomposites, such as magnetic liposomes, magnetic hydrogels, and magnetic dendrimers, have improved the premature burst release of loaded drugs and increased targeting efficacy. Even the most dreaded disease, cancer, could be fought by making use of the principles of 'Magnetic hyperthermia-assisted apoptosis and necrosis. These potent magnetically driven delivery systems can facilitate quick and versatile site-specific delivery of biotherapeutic interventions, which include therapeutic viruses, cell-based therapies, and nucleic acid and protein delivery. Many stringent maladies could be grappled with the synergistic effect of localized magnetic hyperthermia and can pinpoint targeted drug delivery to diseased sites, entitling it as a cutting-edge technology.

13.1 INTRODUCTION

Nanomedicine has adopted the principle of induction of kinetic motion through magnetism, a theory that had a profound stance in physics and chemistry. It also has found a vehement role in biology that enabled to stretch conventional biology far beyond towards arenas of nanotechnology due to the multidimensional roles played by the so-called Magnetic nanoparticles (MNPs). These magic particles have greatly impacted the fields of analytical chemistry, biosensing, drug delivery, and biomedical diagnostic research. MNPs or microparticles used in the diagnosis and treatment of diseases are as follows:

- Magnetic segregation of biological entities in diagnostic research;
- Magnetic nano-carriers as drug delivery systems;
- Magnetic nanoparticles in cancer therapy which are RF (radiofrequency controlled);
- "MRI" or "magnetic resonance imaging" applications.

The functionalization along with modification of MNPs with various kinds of ligands of biomolecules have been already exploited. Innate molecular configuration of the novel micro-sized objects introduced in nanomedicine for drug delivery and applications has got its fingerprints in molecular imaging, making it readily detectable for analysis. Proper layering and coating of particles make them biocompatible and less immunogenic. Magnetic particles have found their applications ranging from catalysis to drug delivery and remediation. Also, its activity spectrum ranging from catalysis, drug delivery, and antidotal qualities are warranted.

To achieve a fast progress in this area of study and the prospects into the limelight, it is very essential to work on challenges connected with the design, synthesis, and characterization of the MNP [1].

Previously, MNP synthesis was a pivotal step in all research connected to ameliorating their properties and testing their realistic utilization. Currently, many of these nanomaterials are extensively used commercially. MNPs comprise a magnetic material (such as iron, cobalt, and nickel) with a chemical component. These are the class of nanoparticles that can be modulated and set into a magnetic field. MNPs are composed of magnetic nanobeads of diameter 50–200 nm exhibiting superparamagnetic nature.

13.2 PHYSICAL DESIGN

The size of MNPs is one of the major physical properties that could be applied to tailor other properties such as surface area and magnetism. Control over size synthesis of iron oxide NPs has been examined by many scientists [2, 3].

The fate of a cell can be controlled by manipulating cell signaling at the molecular level which can be achieved by driving a mechanical force inside the cell. The "magnetic force" can be applied for the movement and transportation of biological entities. The decisive magnetic properties harvested as a result of externally controlled hyperthermia or heat-induced by generated magnetic fields were used in:

- Applications of drug release;
- Disease treatment; and
- Remote of single-cell function control [5, 6].

MNPs are steadier and more flexible as single-domain structures deployed in conditions that pass the "Curie" temperature and are therefore super-paramagnetic (SPNs) or SPNs nanoparticles show several desirable characteristics, such as "low remanence" and "coercivity," coupled with "high magnetic susceptibility."

The particles are coated to produce a stable system. The "surface charge" plays an important role in maintaining repulsion between particles; however, it is very important to optimize the ratio of inert to reactive compounds on the surface to ensure the colloidal stability of the NPs. For example, despite iron fabricated MNPs showing a higher level of magnetic characteristics, they must be coated to protect themselves from oxidation. Gold fabricated MNPs show desired magnetic and optical characteristics. Platinum fabricated MNP acts as a great contrast agent for "magnetic resonance imaging (MRI)" and "X-ray" computed tomography or "CT." Meanwhile, porous MNPs have obviously the same characteristics as solid NPs, yet they offer the additional opportunity to store and release drugs.

13.3 CHEMICAL DESIGN

The stratagem involved in the functionalization of the MNP reports the use of oligonucleotides and antibodies, for example, in the treatment of cancer cells by hyperthermia and magneto-liposomes or by the integration of inorganic materials with other nanocomponents like quantum dots [4].

Covalent and non-covalent modification opens up a modern view of addressing the chemical designing of a particle's surface, with the objective of drug delivery [5]. The former is functional to avoid "cancer drug resistance" and "therapeutic side effects." Non-covalent modifications have been explored but successful drug delivery was reported using conjugate drugs to "magnetic nanoparticles" via hydrophobic interaction [6], electrostatic interactions [7], and coordination chemistry [8]. During the design process, it is quite important to consider the strength of the "binding interactions" and the "collective strength" of these compounds with various other interactions at the same time. Much of the current works on "MNPs" is focused on studying such interactions for the design and "innovation" of novel and desirable "functions" [9] (Figure 13.1).

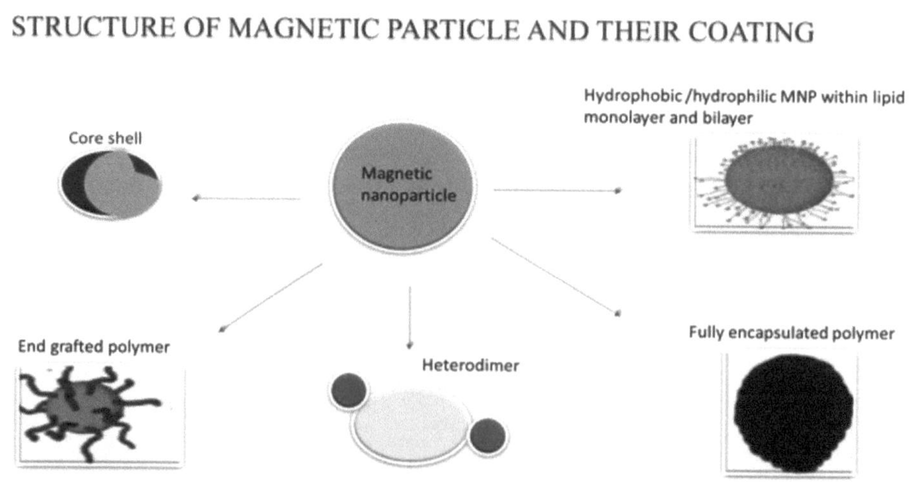

FIGURE 13.1 Structure of magnetic particle and their coating.

Any drug delivery system (DDS) whether polymeric or based on magnetic iron oxide nanoparticles (MIONs), must satisfy several common criteria. Specifically, a DDS must:

- Avoid nonspecific interactions with the body cells or the induction of unpropitious reactions and avoid capture by macrophages.
- Facilitate the biologically active molecules-BAM's (like genes, enzymes, drugs, proteins, or nucleotides) transport to the site of action (organ, tissue, cell, or organelle) from the site of administration in a high yield while keeping the BAM in a safe (inactive) state during transport.
- Protect the BAM from detrimental effects during transport like enzymatic degradation and hydrolysis in the body.
- Release effective quantities of active BAM in or around the target in a controlled fashion such that a desired tissue/cell concentration is achieved.
- Total elimination of all components of the DDS from an organism after the carrier function is fulfilled [10].

Each DDS offers its own set of specific solutions. Some of these are routine, but others are quite unique and depend on the details of the carrier's structure and architecture. Some DDS are better suited to covalent drug conjugation (e.g., polymer-drug conjugates or single magnetic nanocrystals) and some to noncovalent approaches (e.g., polymer NPs or magnetic nanoclusters). Broadly speaking, these approaches apply to diverse kinds of DDS for cancer treatment including MNPs and nanoclusters as well as polymer-drug conjugates, polymer micelles, and polymer NPs. Other polymer-based drug delivery systems such as nonmodified and polymer-modified liposomes (one of a few DDS to have been extensively clinically tested and approved for clinical use) [11].

MNPs can be synthesized in various sizes with a "submicron diameter" and with the same composition, e.g., according to the work published by Grasset et al. [12] (Figure 13.2).

Mn–Zn ferrite MNPs were prepared through coprecipitation and are used for the heat-inducible gene expression method and the surface of particles was covered by polyethyleneimine (PEI) (polyaziridine) [13].

Zinc-substituted ferrite nanoparticles $Zn_{0.9}Fe_{0.1}Fe_2O_4$ synthesized by a polyol method were used for heating glioma cells on hyperthermia assay [14].

The potential "toxicity" and their "chemical instability" have partially limited the biological applications of metallic iron NPs. Metal alloy based or Bimetallic MNPs is another propitious nanomaterial with superparamagnetic properties attractive for MRI.

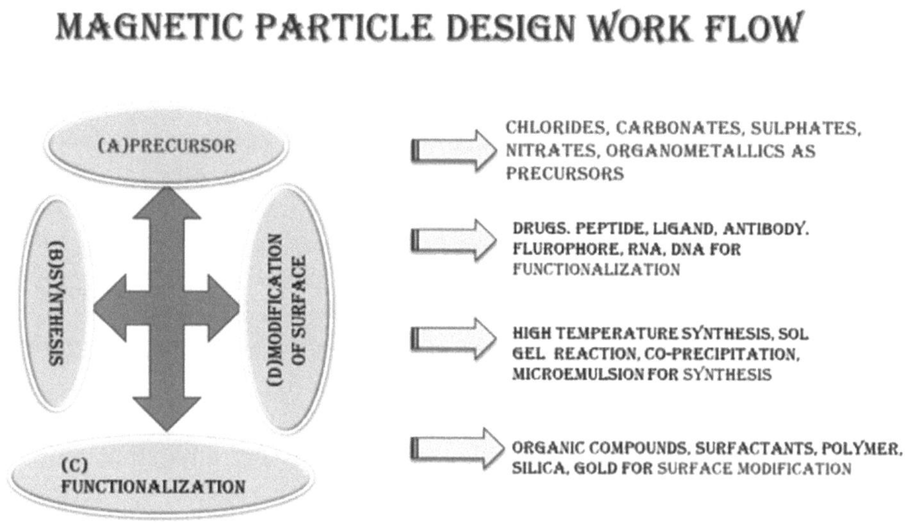

FIGURE 13.2 Outline of synthesis of magnetic nanoparticle.

Iron-platinum (FePt) NPs useful for bio-medical applications have been prepared by different methods, such as solution-phase synthesis or vacuum deposition [15].

The "reduction" of platinum acetylacetonate and "decomposition" of $Fe(CO)_5$ while using oleic acid and oleyl amine as stabilizers enhances the preparation of monodispersed and size tenable Fe–Pt NPs [16].

Such NPs were stable in a "cell culture medium" or phosphate-buffered saline and exhibit the ability of DNA and protein binding.

Carboxylate- and amine-based surfactants improve the water solubility of Fe–Pt NPs by aiding in surface modification [17].

Furthermore, FePt NPs can be encapsulated with shells based on cobalt sulfide ($FePt@CoS_2$). Such NPs exhibit cytotoxicity towards cancer cells [18, 19].

Biocompatibility of FePt NPs can be increased by covering it with a gold shell (FePt/Au) [20]. There are references for works coating gold with FePt NPs that were synthesized from a "High-Temperature" solution phase by the simultaneous reduction of platinum(II) acetylacetonate and decomposition of $Fe(CO)_5$ in octyl ether solvent. Other types of binary metallic nano-alloys with advantageous "magnetic properties" are the NPs containing iron and cobalt [21].

These NPs are formed by a "physical" or "chemical" vapor deposition process ($Fe_{12}Co_{88}$, $Fe_{40}Co_{60}$, and $Fe_{60}Co_{40}$). These require gold, silver, or graphitic protective coating to impede oxidation based on studies presented in [22, 23].

Gadolinium iron garnet particles were reported to be prepared from two techniques: '$Gd_3Fe_5O_{12}$ preparation.'

i. Pyrolysis of aerosol spray; and
ii. Drying the precursor solutions gives a crispy product that is mechanically grounded then dried and this was further subjected to pyrolysis [24].

13.4 BIOMEDICINE FIELD

As therapeutic agent, magnetic particles possess several applications—hyperthermia, drug, and "DNA" delivery exploiting magnetic field guiding and MRI. In modern medicine, the local increase of tumor tissue temperature (hyperthermia) is extensively accepted as an effective adjunctive cancer therapy as indicated in Ref. [25].

Prevalent ways to increase tissue temperature to the required level using "RFM" (radio frequency microwave) or "laser wavelengths" were proposed. Hyperthermia is in agreement with the fact that the cells heated to a temperature greater than 42°C show signs of apoptosis and cells heated above 50C show necrosis [26]. Despite this, there is an ambiguous understanding of the optimum temperature that should be switched on for acquiring temperature homogeneity in the target tissue. Physiological conditions like local perfusion variations render the situation more challenging [27]. A higher rate of metabolism greater will be the cancer cell's susceptibility to heat than normal cells [28]. On a tissue level, disorganized vasculature and decreased ability to dissipate heat are observed in tumor cells.

Radiation therapy or chemotherapy finds its effectiveness of cellular sensitivity amidst elevated temperatures. Regarding other methods, magnetic particle hyperthermia enables local heating of the target tissue by embedding "Magnetic Particles" to the target tissue and by using a "Magnetic Field" externally to heat it. "Magnetic Particles" are directly injected within the tumor body or in the artery furnishing the tumor. "Magnetic Particles" can now be visualized using MRI, as a result of which both diagnosis and combination therapies could be possible.

The potential for using MNPs as contrast enhancers in MR imaging is now a topic of research due to the detailed results and minimal irradiation impact [63, 64].

Visualizing the uptake of the cell line (head and neck cell lines) is made especially easy by magnetic iron nanoparticles called as 'SPIONS' and, both MRI and MPI (magnetic particle imaging) began to employ them extensively.

Surface modification of such particles is vital to hamper them from aggregation. Surface modification of "Magnetic Particles" effects their magnetization and "Specific Absorption Rate" (which measured in $W \cdot kg^{-1}$), the most important property for practical use ("the higher the specific absorption rate, the lower the injected dose to the patient"). It has been seen that the specific absorption rate of "Magnetic Particles" decreases by several tens of percent when it was modified with SiO_2 according to the studies reported in Ref. [29].

The shape of "Magnetic Particles" also plays an important role during hyperthermia. In the case of ellipsoidal NPs, heat release is increased due to the additional anisotropy shape and dynamic reorientation of rods [30].

It was shown that using a "micro-tumor-like" environment that in comparison with exogenous hyperthermia magnetic hyperthermia requires approximately 6C lower target temperature to produce same the cell death effect and exhibit more significant cytotoxic effects according to the results observed in Ref. [31]. Meanwhile, it should be noted that the heat is generated because of the friction of particles due to their oscillation and due to the particle's magnetic moment rotation owing to an alternating magnetic field.

On the other hand, "photo-thermal therapy" based on NPs with strong plasmonic properties was explored as an alternative to "Magnetic Hyperthermia." The enhanced heating ability is reached by a successful combination of plasmonic, and "magnetic properties" called "hybrid multifunctional," and MPPs or magnetic/photonic particles are based on "magnetic particles" decorated with Au or Ag NPs (or vice versa) based on the research studies presented in Refs. [32, 33].

The absorption spectrum of Au and Ag NPs absorbs strongly in the Vis-NIR region and heating is induced by SPR. Meanwhile, Au provides promising surface chemistry for modifications, and in the case of Au presence in shells protects NPs against external agents. On the other hand, the synergic effects of heating can be easily enhanced using the loading of MMPPs with chemotherapeutic agents according to the research studies presented in detail in Ref. [34].

Systemic drug administration is troublesome in current medicine, especially in the case of drugs exhibiting cytotoxic effects like "cytostatic drugs." The large doses needed to obtain a sufficient drug concentration within the pathological site and low drug specificity towards the target causes objectionable treatment side effects based on the report presented in Ref. [35]. Meanwhile, "site-specific drug delivery" is a commonly accepted mechanism to obviate such complications. "Magnetic" targeting (drug immobilized on "Magnetic" materials and targeting it using a "Magnetic" field) is one possibility in this field. For such applications, the size, charge, and surface chemistry of particles are important since they influence blood circulation time and bioavailability. "Magnetic" particles (10–100 nm) are the most suitable for intravenous injection and perform the most prolonged bloodstream circulation time as it is reported in Ref. [36].

It is worth mentioning that particles of size higher than 200 nm are removed from the bloodstream by mechanical filtration in the spleen and eventually by the phagocyte system. Smaller particles of size <10 nm is rapidly removed by extravasation and renal clearance. Not only the surface of "Magnetic," but the surface of all colloidal particles influences their fate within living organisms. The stability of materials in physiological conditions is complicated by the presence of proteins and high ion concentration based on studies presented in Ref. [37]. For example, the adsorption of counterions can screen the repulsive surface charge of particles and decrease their stability. In addition, it is well known that proteins adsorb to solid materials. According to Ref. [37], in the case of colloidal particles, the title protein corona is being used.

However, the biological fluids are so complicated that the reason for particles instability can hardly be ascertained. Particle aggregation can make treatment ineffective or can cause negative health effects, so surface modification which makes magnetic particles stable in physiological conditions is a highly researched area. Innumerable preparation techniques of magnetic particle and their functionalization procedures were reported, but particles coated with polysaccharides or silica were only the two types of particles that entered clinical trials readily [38, 39].

Further, polymeric compounds should be sequestered from the immune system. PEG is one among them since it is biocompatible, uncharged, and hydrophilic, possesses low toxicity and immunogenicity, and can be prepared with a wide range of terminal functional groups and sizes. Other polymers like chitosan (CS) or PEI were also used [40, 41].

Sufficiently modified magnetic particles mostly exhibit no side effects on cell viability [42]. Targeting drug delivery can be based on active-passive or physical targeting. Magnetic particles can be used in all approaches. In the case of physical targeting, magnetic particles can be localized in the desired place by an external magnetic field; however, a low ability to recognize specific tissues is evident.

Passive targeting is based on specific properties of a drug or its carrier and differences between a target site and a healthy site. Typically, enhanced permeability and retention effect (EPR effect) can enhance localization of drug/carrier complex in tumors due to disrupted vasculature. Active targeting is based on transfiguring target material with target-specific biomolecules, which increase the probability of its uptake by specific cells.

- Targeted drug delivery based on magnetic particles was reported by Huang et al. [43]. The proposed magnetic particles were modified with two polymers PEG and PEI. These could conjugate targeting ligand for receptors overexpressed on cancer cells—folic acid in this case—and to interact with negatively charged cell membranes.
- The acidic environment of tumors caused polymer's protonation and drug release, so drugs modified with doxorubicin by manipulating the electrostatic attraction and hydrogen bonds find their application.
- Reports are suggesting pH-responsive behavior for γ-Fe_2O_3 particles modified with porous silica using rhodamine B as a reporter [44].
- Magnetic targeting of Paclitaxel, an anti-cancer drug sparingly soluble in water can be stabilized by a carrier based on L-arginine and human serum albumin [42].
- Therapeutic delivery of nucleic acids into a patient's cells as an alternative to drugs is of importance currently. However, low cell membrane translocation efficiency, short *in vivo* half-time, and poor cell-specific targeting complicate its wider application.
- miRNAs are short non-coding RNA molecules that interact with target mRNAs to down-regulate or inhibit translation. This field is highly investigated for the delivery of microRNA (miRNA). Enhanced apoptosis of cancer cells is induced by the combined effect of hyperthermia and miRNA delivery by magnetic particles. They target heat shock proteins Hsp70 and Hsp90 which protect cellular proteins from degradation and hinder hyperthermia-induced apoptosis [45].

13.5 USE OF MNPS IN MOLECULAR IMAGING

It is observed that "magnetic resonance imaging (MRI)" is one of the most important diagnostic tools in medical science due to its soft tissue contrast and high spatial resolution. It mostly uses the "Magnetic Properties" of hydrogen and its interaction with "Magnetic Fields" and radio waves (like nuclear magnetic resonance or NMR) to produce images of living organisms.

There is a high abundance of hydrogen in tissues (approximately 63%) signifies "Magnetic Moments." The contrast in images is obtained due to different molecular structures and hydrogen densities in tissues which are responsible for specific proton behavior in "Magnetic Fields" that enable observation of it. It is noteworthy that without contrast agents in most tissues, it is difficult to obtain information-rich images. There are two main groups of contrast agents—chelated lanthanide or transition metal ions (Gd^{3+}, Mn^{2+}) and "Paramagnetic Particles." Iron oxide "Magnetic Particles" are the most frequently used "magnetic-particle-based" contrast agents. Requirements for MRI contrast agents are:

- they need to improve contrast (induce large fluctuating magnetic fields);
- should be biocompatible;
- should provide a desirable surface for targeting therapeutic molecules.

The advantage of "Magnetic Particles" is that their biocompatibility has been highly explored based on the results reported in Ref. [45].

Physiologic iron metabolism is responsible for the degradation of MNPs. In addition, due to their 'superparamagnetic' properties, iron oxide MNPs strongly distort magnetic fields and create contrast when exceeding their physical size. The current trend in medical research is to bridge desired properties for interventions to obtain diagnostic and therapeutic (shortened as theranostic) tools.

A convenient example is magnetic particles where drug delivery combined with hyperthermia or MRI is customary. The phenomenon of magnetism provides easy, fast, and cost-effective separation of targets. The list of targets is broad and is not limited since several biological and physics-chemical approaches for target binding on magnetic materials are available. Beyond this magnetic material can serve as therapeutic agent carriers, magnetic hyperthermia treatment (MHT) of malignancies, and MRI contrast agents for diagnosis.

13.6 MNPs AND TARGETED DRUG DELIVERY

Innate specificities of MNPs make these procedures significant practices in modern clinical interventions [46].

13.6.1 USE OF MNPS WITH THERAPEUTIC VIRUSES

Therapeutic viruses especially the adenoviruses had played a profound role in cancer therapy as a powerful gene-delivery system [47]. The hurdle faced with the use of adenoviruses was their low transfection efficiency and non-specific immune responses of the body which impedes the viable virus-based interventions in the clinical arena. The main problem encountered with such treatments is the degradation of these vector viruses by the body's immune cells. Conjugation of adenoviral particles to MNPs employing electrostatic interactions guards the virus against inactivation against the defender cells of the body.

Magnetofection is the process by which the application of an external magnetic field aids the virus conjugated MNPs to site targeted gene transfection [48]. SPNs in complex with adenovirus particles have a higher transfection rate than compared adenovirus particles lacking SPN functionalization [49]. The potential use of therapeutic viruses has given attractive results authenticating paving the way to developing techniques that can use viruses' clinical interventions [50].

13.6.2 USE OF MNPS WITH NUCLEAR AND PROTEIN DELIVERY

Cellular uptake is the key factor that plays a major role in Nucleic acid and protein-based treatments. In the light of issues like viral safety, the requirement of high-end machinery, higher cost of production, inconsistent use of non-viral vectors, MNPs are resourceful and highly sought after. *In-vitro* studies using silencing RNA molecules (siRNA) and functional MNPs are gaining interest due to their suppressing gene regulation action [51]. Magnectofection was used by Ref. [52] had conjugated polyethylenimine-coated MNPs and plasmid DNA expressing short-hairpin RNA with lipofectamine targeting Type 1 insulin-like growth factor receptor (IGF-1R), that recorded 2-fold higher gene suppression rate after 72 hrs. incubation time.

SPNs are also promising vectors for nucleic acid-based vaccines [53]. A nucleic acid–base vaccine that encoded for the *Plasmodium yoelii* merozoite surface protein ($MSP1_{19}$) was able to transfect African green monkey kidney cells at a higher rate than comparable chemical-based methods *in vitro*. Moreover, the delivery of therapeutic proteins promoting their deposition at target sites has been mediated SPNs [54, 56, 57].

Cancer cells are inhibited by two peptides named Cathelicidin LL-37 (CLL-37) and chlorotoxin (from *Leiurus quinquestriatus*) [58]. In vitro experiments showed that chlorotoxin-coated SPNs significantly reduced the proliferation of glioma cells by 98% compared to free chlorotoxin, which predominantly inhibited cell growth by 48%. The magnetic driven technology will soon be deployed in near future.

13.6.3 USE OF MNPS WITH CELL-BASED THERAPIES

Cell-based therapies along with stem cell research are great interventions for life-threatening conditions, such as ischemia [59]. Cell transplantation faces a serious drawback of poor cell retention that comes in the way of cell transplantation procedures. Graft retention and subsequent gene transfection have been improved by conjugating the MNPs on the surface of stem cells and immune cells.

A research work done by Kyrtatos et al. [60] in an *in-vivo* demonstrated the delivery of human $CD133^+$ endothelial progenitor cells to sites of vascular injury to male Sprague-Dawley rats Human $CD133^+$ were successfully delivered to sites of vascular injury to male Sprague-Dawley rats $CD133^+$ cell engraftment was increased to 5.4 fold, reduced restenosis, and fewer incidences of scar tissue formation was observed. This was done by the application of an external magnetic field accumulating MNPs at sites of catheterization. More recent works were done on female Wistar Kyoto rats that were injected with nanoparticle labeled cardiospheres. The results showed 4-fold higher cell retention in 24 hr. period and 3-week period [61]. Host cells can be made super magnetic as they can enhance molecular mimesis by adhering SNPs to immune cells, thereby stimulating the endocytosis of MNPs by cells of interest [62].

13.7 CONCLUSION

The emergence of therapeutics based on nanoparticulate polymer and iron oxide drug carriers is currently evolving. These comprise magnetic nanoclusters, single-crystal NPs, polymer-drug conjugates, sophisticated stimuli-responsive systems, polymerosomes, micelles, dendritic polymer-carriers, polymer-coated liposomes, magnetoliposomes, and niosomes, nanocapsules, and nanoemulsions, nanorobots, and other diagnostic devices. Clinical development of polymer conjugates and other organic or inorganic Drug Delivery systems and their evaluation are augmenting biomedical interventions. PEG conjugates and surface-modified nanomedicines in combination with other hydrophilic polymers and novel biodegradable polymer systems. Together, these advancing technologies have consummated the development and exploration of new polymer-based therapeutic concepts for cancer treatment including combination therapies and new vectors for the delivery of siRNA, genes, and other oligonucleotides. As of now, there are still limitations to the use of "MNPs" as a therapeutic agent. But the extensive research data demonstrated in myriad studies show the versatility of MNPs as a profound agent for clinical interventions and diagnosis in the near future.

KEYWORDS

- **biocompatibility**
- **drug delivery system**
- **enhanced permeability**
- **hyperthermia**
- **magnetic nanoparticles**
- **molecular imaging**
- **retention effect**

REFERENCES

1. Mahmoudi, M., Sant, S., Wang, B., Laurent, S., & Sen, T., (2011). Superparamagnetic iron oxide nanoparticles (SPIONs): Development, surface modification and applications in chemotherapy. *Advanced Drug Delivery Reviews, 63*(1, 2), 24–46.
2. Sun, S., & Zeng, H., (2002). Size-controlled synthesis of magnetite nanoparticles. *Journal of the American Chemical Society, 124*(28), 8204, 8205.
3. Sun, S., Zeng, H., Robinson, D. B., Raoux, S., Rice, P. M., Wang, S. X., & Li, G., (2004). Monodisperse MFe_2O_4 (M= Fe, Co, Mn) nanoparticles. *Journal of the American Chemical Society, 126*(1), 273–279.
4. Lee, J. H., Jang, J. T., Choi, J. S., Moon, S. H., Noh, S. H., Kim, J. W., Kim, J. G., et al., (2011). Exchange-coupled magnetic nanoparticles for efficient heat induction. *Nature Nanotechnology, 6*(7), 418.
5. Yoo, D., Lee, J. H., Shin, T. H., & Cheon, J., (2011). Theranostic magnetic nanoparticles. *Accounts of Chemical Research, 44*(10), 863–874.
6. Gao, J., Gu, H., & Xu, B., (2009). Multifunctional magnetic nanoparticles: Design, synthesis, and biomedical applications. *Accounts of Chemical Research, 42*(8), 1097–1107.
7. Ulbrich, K., Hola, K., Subr, V., Bakandritsos, A., Tucek, J., & Zboril, R., (2016). Targeted drug delivery with polymers and magnetic nanoparticles: Covalent and noncovalent approaches, release control, and clinical studies. *Chemical Reviews, 116*(9), 5338–5431.
8. Tietze, R., Lyer, S., Dürr, S., Struffert, T., Engelhorn, T., Schwarz, M., Eckert, E., et al., (2013). Efficient drug-delivery using magnetic nanoparticles—Biodistribution and therapeutic effects in tumor bearing rabbits. *Nanomedicine: Nanotechnology, Biology and Medicine, 9*(7), 961–971.

9. Yu, M. K., Jeong, Y. Y., Park, J., Park, S., Kim, J. W., Min, J. J., Kim, K., & Jon, S., (2008). Drug-loaded superparamagnetic iron oxide nanoparticles for combined cancer imaging and therapy *in vivo*. *Angewandte Chemie International Edition, 47*(29), 5362–5365.
10. Gautier, J., Allard-Vannier, E., Burlaud-Gaillard, J., Domenech, J., & Chourpa, I., (2015). Efficacy and hemotoxicity of stealth doxorubicin-loaded magnetic nanovectors on breast cancer xenografts. *Journal of Biomedical Nanotechnology, 11*(1), 177–189.
11. Pillai, G., (2014). Nanomedicines for cancer therapy: An update of FDA approved and those under various stages of development. *SOJ Pharm. Pharm. Sci., 1*(2), 13. Nanomedicines for cancer therapy: An update of FDA approved and those under various stages of development.
12. Grasset, F., Mornet, S., Demourgues, A., Portier, J., Bonnet, J., Vekris, A., & Duguet, E., (2001). Synthesis, magnetic properties, surface modification and cytotoxicity evaluation of $Y_3Fe_5-{}_xAl_xO_{12}$ ($0\leqslant x \leqslant 2$) garnet submicron particles for biomedical applications. *Journal of Magnetism and Magnetic Materials, 234*(3), 409–418.
13. Tang, Q. S., Zhang, D. S., Cong, X. M., Wan, M. L., & Jin, L. Q., (2008). Using thermal energy produced by irradiation of Mn–Zn ferrite magnetic nanoparticles (MZF-NPs) for heat-inducible gene expression. *Biomaterials, 29*(17), 2673–2679.
14. Hanini, A., Lartigue, L., Gavard, J., Kacem, K., Wilhelm, C., Gazeau, F., Chau, F., & Ammar, S., (2016). Zinc substituted ferrite nanoparticles with $Zn_{0.9}Fe_{2.1}O_4$ formula used as heating agents for *in vitro* hyperthermia assay on glioma cells. *Journal of Magnetism and Magnetic Materials, 416*, 315–320.
15. Sun, S., (2006). Recent advances in chemical synthesis, self-assembly, and applications of FePt nanoparticles. *Advanced Materials, 18*(4), 393–403.
16. Sun, S., Murray, C. B., Weller, D., Folks, L., & Moser, A., (2000). Monodisperse FePt nanoparticles and ferromagnetic FePt nanocrystal superlattices. *Science, 287*(5460), 1989–1992.
17. Hong, R., Fischer, N. O., Emrick, T., & Rotello, V. M., (2005). Surface PEGylation and ligand exchange chemistry of FePt nanoparticles for biological applications. *Chemistry of Materials, 17*(18), 4617–4621.
18. Gao, J., Zhang, B., Gao, Y., Pan, Y., Zhang, X., & Xu, B., (2007). Fluorescent magnetic nanocrystals by sequential addition of reagents in a one-pot reaction: A simple preparation for multifunctional nanostructures. *Journal of the American Chemical Society, 129*(39), 11928–11935.
19. Gao, J., Liang, G., Zhang, B., Kuang, Y., Zhang, X., & Xu, B., (2007). FePt@ CoS_2 yolk– shell nanocrystals as a potent agent to kill HeLa cells. *Journal of the American Chemical Society, 129*(5), 1428–1433.
20. De La Presa, P., Multigner, M., Morales, M. P., Rueda, T., Fernandez-Pinel, E., & Hernando, A., (2007). Synthesis and characterization of FePt/Au core-shell nanoparticles. *Journal of Magnetism and Magnetic Materials, 316*(2), e753–755.
21. Reiss, G., Brueckl, H., Huetten, A., Schotter, J., Brzeska, M., Panhorst, M., Sudfeld, D., et al., (2005). Magnetoresistive sensors and magnetic nanoparticles for biotechnology. *Journal of Materials Research, 20*(12), 3294–3302.
22. Bai, J., & Wang, J. P., (2005). High-magnetic-moment core-shell-type FeCo–Au/ Ag nanoparticles. *Applied Physics Letters, 87*(15), 152502.
23. Seo, W. S., Lee, J. H., Sun, X., Suzuki, Y., Mann, D., Liu, Z., Terashima, M., et al., (2006). FeCo/graphitic-shell nanocrystals as advanced magnetic-resonance-imaging and near-infrared agents. *Nature Materials, 5*(12), 971.

24. Xu, H. K., Sorensen, C. M., Klabunde, K. J., & Hadjipanayis, G. C., (1992). Aerosol synthesis of gadolinium iron garnet particles. *Journal of Materials Research, 7*(3), 712–716.
25. Issa, B., Obaidat, I. M., Albiss, B. A., & Haik, Y., (2013). Magnetic nanoparticles: Surface effects and properties related to biomedicine applications. *International Journal of Molecular Sciences, 14*(11), 21266–21305.
26. Milleron, R. S., & Bratton, S. B., (2007). 'Heated' debates in apoptosis. *Cellular and Molecular Life Sciences, 64*(18), 2329–2333.
27. Wust, P., Hildebrandt, B., Sreenivasa, G., Rau, B., Gellermann, J., Riess, H., Felix, R., & Schlag, P. M., (2002). Hyperthermia in combined treatment of cancer. *The Lancet Oncology, 3*(8), 487–497.
28. Huff, T. B., Tong, L., Zhao, Y., Hansen, M. N., Cheng, J. X., & Wei, A., (2007). *Hyperthermic Effects of Gold Nanorods on Tumor Cells*, 125–132.
29. Larumbe, S., Gomez-Polo, C., Pérez-Landazábal, J. I., & Pastor, J. M., (2012). Effect of a SiO_2 coating on the magnetic properties of Fe_3O_4 nanoparticles. *Journal of Physics: Condensed Matter, 24*(26), 266007.
30. Simeonidis, K., Morales, M. P., Marciello, M., Angelakeris, M., De La Presa, P., Lazaro-Carrillo, A., Tabero, A., et al., (2016). In-situ particles reorientation during magnetic hyperthermia application: Shape matters twice. *Scientific Reports, 6*, 38382.
31. Sanz, B., Calatayud, M. P., Torres, T. E., Fanarraga, M. L., Ibarra, M. R., & Goya, G. F., (2017). Magnetic hyperthermia enhances cell toxicity with respect to exogenous heating. *Biomaterials, 114*, 62–70.
32. Das, R., Rinaldi-Montes, N., Alonso, J., Amghouz, Z., Garaio, E., García, J. A., Gorria, P., et al., (2016). Boosted hyperthermia therapy by combined AC magnetic and photothermal exposures in Ag/Fe_3O_4 nanoflowers. *ACS Applied Materials & Interfaces, 8*(38), 25162–25169.
33. Lai, J. J., Lai, W. R., Chen, C. Y., Chen, S. W., & Chiang, C. L., (2013). Multifunctional magnetic plasmonic nanoparticles for applications of magnetic/photo-thermal hyperthermia and surface-enhanced Raman spectroscopy. *Journal of Magnetism and Magnetic Materials, 331*, 204–207.
34. Balasubramanian, S., Girija, A. R., Nagaoka, Y., Fukuda, T., Iwai, S., Kizhikkilot, V., Kato, K., et al., (2015). An 'all in one' approach for simultaneous chemotherapeutic, photothermal and magnetic hyperthermia mediated by hybrid magnetic nanoparticles. *RSC Advances, 5*(32), 25066–25078.
35. Verstappen, C. C., Heimans, J. J., Hoekman, K., & Postma, T. J., (2003). Neurotoxic complications of chemotherapy in patients with cancer. *Drugs, 63*(15), 1549–1563.
36. Gupta, A. K., & Gupta, M., (2005). Synthesis and surface engineering of iron oxide nanoparticles for biomedical applications. *Biomaterials, 26*(18), 3995–4021.
37. Nissinen, T., Nä kki, S., Laakso, H., Kuč iauskas, D., Kaupinis, A., Kettunen, M. I., Liimatainen, T., et al., (2016). Tailored dual PEGylation of inorganic porous nanocarriers for extremely long blood circulation *in vivo*. *ACS Applied Materials & Interfaces, 8*(48), 32723–32731.
38. Karimi, Z., Karimi, L., & Shokrollahi, H., (2013). Nano-magnetic particles used in biomedicine: Core and coating materials. *Materials Science and Engineering: C, 33*(5), 2465–2475.
39. McBain, S. C., Griesenbach, U., Xenariou, S., Keramane, A., Batich, C. D., Alton, E. W., & Dobson, J., (2008). Magnetic nanoparticles as gene delivery agents: Enhanced transfection in the presence of oscillating magnet arrays. *Nanotechnology, 19*(40), 405102.

40. Patil, R. M., Shete, P. B., Thorat, N. D., Otari, S. V., Barick, K. C., Prasad, A., Ningthoujam, R. S., et al., (2014). Superparamagnetic iron oxide/chitosan core/shells for hyperthermia application: Improved colloidal stability and biocompatibility. *Journal of Magnetism and Magnetic Materials, 355*, 22–30.
41. Sattarahmady, N., Azarpira, N., Hosseinpour, A., Heli, H., & Zare, T., (2016). Albumin coated arginine-capped magnetite nanoparticles as a paclitaxel vehicle: Physicochemical characterizations and *in vitro* evaluation. *Journal of Drug Delivery Science and Technology, 36*, 68–74.
42. Huang, Y., Mao, K., Zhang, B., & Zhao, Y., (2017). Superparamagnetic iron oxide nanoparticles conjugated with folic acid for dual target-specific drug delivery and MRI in cancer theranostics. *Materials Science and Engineering: C., 70*, 763–771.
43. Li, L., Zhang, R., Guo, Y., Zhang, C., Zhao, W., Xu, Z., & Whittaker, A. K., (2016). Functional magnetic porous silica for T 1–T 2 dual-modal magnetic resonance imaging and pH-responsive drug delivery of basic drugs. *Nanotechnology, 27*(48), 485702.
44. Yin, P. T., Shah, B. P., & Lee, K. B., (2014). Combined magnetic nanoparticle-based microRNA and hyperthermia therapy to enhance apoptosis in brain cancer cells. *Small, 10*(20), 4106–4112.
45. Miyagawa, T., Saito, H., Minamiya, Y., Mitobe, K., Takashima, S., Takahashi, N., Ito, A., et al., (2014). Inhibition of Hsp90 and 70 sensitizes melanoma cells to hyperthermia using ferromagnetic particles with a low curie temperature. I*nternational Journal of Clinical Oncology, 19*(4), 722–730.
46. Arami, H., Khandhar, A., Liggitt, D., & Krishnan, K. M., (2015). *In vivo* delivery, pharmacokinetics, biodistribution and toxicity of iron oxide nanoparticles. *Chemical Society Reviews, 44*(23), 8576–8607.
47. Mok, H., & Zhang, M., (2013). Superparamagnetic iron oxide nanoparticle-based delivery systems for biotherapeutics. *Expert Opinion on Drug Delivery, 10*(1), 73–87.
48. Wold, W. S. M., & Toth, K., (2013). Adenovirus vectors for gene therapy, vaccination cancer gene therapy. *Current Gene Therapy, 13*(6), 421–433.
49. Choi, J. W., Park, J. W., Na, Y., et al., (2015). Using a magnetic field to redirect an oncolytic adenovirus complexed with iron oxide augments gene therapy efficacy. *Biomaterials, 65*, 163–174.
50. Sapet, C., Pellegrino, C., Laurent, N., et al., (2012). Magnetic nanoparticles enhance adenovirus transduction *in vitro* and *in vivo*. *Pharmaceutical Research, 29*(5), 1203–1218.
51. Singh, D., McMillan, J. M., Liu, X. M., et al., (2014). Formulation design facilitates magnetic nanoparticle delivery to diseased cells and tissues. *Nanomedicine: Nanotechnology, Biology, and Medicine, 9*(3), 469–485.
52. Lee, S., Shim, G., Kim, S., et al., (2011). Enhanced transfection rates of small-interfering RNA using dioleylglutamide-based magnetic lipoplexes. *Nucleic Acid Therapeutics, 21*(3), 165–172.
53. Wang, C., Ding, C., Kong, M., et al., (2011). Tumor-targeting magnetic lipoplex delivery of short hairpin RNA suppresses IGF-1R overexpression of lung adenocarcinoma A549 cells *in vitro* and *in vivo*. *Biochemical and Biophysical Research Communications, 410*(3), 537–542.
54. Nawwab Al-Deen, F., Ho, J., Selomulya, C., et al., (2011). Superparamagnetic nanoparticles for effective delivery of malaria DNA vaccine. *Langmuir: The ACS Journal of Surfaces and Colloids, 27*(7), 3703–3712.

55. Veiseh, O., Gunn, J. W., Kievit, F. M., et al., (2009). Inhibition of tumor-cell invasion with chlorotoxin-bound superparamagnetic nanoparticles. *Small (Weinheim and Der Bergstrasse, Germany), 5*(2), 256–264.
56. Kim, C., Lee, Y., Kim, J. S., et al., (2010). Thermally triggered cellular uptake of quantum dots immobilized with poly(N-isopropylacrylamide) and cell penetrating peptide. *Langmuir: The ACS Journal of Surfaces and Colloids, 26*(18), 14965–14969.
57. Niemirowicz, K., Prokop, I., Wilczewska, A., et al., (2015). Magnetic nanoparticles enhance the anticancer activity of cathelicidin LL-37 peptide against colon cancer cells. *International Journal of Nanomedicine, 10*, 3843–3853.
58. Ojeda, P. G., Wang, C. K., & Craik, D. J., (2016). Chlorotoxin: Structure, activity, and potential uses in cancer therapy. *Biopolymers, 106*(1), 25–36.
59. Cores, J., Caranasos, T. G., & Cheng, K., (2015). Magnetically targeted stem cell delivery for regenerative medicine. *Journal of Functional Biomaterials, 6*(3), 526–546.
60. Kyrtatos, P. G., Lehtolainen, P., Junemann-Ramirez, M., et al., (2009). Magnetic tagging increases the delivery of circulating progenitors in vascular injury. *JACC: Cardiovascular Interventions, 2*(8), 794–802.
61. Cheng, K., Malliaras, K., Li, T. S. S., et al., (2012). Magnetic enhancement of cell retention, engraftment, and functional benefit after intracoronary delivery of cardiac-derived stem cells in a rat model of ischemia/reperfusion. *Cell Transplantation, 21*(6), 1121–1135.
62. Riemer, J., Hoepken, H. H., Czerwinska, H., et al., (2004). Colorimetric ferrozine-based assay for the quantitation of iron in cultured cells. *Analytical Biochemistry, 331*(2), 370–375.
63. Comanescu, C. (2022). Magnetic Nanoparticles: Current Advances in Nanomedicine, *Drug Delivery and MRI. Chemistry. 4*(3), 872–930.
64. Balk, M., Haus, T., Band, J., Unterweger, H., Schreiber, E., Friedrich, R. P., Alexiou, C., & Gostian, A. O. (2021). Cellular SPION uptake and toxicity in various head and neck cancer cell lines. *Nanomaterials. 11*(3), 726.

CHAPTER 14

A Technical Note on Metal-Carbon Mesocomposites Magnetic Characteristics Growth

V. I. KODOLOV[1,2] and V. V. KODOLOVA-CHUKHONTZEVA[2,3]

[1]*M.T. Kalashnikov Izhevsk State Technical University, Izhevsk, Russia*

[2]*Basic Research–High Educational Center of Chemical Physics and Mesoscopic, UD RAS, Izhevsk, Russia*

[3]*Peter Great St. Petersburg Technical University, St. Petersburg, Russia*

ABSTRACT

This information is dedicated to the reasons for the metal-carbon mesocomposites (Me–C MCs) magnetic characteristics growth. The magnetic properties of substances and materials are usually caused by the presence of the unpaired electron. Among these substances, the series of metals and carbon radicals may be considered. The possibility of magnetic characteristics growth with simultaneous increasing of unpaired electrons quantity may be realized by reduction-oxidation processes. These possibilities are considered in the present communication.

14.1 STRUCTURE OF METAL-CARBON MESOCOMPOSITES (ME–C MCS)

The mechanochemical process of metal-carbon mesocomposite (Me–C MC) obtaining consists in the joint grinding of polyvinyl alcohol (PVA) with metal oxides in the mechanical mortars or in the paddle mills. At this red-ox

Modern Magnetic Materials: Properties and Applications. Iuliana Stoica, Ann Rose Abraham, & A. K. Haghi (Eds.)

process, the metal from its oxide is reduced with the metallic cluster formation and simultaneous water vaporization at the PVA destruction that leads to the carbon fibers formation. These fibers contain the polyacetylene and carbine fragments with coordinate bonds which are created between metal cluster and unsaturated bonds of carbon fibers. As a result, the creation of a common electron system which includes nucleus-metal cluster with carbon cover which consists of polyacetylene and carbine fragments.

14.2 MECHANISM OF REDUCTION OXIDATION REACTIONS OF METAL-CARBON MESOCOMPOSITES

The Me–C MC red-ox processes consist in the grinding of copper–carbon or nickel–carbon nanocomposites with the corresponding substances containing P or Si. As the mechanochemical process result – the xerogel is formed which is dried at 80°C. At the interaction of oxidizers with Me–C MC the common electron system is organized and, in this system, the red-ox process flows owing to the interaction of negative charged quants on the positive charged quants. The annihilation phenomenon is appeared because of this interaction which leads to the creation of inner electromagnetic (EM) radiation rousing the unpaired electrons formation with simultaneous shift on the high energetic orbitals. The presence of coordinate bonds between metal cluster and unsaturated bonds of carbon cover rouses the cover change with the sizes carbine fragments decreasing, that correspondent to the fragments number increasing and the unpaired electrons growth. In other words, the growth of fragments alternation number takes place at which the unpaired electrons quantity grows.

14.3 ABOUT THE REASONS OF THE METAL-CARBON MESOCOMPOSITES (ME–C MCS) MAGNETIC CHARACTERISTICS GROWTH AT RED-OX PROCESSES

The first reason is caused by the magnetic characteristics' growth because of the formation of unpaired electrons in metal cluster electron shell. This effect is explained by the annihilation phenomenon at the reduction oxidation process.

The second reason is associated with the electron structure carbon cover change at the alternation fragments order change. It's noted that the unsaturated bonds (double bonds) of polyacetylene and carbine are alternative. On the boundary of these bonds the unpaired electrons are appeared.

14.4 CONCLUSION

All above statement is confirmed by the theoretical investigations and authors experimental results presented in Refs. [1–35].

KEYWORDS

- **electromagnetic radiation**
- **magnetic characteristics growth**
- **magnetic properties**
- **mechanochemical process**
- **nanocomposites**
- **reduction-oxidation processes**

REFERENCES

1. Imri, (2009). *Introduction in Mesoscopic Physics* (p. 304). M.: Physmatlit.
2. Moskalets, M. V., (2010). *Fundamentals of Mesoscopic Physics*. Khar'kov: NTU KhPI.
3. Morozov, A. D., (2002). *Introduction in Fractal Theory* (p. 160). M.-Izhevsk: ICT.
4. Kolmogorov, A. N., & Fomin, S. V., (2009). *Introductory Real Analysis* (p. 403). USA, Portland: Prentice Hall.
5. Wunderlikh, B., (1979). *Physics of Macromolecules* (Vol. 2, p. 422). M.: Mir.
6. Brown, J. F. Jr., & White, D. M., (1960). *J. Am. Chem. Soc., 82*, 5671.
7. Morgan, P. W., (1970). Condensation polymers by interfacial and solution methods (translated) L.; *Chemistry*, p. 448.
8. Buchachenko, A. L., (2003). *Nanochemistry – Direct Way to High Technologies* (Vol. 72, No 5, pp. 419–437). Uspechi Chimii.
9. Kodolov, V. I., & Khokhriakov, N. V., (2009). *Chemical Physics of Formation and Transformation Processes of Nanostructures and Nanosystems* (Vol. 1, 2, pp. 361, 415). Izhevsk: Publ. IzhSACA.
10. Shabanova, I. N., Kodolov, V. I., Terebova, N. S., & Trineeva, V. V., (2012). *X-Ray Photoelectron Spectroscopy Investigations of Metal/Carbon Nanosystems and Nanostructured Materials* (p. 252). M.: Izhevsk: Publ. "Udmurt university."

11. Kodolov, V. I., Khokhriakov, N. V., Trineeva, V. V., & Blagodatskikh, I. I., (2008). Nanostructure activity and its display in nanoreactors of polymeric matrices and in active media. *Chemical Physics and Mesoscopy, 10*(4), 448–460.
12. Kodolov, V. I., (2009). The addition to previous paper. *Chemical Physics and Mesoscopy, 11*(1), 134–136.
13. Trineeva, V. V., Vakhrushina, M. A., Bulatov, D. I., & Kodolov, V. I., (2012). The obtaining of metal/carbon nanocomposites and investigation of their structure phenomena. *Nanotechnics*, (4), 50–55.
14. Trineeva, V. V., Lyakkhovich, A. M., & Kodolov, V. I., (2009). Forecasting of the formation processes of carbon metal containing nanostructures using the method of atomic force microscopy. *Nanotechnics, 4*(20), 87–90.
15. Kodolov, V. I., Blagodatskikh, I. I., Lyakhovich, A. M., et al., (2007). Investigation of the formation processes of metal containing carbon nanostructures in nanoreactors of polyvinyl alcohol at early stages. *Chemical Physics and Mesoscopy, 9*(4), 422–429.
16. Kodolov, V. I., Khokhriakov, N. V., & Kuznetsov, A. P., (2006). To the issue of the mechanism of the influence of nanostructures on structurally changing media at the formation of "intellectual" composites. *Nanotechnics, 3*(7), 27–35.
17. Kodolov, V. I., Khokhriakov, N. V., Trineeva, V. V., & Blagodatskikh, I. I., (2010). Problems of nanostructure activity estimation, nanostructures directed production and application. *Nanomaterials Yearbook–2009: From Nanostructures, Nanomaterials and Nanotechnologies to Nanoindustry* (pp. 1–18). N.Y.: Nova Science Publishers, Inc.
18. Fedorov, V. B., Khakimova, D. K., Shipkov, N. N., & Avdeenko, M. A., (1974). To thermodynamics of carbon materials (Vol. 219. No. 3. pp. 596–599). Doklady AS USSR.
19. Fedorov, V. B., Khakimova, D. K., Shorshorov, M. H., et al., (1975). *To Kinetics of Graphitation* (Vol. 222. No. 2. pp. 399–402). Doklady AS USSR.
20. Khokhriakov, N. V., & Kodolov, V. I., (2005). Quantum-chemical modeling of nanostructure formation. *Nanotechnics, 2*, 108–112.
21. Lipanov, A. M., Kodolov, V. I., Khokhriakov, N. V., et al., (2005). Challenges in creating nanoreactors for the synthesis of metal nanoparticles in carbon shells. *Alternative Energy and Ecology, 2*(22), 58–63.
22. Kodolov, V. I., Didik, A. A., Volkov, Yu. A., & Volkova, E. G., (2004). Low-temperature synthesis of copper nanoparticles in carbon shells. HEIs' news. *Chemistry and Chemical Engineering, 47*(1), 27–30.
23. Serkov, A. T., (1975). *Theory of Chemical Fiber Formation* (p. 548). edited by– M.: Himiya.
24. Palm, V. A., (1967). *Basics of Quantitative Theory of Organic Reactions* (p. 356). L: Himiya.
25. Kodolov, V. I., (1965). On modeling possibility in organic chemistry. *Organic Reactivity, 2*(4), 11–18. Tartu: TSU.
26. Chashkin, M. A., Kodolov, V. I., Zakharov, A. I., et al., (2011). Metal/carbon nanocomposites for epoxy compositions: Quantum-chemical investigations and experimental modeling. *Polymer Research Journal, 5*(1), 5–19.
27. Kodolov, V. I., & Trineeva, V. V., (2013). Fundamental definitions for domain of nanostructures and metal/carbon nanocomposites. In: *Nanostructure, Nanosystems and Nanostructured Materials: Theory, Production and Development* (pp. 1–42). Toronto-New Jersey: Apple Academic Press.
28. Kodolov, V. I., Trineeva, V. V., Blagodatskikh, I. I., Vasil'chenko, Yu. M., Vakhrushina, M. A., & Bondar, A. Yu., (2013). The nanostructures obtaining and the synthesis of metal/carbon nanocomposites in nanoreactors. In: *Nanostructure, Nanosystems and*

Nanostructured Materials: Theory, Production and Development (pp. 101–145). Toronto-New Jersey: Apple Academic Press.
29. Shabanova, I. N., & Terebova, N. S., (2012). Dependence of the value of the atomic magnetic moment of d metals on the chemical structure of nanoforms. In: *Problems of Nanochemistry for the Creation of New Materials* (pp. 123–131). Torun, Poland: IEPMD.
30. Mustakimov, R. V., Kodolov, V. I., Shabanova, I. N., & Terebova, N. S., (2017). Modification of copper nanocomposites with the use of ammonium polyphosphate for the application as modifiers of epoxy resins. *Chemical Physics & Mesoscopics, 19*(1), 50–57.
31. Kodolov, V. I., Trineeva, V. V., Kopylova, A. A., et al., (2017). Mechanochemical modification of metal carbon nanocomposites. *Chemical Physics & Mesoscopics, 19*(4), 569–580.
32. Kodolov, V. I., Trineeva, V. V., Terebova, N. S., et al., (2018). The change of electron structure and magnetic characteristics of modified copper carbon nanocomposites. *Chemical Physics & Mesoscopics, 20*(1), 72–79.
33. Kodolov, V. I., et al., (2019). Modified magnetic metal carbon mesoscopic composites with bio active substances. *Chemical Physics & Mesoscopics, 21*(3), 446–454.
34. Mustakimov, R. V., (2019). *The Method of Modification of Metal Carbon Nanostructures by Ammonium Poly Phosphate.* Pat. N 2694092.
35. Kodolov, V. I., & Kodolova–Chukhontzeva, V. V., (2019). Fundamentals of chemical mesoscopics. *Monograph* (p. 218). Izhevsk: Publisher–M.T. Kalashnikov Izhevsk State Technical University.

CHAPTER 15

Synthesis and Applications of Magnetic Nanoparticles: A Review

VANDANA CHAUHAN,[1] RAJPREET KAUR,[1,2] ANITA GUPTA,[1] and POONAM KHULLAR[2]

[1]*Department of Chemistry, Amity University, Noida, Uttar Pradesh, India*

[2]*Department of Chemistry, BBK DAV College for Women, Amritsar, Punjab, India*

ABSTRACT

Magnetic nanoparticles (MNPs) have gained tremendous interest in the past few decades due to their superparamagnetism property. To produce MNPs with desired structures and properties, various routes for its preparation have been explored. To produce the next generation of advanced magnets, it is important to synthesize MNPs that have controllable size, shape, and chemical composition. So here, some main methods are covered to synthesize MNPs. As MNPs show unique characteristics which have increased their demand in the field of industries and biomedical. Therefore, here the application of MNPs is covered in three main areas that represent them are biomedical, ferrofluids, and data storage. In the case of biomedical MNPs can target or detect biological entities with the help of modification of specific biomolecules and they can be used to apply in photothermal therapy, magnetic resonance imaging (MRI), biosensing, and drug delivery.

Modern Magnetic Materials: Properties and Applications. Iuliana Stoica, Ann Rose Abraham, & A. K. Haghi (Eds.)

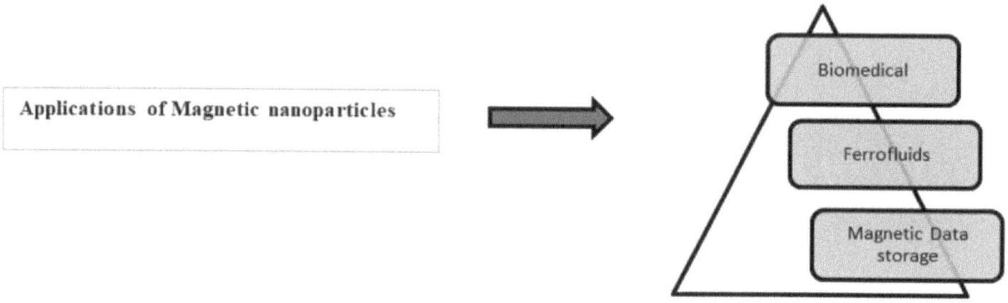

15.1 INTRODUCTION

Nanomaterials are really small particles ranges between 1 and 100 nm in diameter and also, they are about 1,000 times smaller than a single cell. Sometimes this term is also used for the larger particles up to 500 nm or tubes and fibers that are less than 100 nm in only two directions. Nanoparticles (NPs) are classified into different types depending on their morphology and size [1].

Some of the main NPs are lipid-based NPs, ceramic NPs, polymeric NPs, semiconductor NPs, carbon-based NPs, and metal NPs. In recent times NPs are being used in the manufacturing of various things like crack-resistant paints, coating for walls, stain-repellent fabrics, ceramic coating for solar cells, windows that can be easily self-cleaned, and transparent sunscreen. When these are compared with bulk material then they show some unique characteristics which make them interesting in the field of industries and for medical applications [2].

Magnetic nanoparticles (MNPs) are a group of NPs which in presence of an external magnetic field show some responses to an applying magnetic field. These particles consist of magnetic elements like chromium, iron, cobalt, manganese, gadolinium, nickel, and their chemical compounds. MNPs are typically classified into metal oxide NPs, metal alloy NPs, and metal NPs. The common NPs used are silver (Ag), gold (Au), iron (Fe), nickel (Ni), and cobalt (Co). Metal oxide NPs mostly consist of ferrites ($CoFe_2O_4$ and $MnO.6ZnO.4Fe_2O_4$) and iron oxides (γ-Fe_2O_3 and Fe_3O_4) and metal alloy NPs include FePt, FeCo, and so on. Among all of them, the most widely use MNPs are metal oxide magnetite $(Fe_3O_4)_2$ and maghemite (γ-Fe_2O_3) MNPs these particles can be easily controlled in particle shape and size also can be conveniently prepared and have low toxicity. Figure 15.1 shows the general view and selected area electron diffraction (SAED) of β-magnetite by TEM analysis. By adding zinc (Zn) and manganese (Mn) to nanosized iron oxide structure variety of ferrite NPs like $Mn_3Zn_7Fe_2O_4$, MnO.6ZnO.4Fe_2O_4, and many more can be prepared. So, when these ferrite NPs get prepared then these have stronger magnetism and higher relaxation rate [3–6].

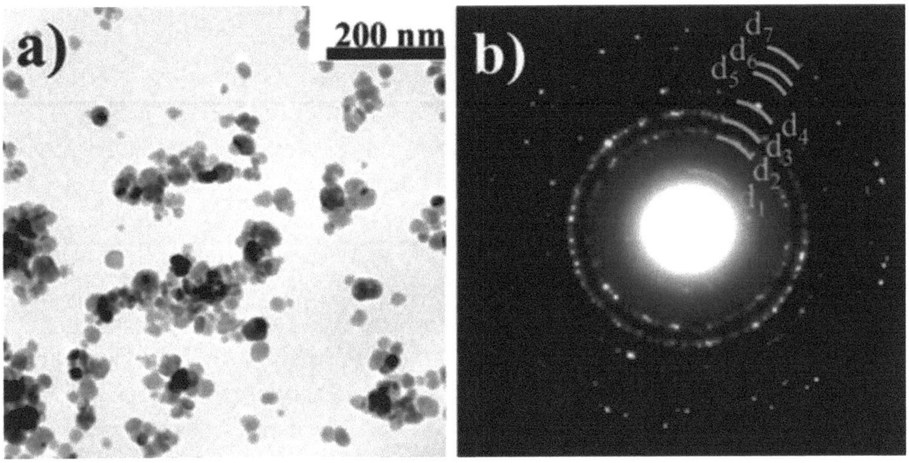

FIGURE 15.1 Transmission electron microscope (TEM) analysis of β-magnetite NPs (a) general view; and (b) corresponding SAED pattern.

In different types of NPs, MNPs have gained a lot of attention among researchers because of their unique magnetic properties like superparamagnetic, low Curie temperature, high magnetic susceptibility, etc. Superparamagnetic is observed in small materials (10 nm–150 nm) and MNPs are of nano-size, so they show this feature [4]. As MNPs show superparamagnetic feature, this, when the field is applied than the particles, get attracted towards the magnetic field but after removal of the field, they retain no residual magnetism. Thus, this characteristic makes these MNPs easily separated with target compounds from complex matrices like biological samples, environment, and food by just applying an external magnetic field [5, 6].

Synthesizing MNPs is an important part, over the years several methods have been developed to obtain MNPs with magnetic characteristics, controllable size, shape, reproducibility, surface chemistry, and size distribution. For a long period for synthesizing MNPs by chemical method has been commonly used, as it is one of the cheapest ways of preparing large quantities of the desired MNPs. There are many techniques to synthesize MNPs such as combustion, chemical vapor deposition, lab on a chip system, carbon arc, laser pyrolysis techniques, electrochemical synthesis, and microbial synthesis. For obtaining stable, monodisperse, and shape controllable MNPs various popular routes are present such as co-precipitation technique, thermal decomposition technique, hydrothermal synthesis, sol-gel synthesis, aero-sol/vapor-phase based synthesis techniques, flow injection, microemulsion, and hydrothermal synthesis. The simplest and efficient chemical pathway among all of them to synthesize MNPs is the co-precipitation technique, then for obtaining monodispersed NPs with high crystallinity, size, and shape control high-temperature thermal decomposition seems to be an interesting technique. The hydrothermal process is used for the formation of small-sized MNPs whereas the microemulsion method is used to synthesize MNPs of size and shape-controlled. MNPs obtained by using co-precipitation and decomposition approaches are the finest studied types of the synthesis techniques of the MNPs. According to the work of Grassel et al., he explained that $\gamma 3Fe_5O_{12}$ particles were prepared by different synthetic approaches – hydroxide coprecipitation, hydrolysis, citrate gel process, co-precipitation in microemulsion, glass crystallization or glycothermal synthesis, and some other related method.

MNPs are used in several technological applications in the field of information storage, high-frequency electronics, ferrofluids, magnetic refrigerants, and high-performance permanent magnets. MNPs are also widely used in biomedical and industrial applications. The main four applications of MNPs in biomedical are: (i) for magnetic separation of biological entities; (ii) radio frequency controlled MNPs; (iii) drug delivery; and (iv) magnetic resonance imaging (MRI). Magnetic iron oxide nanoparticles (MIONs) are mostly used for biomedical applications. For biomedical applications, magnetite Fe_3O_4 or maghemite γ-Fe_2O_3 are the most widely used oxides. The main factor for using magnetite and maghemite is their good biocompatibility and chemical stability. When magnetic properties are taken into consideration in that case metal NPs are more powerful, but their downfall is that they are toxic, so cannot be used for clinical applications. Also, MNPs are used in the treatment of infectious diseases and cancer. For medical applications, the important attributes of MNPs which make them a great choice are their high-level aggregation in the desired tissue, non-toxicity, and biocompatibility. Some real-life application of MNPs is in Analytical chemistry, separation of heavy metals from a contaminant, preconcentration, cells, and biomolecules, inorganic compounds and therapy. MNPs can also be used in the field of environmental purification and catalysis [7–9].

15.2 SYNTHESIS OF MNPS

There are numerous methods developed to synthesize MNPs, a lot of studies have been done by researchers to synthesize MNPs. The most common techniques to synthesize controllable shape, size, high crystallinity, stable, and monodisperse MNPs are sol-gel synthesis, thermal decomposition, co-precipitation, microemulsion, hydrothermal synthesis, aerosol/vapor-phase base synthesis, and flow injection technique. For synthesizing MNPs, there are two approaches: (i) bottom-up; and (ii) top-down approaches. In both approaches, there are many methods present to prepare MNPs. The bottom-up approach contains methods like plasma spraying synthesis, electrochemical reactions, bio-assisted synthesis, chemical vapor deposition, thermal evaporation, molecular condensation, and sol-gel processing synthesis to prepare MNPs while in top-down technique it employs microfabrication processes like lit off, lithographical patterning, wet etching or ion milling. Basically, for representing MNPs, Maghemite (γ-Fe_2O_3) and Magnetite (Fe_3O_4) ferrite colloids are the main representatives of MNPs. They have received so much attention from researchers due to their characteristics of biocompatibility and biodegradability. Mainly there are three main pathways to synthesize Fe_3O_4 NPs: (i) wet chemical preparation method; (ii) microbial methods; and (iii) physical methods [10]. Some of the main syntheses are discussed in subsections.

15.2.1 THERMAL DECOMPOSITION TECHNIQUE

MNPs can be synthesized by thermal decomposition firstly thermal decomposition of metal precursors takes place which includes metal oleates and metal-carbonyls like $Ni(CO)_4$, $Co_2(CO)_8$, $Fe(CO)_5$, and then high temperature and pressure is applied to reactors or autoclaves during the synthesis process. The products yield by the thermal decomposition technique have excellent crystallinity of dispersible iron oxide MNPs, great monodispersity, narrow size distribution, and excellent controllable particle size and shape. By using metal-carbonyl precursors (i.e., $Fe(CO)_5$ and $Co_2(CO)_8$) a number of magnetic metal NPs were synthesized with the help of the Thermal decomposition approach. By Decomposition of polyene metal complexes or low valent alkene with H_2 at room temperature magnetic metal NPs (i.e., Co and Ni) can be easily synthesized. Recently water-soluble MNPs were successfully synthesized, which yields a product having narrow size distribution and controllable particle size.

Three main factors control the shape and size of the obtained MNPs by thermal decomposition technique: (i) duration of the reaction when it reaches the boiling point; (ii) decomposition temperature; and (iii) the precursor. Lee & co-workers carried out a study in which at high temperature in the organic medium high-quality magnetism-engineered Fe_2O_3 NPs as contrast agents for the application of magnetic resonance and also, they tested out their applicability for the sensitive recognition of target biomolecules as molecular imaging probes. Some main drawbacks of this technique are high cost, need for high temperature, use of toxic chemicals/organic solvents, and need for modification to increase biocompatibility during the synthesis process of MNPs [11–15].

15.2.2 COPRECIPITATION TECHNIQUE

To synthesize MNPs below 50 nm diameter, coprecipitation is the most efficient and simplest pathway. So, the synthesis process takes place in an aqueous solution of a mixture of ferrous and ferric salts base is added in it, following chemical reaction takes place:

$$Fe^{2+} + 2Fe^{3+} + 8OH^{-} \rightarrow Fe_3O_4 + 4H_2O$$

After the Massart method which was developed in 1981, a lot of studies have been done to understand the effects of parameters of synthesis like concentration and nature of salts, reaction temperature, ionic strength and pH of the medium, injection flux rates, presence of oxygen. In all these studies not only the synthesis parameter is taken into consideration but also along with it the whole procedure is adopted so that the shape, size, composition, magnetic behavior, and morphology of the MNPs can be controlled. To avoid aggregation of NPs polymers or surfactants are added in order to stabilize and control the surface of MNPs, also the addition of surfactants in the reaction increases biocompatibility. Figure 15.2 shows the TEM analysis of supermagnetic iron oxide NPs synthesized by the coprecipitation method [10–15]. However, there are many advantages of the co-precipitation method, but some disadvantages are also there like it produces NPs which ranges between 4 and 20 nm, control of size distribution is poor, irregular morphologies [14–18].

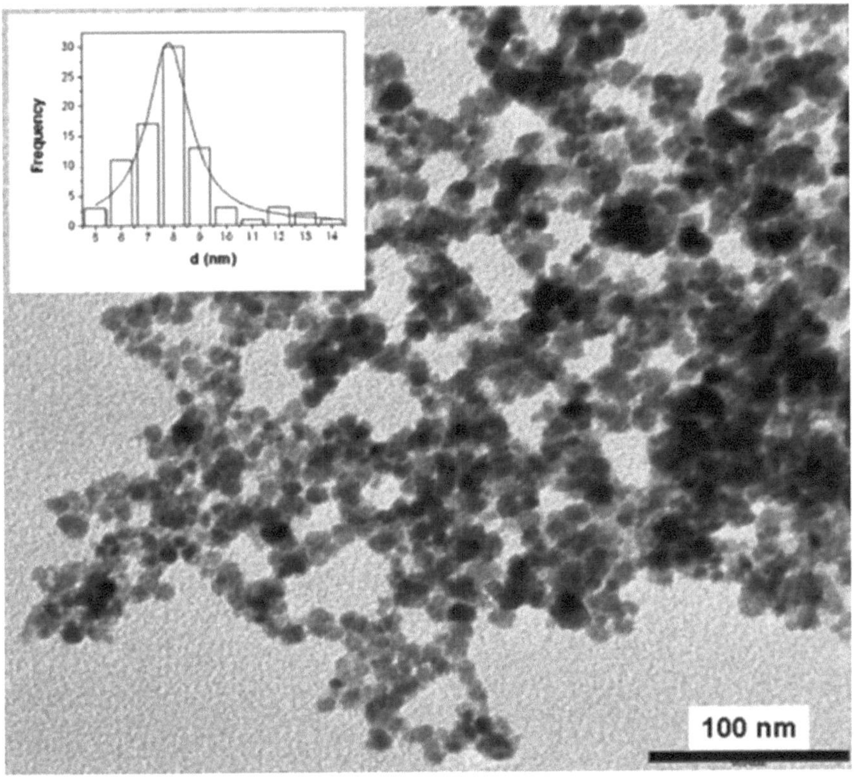

FIGURE 15.2 TEM analysis of Fe_3O_4 particles that are obtained by coprecipitation method.

15.2.3 HYDROTHERMAL METHOD

The hydrothermal process is used to yield MNPs at high pressure and high temperature in autoclaves or reactors in aqueous media which results in fast growth and rapid nucleation of freshly formed MNPs which later on forms in MNPs whose size is really small. When this reaction gets completed then autoclaves are left to cool down to room temperature, after all, this is done then the MNPs are washed numerous times using different solvents so that the impurities present in them can be eliminated. For obtaining hollow iron oxide NPs hydrothermal process is a conventional and facile method. For this the ion which is used as ion source is Fe^{3+} then sodium citrate, acetate, and urea under continuous stirring mixed in ethylene glycol after this the solution is transferred to properly sealed Teflon, these Teflons are tightly sealed because these are to be kept at a high temperature which is approx. 200°C for a time of 8–24 hrs. Figure 15.3 shows the formation of MNPs by the hydrothermal process by using Teflon. There are some other different architectures possible with the help of this method for producing along with nanoring's, short nanorods, and nanospheres [10, 15, 19, 20]. The following reaction is expected to occur in order to produce MNPs:

$$Fe^{2+} + 2Fe^{3+} + 8OH^{-} \rightarrow Fe_3O_4 + 4H_2O$$

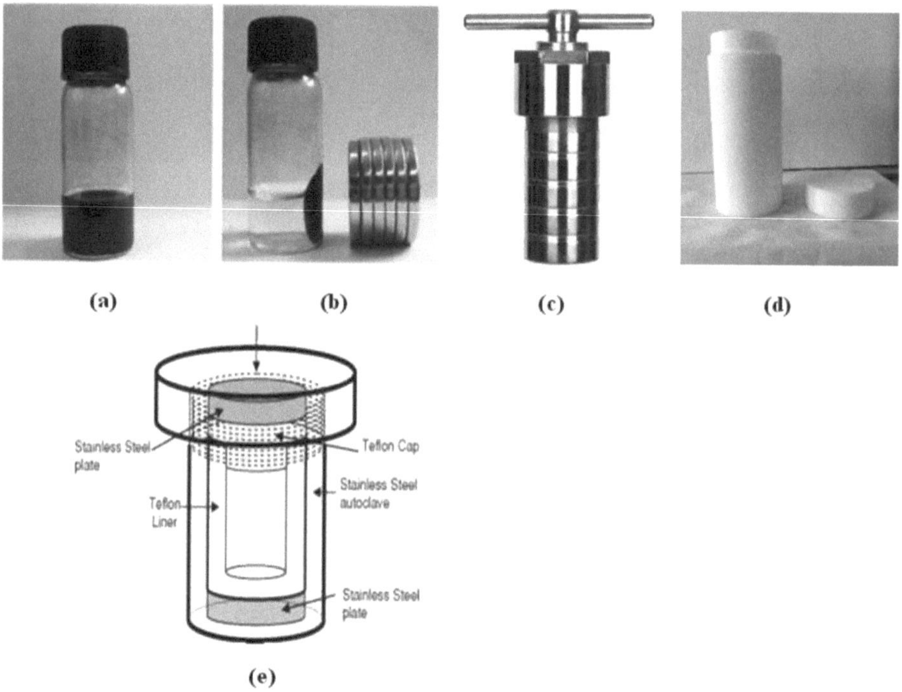

FIGURE 15.3 Synthesis of the magnetic nanoparticle by hydrothermal method. (a) Formation of MNPs; (b) when magnets placed near MNPs then the MNPs gets attracted towards it and moves to the surface; (c) this is the steel jacket and inside it, Teflon is placed at very high temperature; (d) Teflon; and (e) diagrammatic representation of Teflon.

15.2.4 SOL-GEL METHOD

The sol-gel method is used to yield MNPs of controllable size and particle structure. The other name used for the sol-gel method is chemical solution deposition. This method is used to produce nanostructured metal oxides, So, it is an appropriate wet route. This synthesis process is based on the solution of molecular precursors, on its condensation and hydroxylation from this, a sol is formed by nanometric particles. In addition, inorganic polymerization and condensation lead to the formation of a three-dimensional (3D) network of metal oxide-dominated wet gel. As this reaction occurs at room temperature than to reach the final state of crystalline more heat treatments are done. The factor on which gel depends is the sol-stage when the structure is formed during the synthesis of sol-gel. The main parameters that affect the hydrolysis, kinetics, properties, structure, growth reaction of the gel are nature, agitation, salt precursors concentration, and temperature. From an experiment conducted for the synthesis of MNPs it has been found out that maghemite can be synthesized at high temperature (approx. 400°C) by direct heat treatment of gels from all this procedure it is concluded that the NPs form through this procedure ranges between 6 and 15 nm. This approach has many advantages like low process temperature, good mixing uniformity, and high reaction activity but there is a disadvantage of this process that while MNPs are synthesizing during the posttreatment process MNPs get coagulated [15, 21, 22].

15.2.5 MICROEMULSION-BASED SYNTHESIS

The microemulsion method is used to synthesize MNPs of controllable size and shape. This method is based on the isotropic stable system which consists of two immiscible phases a polar (usually water) and a nonpolar (commonly oil) phase in the presence of surfactant, surfactant molecules can be either poly-vinylpyrrolidone, bis(2-ethylhexyl)sulfosuccinate, cetyltrimethylammonium bromide. The microemulsion method for preparing MNPs in a way that the size of MNPs can be controlled is done by a chemical reaction like reduction or co-precipitation in microemulsions. This is accomplished with the help of thermal initiations in which two reactants are placed in a single micelle system or two separate micelle systems. This method is also used to synthesize MNPs which are monodisperse and have various morphologies. There are some disadvantages of this method are also present, some of its main disadvantages are yield produce through this method are low and during the synthesis process, the consumption of organic solvents is high [10, 14, 15, 23, 24].

15.2.6 FLOW INJECTION METHOD

MNPs that have narrow size distribution can be efficiently synthesized through the flow injection method. This method "consists of continuous or segmented mixing of reagents under a laminar flow regime in a capillary reactor." From a study, it is noticed that Fe_3O_4 MNPs of narrow size distribution can be synthesized using this technique. Figure 15.4 shows using flow injection method Fe_3O_4 NPs were obtained having a narrow size distribution range between 2 and 7 nm. Some of the main advantages of this technique are its high mixing homogeneity, precise control of the process, and have good reproducibility [10, 15, 16].

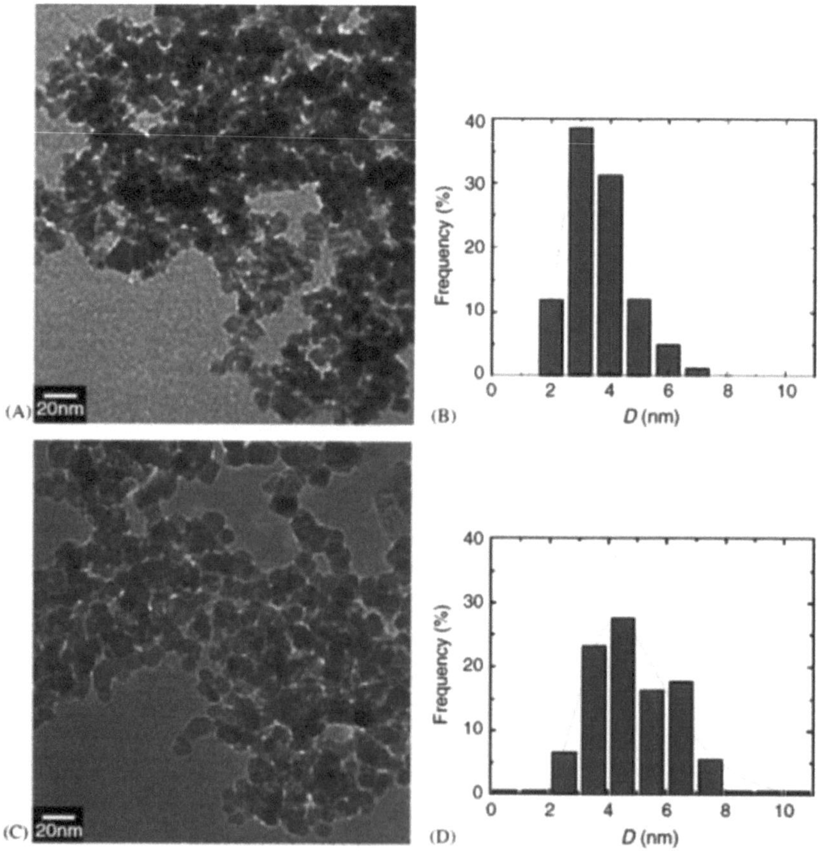

FIGURE 15.4 TEM analysis and diagrams of particle size distribution for iron oxide (Fe_3O_4) nanoparticles synthesized by continuous flow injection (A and B); and batch methods (C and D).

15.2.7 AEROSOL/VAPOR METHODS

Aerosol technologies like laser and spray pyrolysis are efficient techniques because these techniques allow high-rate production in a continuous chemical process. In the case of the spray pyrolysis technique, in an organic solvent solution of reducing agent and ferric salts are sprayed into a series of reactor, then aerosol is condensed, and the solvent is evaporated. After all, these dried residues are obtained which are then changed into MNPs where all matters are the initial size of original droplets of the NPs formed. By using an alcoholic solution that contains "different iron precursors salts" maghemite particles have been obtained whose size ranges from 5 to 60 nm. Now comes laser pyrolysis, which is also a promising approach for synthesizing MNPs. For reducing the volume of reaction, laser pyrolysis can be used. In this technique firstly the laser heats the flowing mixture of gas and iron precursor (like iron pentacarbonyl) mixture which then produces non-aggregated, narrow, and small size NPs. This whole process of laser pyrolysis is carried out into generators which helps in obtaining high yield MNPs. Under conditions of the optimum process, MNPs can be prepared which have a crystal size with narrow size distribution (between 2 and 7 nm) [21, 25].

15.3 APPLICATIONS

15.3.1 BIOMEDICAL

MNPs have shown great potential in the field of biomedicine, as they show some unique characteristics when the magnetic field is applied to them. MNPs are also used in the treatment of cancer. They can detect the biological entity present inside the body and can be used to target that biological entity. Some of the main applications in the field of biomedicine are magnetic resonance imagining (MRI), drug delivery, and imaging-guided photochemical therapy. So, in recent times biomedical applications have become one of the best and leading fields of research because of their different characteristics and irreplaceable advantages [26].

15.3.1.1 MRI

Magnetic resonance imagining (MRI) is basically a diagnostic tool used for the non-invasive imagining of the internal organ of the body of a human. For diagnostics and medical research MRI has become a very essential tool. It is one of the common methods for the diagnosis of tumors [26]. MRI was founded by the NMR phenomenon. So, in the year 1946 through independent experiment two scientists named Edward Mills Purcell and Felix Bloch, they found that when a strong magnetic field is applied on paraffin or water and then at radio frequencies barraged with magnetic oscillation, after this they would release and absorb energy. This all was noticed at the very starting of the nuclear magnetic resonance (NMR), with its help MRI was developed for imaging of the tissues in the body [25]. So MRI helps in monitoring the decay shown by the nuclei during its magnetization, here nuclei are considered as protons of the water molecule when the strong magnetic field is applied on the nuclei after that by a series of radiofrequency (RF) magnetic pulses they get excited [15]. The rate of signal decays depends on two properties first, the density of the nuclei and second on the local magnetic field. In the water contents when the variations are seen between organs then at that time it provides contrast in MRI images. As it shows changes in the content of water to the diseased area, so it is used for the diagnostics of the brain or neural diseases. MRI has shown some good results in the diagnosis of cancer and also in its treatment [27].

15.3.1.2 DRUG DELIVERY

MNPs can efficiently deliver the drug to a particular site and can also control drug release remotely by real-time monitoring. Drug delivery helps in the treatment of cancer through improved chemotherapy treatments. Mostly in chemotherapy treatments, the drugs are circulated to the whole body so that the healthy tissues can be exposed to deleterious side effects. In this therapy, drugs are delivered magnetically so that they can be attached to NPs present in biocompatible ferrofluids, which are then injected into the body of the patient [8]. To concentrate the drugs added inside the body at a specific target than a high gradient external magnetic field may be used for this purpose. So, in this way, it reduces the chances of side effects that can happen on the healthy tissue and may be optimized the quantity of the dosage in a tumor [28].

15.3.1.3 PHOTOTHERMAL THERAPY

Recently a technique developed for the treatment of cancer in that with laser radiation heat is generated under illumination. Firstly, NPs of light-activated heated transfer into tumors than when they enter in the site tumor it attains high temperature and without doing any damage to the surrounding healthy tissue it kills the tumor cells. This is a very effective technique for treating the tumor. In this technique combination of both magnetic properties and plasmonic is seen which successfully provides materials that show enhanced heating ability. Synergic effects of heating can also be improved using multi-functional (magnetic/photonic) particles with chemotherapeutic agents [26].

15.3.1.4 HYPERTHERMIA

In this, the temperature of tumor tissue is increased which shows effective adjunctive cancer therapy. Several ways are proposed to increase the temperature of the tissue to the level required for it by using laser, microwave, or radio frequency wavelengths. This therapy is based on the fact that above 42°C temperature when the cells are heated then it shows signs of apoptosis and when the temperature is set to above 50°C and cells are heated then it shows necrosis. Despite knowing all this the relevant temperature needed clinically to reach is still unclear and it is challenging to reach homogeneity of temperature in the particular target tissue because of local perfusion variations. Due to the higher rate of metabolism, cancer cells are more susceptible to heat compared to normal cells. Magnetic particles hyperthermia by embedding MNPs to the target tissue and also by applying external alternating magnetic field help in heating the particular target tissue. MNPs in this therapy is injected to target cells within the body. Further with the help of the MRI MNPs can be visualized so both diagnosis and treatment are possible. Magnetic iron oxide particles are mostly used for this therapy because of their higher biocompatibility [3, 25, 28–31].

15.3.1.5 BIOSENSING

Using biosensing applications, the high surface area and magnetic properties of MNPs can be exploited to quantity and detect a variety of biological targets (another name analyte) like pathogens, cells, and molecules. The most effective method used for the detection of improved sensitivity is firstly concentrated all MNPs and then the analyte on the surface of the sensor by using a magnet. The MNPs used in biosensors fall into another category which depends on the signal transduction physical principle [15].

15.3.2 MAGNETIC DATA STORAGE

From the invention by Valdemar Poulsen of magnetic recording, the information stored in magnetic storage has become important in the field of digital magnetic data storage media. The progress done in magnetic data storage has been important for the modification of modern computers. The magnetic storage media involves MNPs of a single domain. When the inductive write head is used then the information needed can be easily written on the medium which then generates the localized magnetic field that shows variation in time this all happens when the medium moves below the head. While the writing process has taken place the magnetization directions of the NPs which are single domain and when the field is applied then in its every bit, they get aligned in the direction of the applied field and after the removal of the field, it retains the memory of the field. And when the medium magnetic field sensor below a read head, the magnetization written state then gives rise to the magnetic field that varies with time and can be detected. In many digital media like hard disks of computer, in them, the stored binary information corresponds to magnetization orientation present in antiparallel of the media in its corresponding areas [15, 28].

15.3.3 FERROFLUIDS

Ferrofluids can be defined as a colloidal suspension of ferrimagnetic or ferromagnetic particles present in a liquid. In the 1960s ferrofluids were discovered in the research center of NASA. At that time the scientists of NASA were investigating methods for controlling liquids in space. Then they notice that in absence of gravity location of a ferrofluid could be controlled with magnetic fields. For preparing a ferrofluid, it is very important to remove sedimentation and agglomeration of the particles. So the size of MNPs plays an important in that case, the size of MNPs should be below a critical limit so that the Brownian motion opposes the sedimentation and agglomeration tendency because of the presence of gravity [28]. Ferrofluids can be obtained with spiking effect characteristics when the magnetic field is applied, so the only thing that needed for it is high density of NPs should be dispersed in the water solvent. This was achieved by magnetic drying and decantation, after this few drops of tetrabutylammonium (TBA) hydroxide solution were added as shown in Figure 15.5 [19–23].

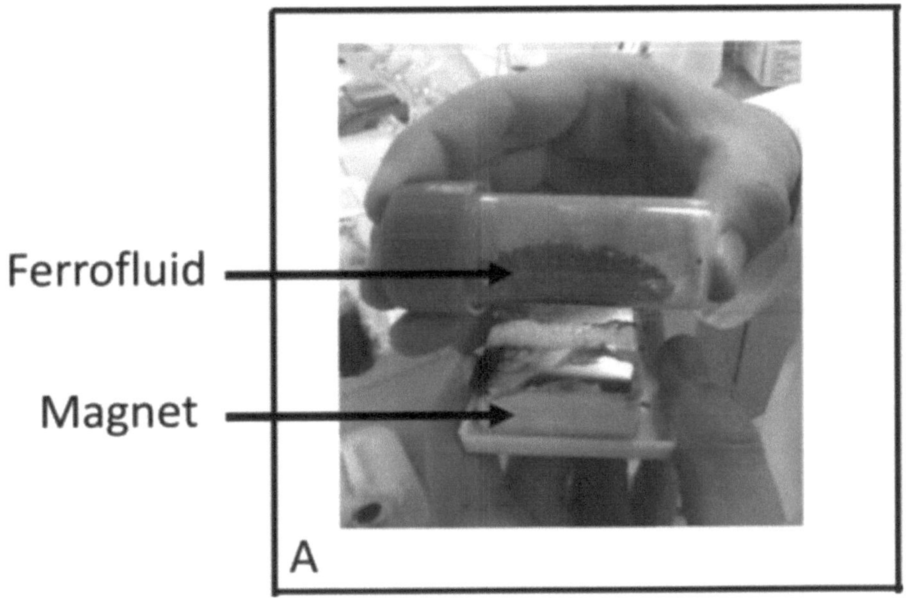

FIGURE 15.5 Tetrabutylammonium (TBA) stabilized aqueous ferrofluids.

15.4 CONCLUSION

MNPs have opened a way to create various magnetic materials that can possess unique properties, especially when their nanoscale effects are considered. As MNPs show some characteristics that have increased their demand in the biomedical and industrial field. MNPs of the desired structure, composition, and sizes can be synthesized by various methods. Different synthesis process gives a different desired result like some gives controllable size or shape while some give great monodispersity and narrow size distribution, etc. The main problem that comes when MNPs are synthesized is aggregation, they can easily get aggregated. MNPs have great potential to play an important role in the progress of future medical applications whereas that is in diagnosis and imaging or through biosensing and drug delivery applications. They can also be used in magnetic data storage and ferrofluids applications. Hence it is not wrong to call MNPs as materials of the future.

KEYWORDS

- **hydrothermal**
- **hyperthermia**
- **maghemite**
- **magnetic nanoparticles**
- **magnetic resonance imaging**
- **magnetite**
- **superparamagnetism**

REFERENCES

1. Yin, L., & Zhong, Z., (2020). Nanoparticles. *Biomater. Sci.*, 453–483. doi: 10.1016/b978-0-12-816137-1.00031-3.
2. Sun, S., & Zeng, H., (2002). Size-controlled synthesis of magnetite nanoparticles. *J. Am. Chem. Soc., 124*, 8204, 8205.
3. Bedanta, S., Barman, A., Kleemann, W., Petracic, O., & Seki, T., (2013). *Magnetic Nanoparticles: A Subject for Both Fundamental Research and Applications, 2013.*
4. Wu, L., Mendoza-Garcia, A., Li, Q., & Sun, S., (2016). Organic phase syntheses of magnetic nanoparticles and their applications. *Chem. Rev., 116*, 10473–10512.
5. Rahim, S., Jan, I. F., & Malik, M. I., (2019). Biomedical applications of magnetic nanoparticles. *Met. Nanoparticles Drug Deliv. Diagnostic Appl.*, 301–328. doi: 10.1016/B978-0-12-816960-5.00016-1.
6. Slimani, Y., & Hannachi, E., (2020). Magnetic nanosensors and their potential applications. *Nanosensors for Smart Cities* (INC, 2020). doi: 10.1016/b978-0-12-819870-4.00009-8.
7. Guo, T., et al., (2018). *Review Article the Recent Advances of Magnetic Nanoparticles in Medicine, 2018.*
8. Sandler, S. E., & Fellows, B. M. © 2019. (2019). doi: 10.1021/acs.analchem.9b03518.
9. Williams, H. M., (2017). *Review Article the Application of Magnetic Nanoparticles in the Treatment and Monitoring of Cancer and Infectious Diseases, 10,* 1–10.
10. Reddy, L. H., Arias, J. L., Nicolas, J., & Couvreur, P., (2012). Magnetic nanoparticles: Design and characterization, toxicity and biocompatibility, pharmaceutical and biomedical applications. *Chem. Rev., 112*, 5818–5878.
11. Bu, S., Hussain, C. M., Science, E., & States, U., (2020). *5.1 Introduction*. doi: 10.1016/B978-0-12-816699-4.00005-0.
12. Odularu, A. T., (2018). Metal nanoparticles: Thermal decomposition, biomedicinal applications to cancer treatment, and future perspectives. *Bioinorg. Chem. Appl., 2018.*
13. Unni, M., et al., (2017). Thermal decomposition synthesis of iron oxide nanoparticles with diminished magnetic dead layer by controlled addition of oxygen. *ACS Nano, 11*, 2284–2303.

14. Cardoso, V. F., et al., (2017). *Advances in Magnetic Nanoparticles for Biomedical Applications, 1700845*, 1–35.
15. Applications, B., (2018). *Multifunctional Magnetic Nanoparticles: Design, Synthesis, and. Comprehensive Nanoscience and Nanotechnology* (2nd edn.). Elsevier Ltd. doi: 10.1016/B978-0-12-803581-8.10462-X.
16. Ahn, T., Kim, J. H., Yang, H. M., Lee, J. W., & Kim, J. D., (2012). Formation pathways of magnetite nanoparticles by coprecipitation method. *J. Phys. Chem. C, 116*, 6069–6076.
17. Pereira, C., et al., (2012). Superparamagnetic MFe_2O_4 (M = Fe, Co, Mn) nanoparticles: Tuning the particle size and magnetic properties through a novel one-step coprecipitation route. *Chem. Mater., 24*, 1496–1504.
18. Suh, S. K., et al., (2012). Synthesis of nonspherical superparamagnetic particles: In situ coprecipitation of magnetic nanoparticles in microgels prepared by stop-flow lithography. *J. Am. Chem. Soc., 134*, 7337–7343.
19. Dalverny, A., et al., (2017). *Comprehensive Approach to Amazing Magnetic Materials*. doi: 10.1021/acs.jchemed.7b00277.
20. Ge, S., et al., (2009). *Facile Hydrothermal Synthesis of Iron Oxide Nanoparticles with Tunable Magnetic Properties*, 13593–13599.
21. Laurent, S., et al., (2008). *Magnetic Iron Oxide Nanoparticles: Synthesis, Stabilization, Vectorization, Physicochemical Characterizations, and Biological Applications*, 2064–2110.
22. Rosa, C., & Segarra, M. (2019). Optimization of the Synthesis of Copper Ferrite Nanoparticles by a Polymer-Assisted Sol – Gel Method. *ACS Omega, 4*(19), 18289.
23. Okoli, C., et al., (2012). *Comparison and Functionalization Study of Microemulsion-Prepared Magnetic Iron Oxide Nanoparticles*. doi: 10.1021/la300599q.
24. Lo, J. A., Lo, M. A., Mira, J., Rivas, J., & Charles, S. W., (1997). *Advances in the Preparation of Magnetic Nanoparticles by the Microemulsion Method, 101*, 4–6.
25. Li, Y., et al., (2020). *Polymer-Assisted Magnetic Nanoparticle Assemblies for Biomedical Applications*. doi: 10.1021/acsabm.9b00896.
26. Zhu, K., et al., (2018). *Magnetic Nanomaterials: Chemical Design, Synthesis, and Potential Applications*. doi: 10.1021/acs.accounts.7b00407.
27. Duguet, E., (2006). *Magnetic Nanoparticles and Their Applications in Medicine, 1*, 157–168.
28. Mørup, S., Hansen, M. F., Frandsen, C. (2011). Magnetic Nanoparticles. *Comp. Nanosci. Technol. 1*, 437–491.
29. Gao, J., Gu, H., & Xu, B., (2009). *Multifunctional Magnetic Nanoparticles: Design, Synthesis, and Biomedical Applications, 42*.
30. Micheletti, G., et al., (2020). *Magnetic Nanoparticles Coated with (R) - 9-Acetoxystearic Acid for Biomedical Applications*. doi: 10.1021/acsomega.0c00163.
31. Kudr, J., Id, Y. H., Richtera, L., Adam, V., & Zitka, O. (2017). *Magnetic Nanoparticles: From Design and Synthesis to Real World Applications*. doi: 10.3390/nano7090243.

CHAPTER 16

Electrospinning Technique and Magnetic Nanofibers

ROOP VARGHESE RUBERT, MARIA MATHEW, and RONY RAJAN PAUL

Department of Chemistry, CMS College, Kottayam, Kerala, India

ABSTRACT

Electrospinning, a type of electrospraying technique that utilizes the electrohydrodynamic phenomenon, emerged as a flexible method to generate nanoscale and microscale objects. Electrospun nanofibers offer a variety of benefits, including a high surface area-to-volume ratio, tunable porosity, adequate malleability to get the desired size, shape, and dimension, and adaptability to achieve the desired physical qualities. Also, exceptional mechanical and thermal properties that are superior to ordinary fibers are observed. The magnetic nanofibers (MNFs) are found to have good electromagnetic (EM) performance, along with obvious magnetism like paramagnetism, ferromagnetism, and diamagnetism, as well as other qualities like superparamagnetic effect, macroscopic tunneling effect, and so on. These special characteristics make them good enough for a wide range of applications, such as ultrahigh-density data storage, sensors, drug delivery systems, catalysis, and biological applications. This work reports updated progress made on electrospinning techniques as well as the recent developments in MNFs and their potential applications.

16.1 INTRODUCTION

Nanofibers are found to have profound applications in the recent years. They have persuasive size dependent chemical, mechanical, thermal, electrical,

Modern Magnetic Materials: Properties and Applications. Iuliana Stoica, Ann Rose Abraham, & A. K. Haghi (Eds.)

and biological properties which are in turn related to their one dimensionality [1, 2]. The widespread applications of nanofibers are evident in everyday implementations including batteries, mobile phones, solar cells, fuel cells, and so on. The fabrication of nanofibers is indeed very fascinating. The fabrication technique can be generally categorized as: (i) physical and chemical techniques; and (ii) electrospinning and non-electrospinning techniques [3].

Electrospinning is a versatile tool that can be used for manufacturing of nanoscale and microscale structures. The process is initiated by the preparation of a viscous solution prepared by dissolving a polymeric substance in a sufficiently volatile solvent which is the fed into the tip of a metallic needle from a syringe. A grounded conducting collector is set at a specific distance from the needle's tip. The needle is then placed at a high potential. Due to the strong electric field applied at the tip of the needle, the solution pop-up and appears in the form of a cone often referred to as Taylor cone [4, 5]. From the peak of this cone, the jet of the viscous solution emerges and gets accelerated to the grounded conducting collector. In most cases, the solvent carrying the polymeric substance evaporates in-flight (before it hits the collector) and the fluid jet solidifies into a fiber. Figure 16.1(a) shows the schematic illustration of an electrospinning set-up Figure 16.1(b) shows the produced fiber mat [6].

Even though, there are some challenges associated with the process, electrospinning has been most widely examined and also applied to fabricate fibers in nanoscale from the early 2000s. Some of the difficulties include the low rate of production and the high capital expenditure in the use of the technology for making fibers over conventional methods [3]. The electrospinning is actually a kind of electrospraying technique which is based on the electro-hydrodynamic phenomenon. The adaptability of the electrospraying/electrospinning technique is accredited to several beneficial characteristics such as:

- The size of the droplet is even smaller than those produced by conventional atomizers. Hence the resulting material can be in the nanoscopic scale.
- Usually, the size distribution of the droplet is thin.
- Due to the mutual repulsion, charged particles are scattered in space. Hence the coalescence of the droplet is absent and consequently lower agglomeration is observed.
- By varying the electric field strength, better control over the motion of charged particle is possible [8].

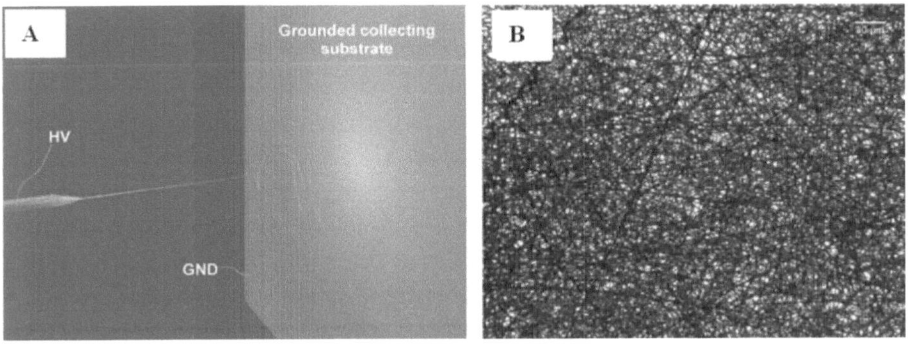

FIGURE 16.1 (a) Illustration of standard electrospinning apparatus; and (b) an example of a fiber mat produced by an electrospinning process. This image is of polyethylene oxide nanofibers electrospun from a water/ethanol solution (20% 1,00,000 M_w polymer in 50/50 water/ethanol by volume).

16.2 HISTORY

Electrospinning is an old method to produce nanofibers. Rayleigh observed the mechanism of hydrodynamic stability of the fluid jet "with" an external electric field and "without" using an external electric field. In 1882, using a charged fluid jet Rayleigh [9] scrutinized the instability mechanism and reported that the electrostatic force could exceed the surface tension, which acts in the opposite direction and the fluid was tossed out in the form of fine jet [10, 11].

In 1914, Zeleny [12] conducted a detailed study of the mechanism. The first patent on the technique of electrospinning was surfaced up in the year of 1934 by Formhals [13], the different experimental set-up, the collection of electrospun fibers and their applications at that time made it particularly attractive. The set-up consisted of an apparatus that can produce polymer filaments by taking advantage of the electrostatic repulsion between the surface charges. In 1964 Taylor [5] put forward the hypothesis of Taylor cone, which mainly dealt with the shape of the liquid droplet due to the electrostatic forces which in turn is explained by some mathematical equations. He also derived the voltage balance equation of the cone and found that the stable Taylor cone is with a semi-vertical angle of 49.3°. Later, in 1969 this was experimentally proven and laid the foundation for the development in electrospinning processes.

The phrase "electrospinning" comes from the term "electrostatic spinning" which had been in use until 1994. Hayati & co-workers [14] in 1987 investigated the factors affecting the process including experimental condition (ambient condition), electric field strength, stability of the fiber and atomization. They found that when it comes to electrostatic disruption of liquid surfaces, fluid conductivity is crucial. In conclusion, fluid which are highly conducting produced streams that are highly unstable, that whipped around various directions in the presence of an increasing applied potential difference. In 1971, Baumgarten [15] and in 2005 Subbiah & co-workers [16] successfully fabricated electrospun microfibers using a solution of polyacrylonitrile and dimethylformamide (DMF). The polymer solution was suspended from a stainless-steel needle when applied with a high potential difference. They observed that initially there is an inverse relationship between the diameter of the jet and the applied voltage. After reaching a minimum value of the diameter, a direct relationship was observed. An investigation of the relationship between the melt temperature and the fiber diameter was conducted by Larrondo & Manley [17] using polyethylene and polypropylene as substrate and found that an inverse relationship exists between the fiber diameter and the melt temperature. In 1995, Chun [18] as well as in 1999, Fong & Reneker [19] fabricated different nanofibers from a number of polymeric solutions including poly(amic acid) and poly(acrylonitrile) using the electrospinning technique. By investigating the bending instability during the flight Reneker & co-workers [20, 21] proposed the experimental phenomenon and theory of electrospinning mechanical process (Figure 16.2). With the help of high-speed cameras, the jet path has been observed and the inception and growth of the bending, whipping, and stretching motions were examined and it was found that it could reduce the cross-sectional area of the liquid up to 105 times. Rutledge & co-workers [22] examined the solution parameters and processing parameters such as surface tension, flow rate, electrical current, etc., which influences the diameter of the liquid jet during electrospinning and an analytical model was developed to foretell the existence of a terminal jet diameter.

In 2003, Larsen & co-workers [23] put forward the idea of fabricating core-shell composite ultrafine fiber using coaxial electrospinning. Chen et al. [24–26] in 2014 suggested the first ever triple coaxial electrospinning technique to produce amorphous carbon nanotubes (CNTs). Jeong & co-workers [27] reported a method of incorporating antimicrobial actives such as silver nanoparticles (NPs) and Mefoxin in electrospun fibers. For the past couple of decades, electrospinning has been an area of notable research works.

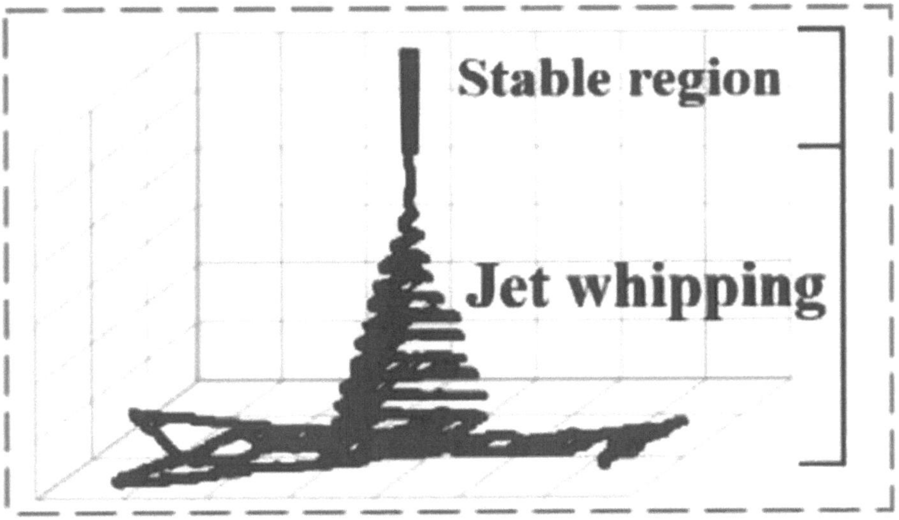

FIGURE 16.2 Mathematical modeling of the electrospinning jet.

16.3 SET-UP AND MECHANISM

The electrospinning technique has been widely explored and used for the fabrication of different nanofibers for the past few decades. Spun nanofibers have a wide range of advantages including high surface area-to-volume ratio, tunable porosity, sufficient malleability to attain the required size, shape, and dimension and the adaptability to attain the required physical properties [3].

An electrospinning apparatus consists of three important components namely: (i) a high voltage power supply; (ii) a spinneret; and (iii) a grounded conducting collector [3]. Figure 16.3 depicts the schematic illustration of the set-up [28]. Though alternating current (AC) potential can be used for the purpose, direct current (DC) power supply is usually used [29–32]. The spinneret is associated with a syringe from which the solution of the polymer is supplied. The syringe pump enabled the controlled and constant delivery of the solution through the spinneret. DC voltage in the range of 10 kV is required in order to meet the electric field for the electrospinning process [7]. The process starts at the tip of the spinneret needle, where the droplet of the polymer solution is applied with a high potential difference. This will induce charges and eventually they will get dispersed over the collector surface. As a result of this, the droplets will experience two major kinds of electrostatic interactions; the electrostatic repulsion between the surface charges and the Coulombic interaction brought to bear by the external electric field [28]. The charged droplets that emerge from the tip of the spinneret are influenced by electrostatic forces and gradually deform into a cone known as the Taylor cone (Figure 16.3) [20, 29–31, 34]. Beyond a certain electric field strength value *viz.*, threshold value, the electrostatic repulsion will exceed the surface tension offered by the polymer solution and hence compel the ejection of a liquid jet from the tip of the spinneret (nozzle). This electrified jet of the liquid then experiences a stretching, looping around and whipping process resulting in the production of long and fine threads [28]. During the flight Coulombic fission between the charge particles takes place and results in the evaporation of the solvent.

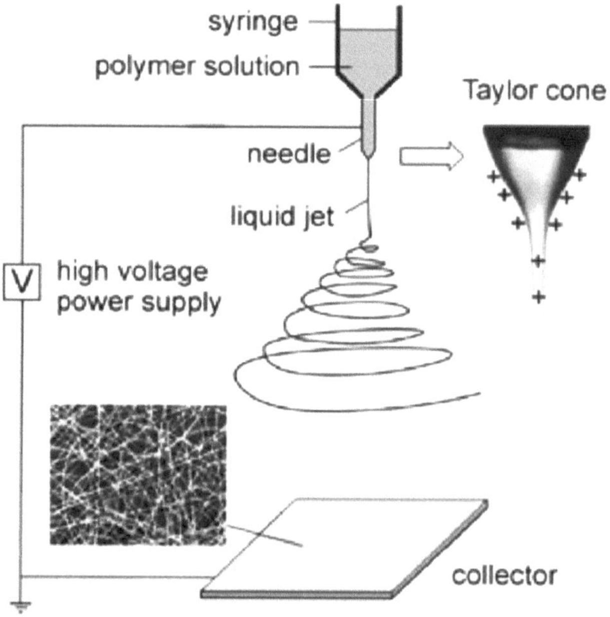

FIGURE 16.3 Schematic illustration of the basic setup for electrospinning. The insets show a drawing of the electrified Taylor cone and a typical SEM image of the nonwoven mat of poly(vinyl pyrrolidone) (PVP) nanofibers deposited on the collector.

The fabricated nanofibers will be amorphous, polycrystalline, dense, porous, hollow aligned or unaligned (random) [35]. The electrospun nanofibers are usually obtained as randomly oriented unaligned or partially aligned fibrous mats. Figure 16.4 portrays the rendering process of these nanofibers.

The diameter of the formed nanofibers was usually in the range of 10 nm to a few micrometers. The length and the diameter of the nanofibers can be generally altered by changing the processing parameters and solution parameters including applied electric field, concentration of the polymer solution, feed rate, etc. [36].

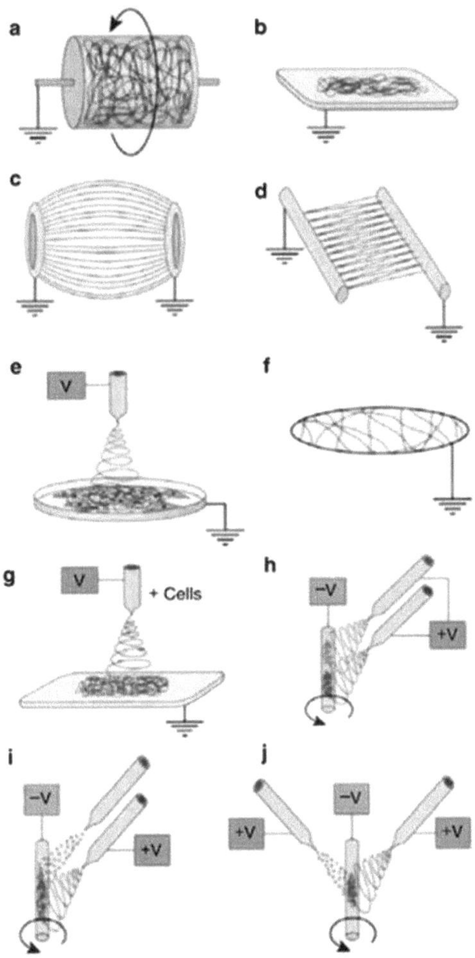

FIGURE 16.4 The setup for electrospinning technique with different collecting methods: (a) with using a rotating single ground cylinder; (b) with using a single ground plate; (c) designing a practical dual ring; (d) designing a flexible dual bar; (e) the technique of electrospinning *in vitro* into cells; (f) using a practical single horizontal ring; (g) designing an electrospinning cells with polymer; (h) using a dual spinneret electrospinning technique (with second pink fibers); (i) electrospinning/electrospraying technique with parallel; and (j) also with a unique perpendicular spinnerets.

16.4 BENDING INSTABILITY THEORY

Positive and negative ions are formed when an external electric field is introduced to a polymer solution, and they migrate in opposite directions. Positive ions are drawn to the negative electrode and vice-versa. The difference in the number of negative and positive ions in the polymer solution is called excess charge. In case of highly insulating polymer solution, salts like KCl, NaCl (1:1 electrolyte) are added which may dissociate into equal number of anions and cations. This can boost the polymer solution's ionic conductivity without changing the excess charge since the addition of electrolyte only increases the ion per unit volume. The higher the conductivity, the faster the movement of excess charge will move in a certain direction under the effect of the applied electric field.

When the polymer solution is encountered with the electrostatic field, it compels the surface of the solution to get charged, thereby causing the flood of ions through the polymer solution where the fluid jet is produced [28, 37–39].

The jet of the fluid emerges out from the peak of the cone resume to narrow down along the path through which it travels towards the collector. This mode of jetting is often referred to as electrohydrodynamic cone jet [40]. The jet of the fluid ejected from the needle of the spinneret initially follows a straight path. But later on, segments of a loop develop back-to-back bending instabilities. There are three stages to each cycle of the bending instability:

- A laminar segment of the fluid jet that was initially straight suddenly starts to develop an array of bends;
- The elongation of the segment of the jet in each bends and the array of bends eventually becomes sequence of spiraling loops with increasing diameter; and
- The cross-sectional diameter of the fluid jet becomes thinner as the circumference of the loops get higher.

The cycle of instability was found to be repeating several times on a smaller scale. It was also found that as time passes increasingly more instability in the cycle takes place, linearly decreasing the diameter of the fluid jet and thereby fabricating nanofibers. The axis of particular segment pointed in any direction (random orientation) occurs after the second cycle. The solvent in the jet evaporates before it hits the collector screen thereby producing electrospun nanofibers [40]. The bending instability was at first considered to be taking place as a result of the splitting of a single jet into a number of thin fiber filaments due to radial charge repulsion, which was known to be splaying. Reneker & Doshi [41] suggested that as a result of the concurrent stretching of the jet and the solvent evaporation the diameter of the fiber decreases. Also, the increased charge density causes the splitting of the fluid jet into smaller ones. This was expected to occur continuously, causing fibers with smaller diameter. But recent studies confirmed that the development of a non-axisymmetric or whipping instability is the reason for the stretching and bending of the fluid jet [34]. Warner & co-workers [42] as well as Shina and co-workers confirmed that in the shape of an inverted cone, the fluid jet's unstable area appeared questioning the occurrence of multiple splitting which is in reality a single rapidly whipping jet (Figures 16.5(a) and (b)). Figure 16.5(c) shows the schematic illustration of different types of instabilities ejected from the Taylor cone.

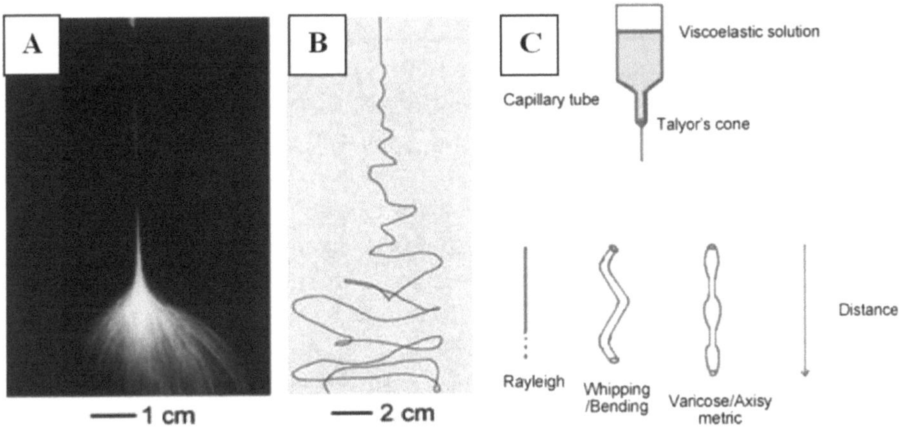

FIGURE 16.5 Photographs illustrating the instability region of a liquid jet electrospun from an aqueous solution of poly(ethylene oxide) (PEO). The capture time was on two different scales: (a) 1/250s; and (b) 18ns. (c) Schematic representation of various types of instabilities emerging from Taylor cone.

16.5 FACTORS AFFECTING ELECTROSPINNING PROCESS

There are two types of parameters that influence the electrospinning process and fiber creation: (i) system parameters; and (ii) process parameters. System parameters are those accompanied with the experimental set-up before the electrospinning process starts. This is usually based on the inference from previous works. Factors such as nature of the solvent, polymer used of the electrospinning, viscosity, conductivity, dielectric constant and surface tension of the solution can be brought under this category. Process parameters are those which are chosen after the experiment is being set-up. Factors including electrostatic field strength, flow/feed rate, the separation between the spinneret's tip and the collector screen, ambient condition such as humidity, temperature, air velocity within the electrospinning chamber, concentration of the polymer solution, etc., falls under this division. These factors are supposed to influence the process and play an important role in determining the quality of the produced nanofibers [10, 43].

16.5.1 SYSTEM PARAMETER/SOLUTION PARAMETER

16.5.1.1 CONCENTRATION

Solution concentration determines the limiting boundaries for the fiber formation to occur since it can alter the viscosity and surface tension as well [44]. For the electrospun fiber formation to take place, a minimum concentration of the solution should be maintained. Several researchers reported that at low concentration of the polymer, a mixture of fibers and beads are formed and as gradually the concentration of the polymer solution increases, the appearance of spherical beads changes to spindle-like and later to uniform fibers with increased diameter due to the increased viscosity resistance [44–49]. Likewise, there is also an optimum concentration beyond which the continuous fiber formation declines. This is due to the difficulty in maintaining the solution flow at the tip of the needle which may further end up in the formation of large fibers [50]. Studies conducted by Tan & co-workers; Rutledge & co-workers; and Hsiao & co-workers [22, 44, 51] reported that at certain concentration level polymers such as PEO, PAN, PDLA show difficulty in forming the fiber *via* electrospinning. Some other studies showed that by increasing the concentration level diameter of the fiber increases and pore size decreases to narrow in case of polystyrene (PS) solution [52]. In case of PEO solution made in water, at elevated concentration a bimodal distribution in the diameter of fiber was observed [44]. The average fiber diameter and the concentration of polymer solution were related as a power law relationship in case of PEO system. Table 16.1 shows a relation between the concentration and fiber diameter for certain polymers. Tan and co-workers also conducted a study in relating the variation in diameter of fiber and the morphology to the shape of the jet-originating surface. The result was concordant with the studies conducted by Hsiao & co-workers [51].

TABLE 16.1 Properties of Some Typical Polymers and Their Corresponding Electrospun Fiber Diameters

Polymer	Structural Formula	Electrospinning Solution (Polymer/Solvent) Molecular Weight, Concentration	Melting Point (°C)	Fiber Diameter (nm)	References
PAN	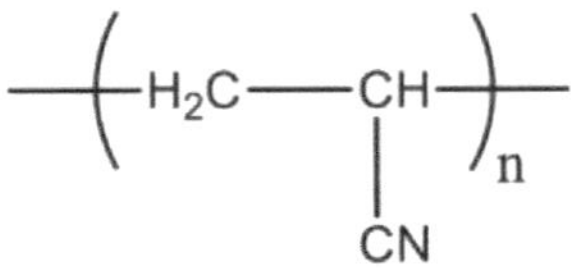	PAN/DMF M_W = 1,00,000, ~10 wt.%	317	~100	[76]
PVP	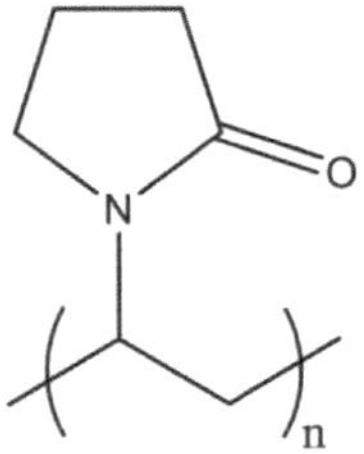	PVP/Ethanol M_W = 13,00,000, ~10 wt.%	130	~1,500	[76]
PS	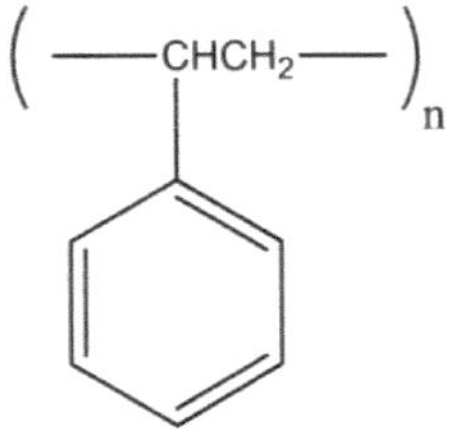	PS/DMF M_W = 2,80,000, ~10 wt.%	240	~600	[35]
PVA	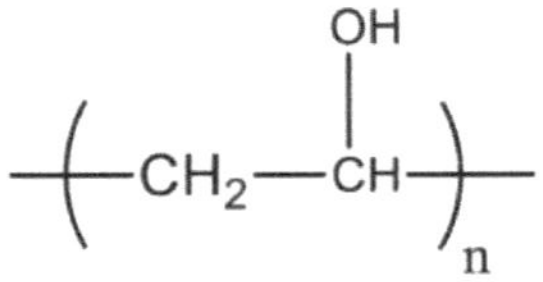	PVA/Water M_W = 78,672, ~10 wt.%	230–240	200–500	[76]

TABLE 16.1 *(Continued)*

Polymer	Structural Formula	Electrospinning Solution (Polymer/Solvent) Molecular Weight, Concentration	Melting Point (°C)	Fiber Diameter (nm)	References
PMMA	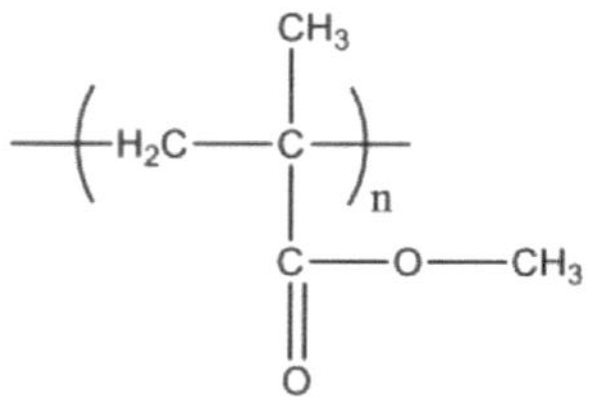	PMMA/DMF M_W = 60,000, ~8 wt.%	130–140	~300	[76]
PEO	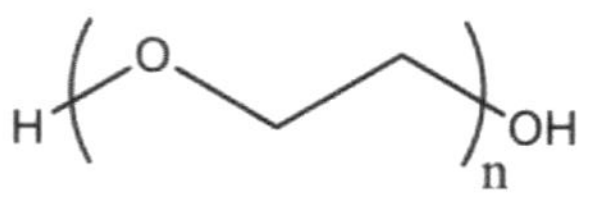	PEO/Water M_W = 3,00,000, ~6 wt.%	62–64	~300	[76]

Surging morphology in fibers were accredited to the delayed drying of the solvent and the stress relaxation nature of the fiber exhibited at lower polymer concentration. It is obvious that the polymer solution concentration controls the spinning of fibers and influences the fiber structure and morphology.

16.5.1.2 MOLECULAR WEIGHT OF THE POLYMER

It is very important to note that the "molecular weight" of the polymer has a direct influence on the morphology of electrospun nanofibers. The molecular weight has showed a linear relationship with the viscosity of the polymer solution and the mixing-up of the polymer chain in solution [53]. Molecular weight also affects the electrical properties as well as other factors like surface tension, conductivity, and dielectric strength [49]. Solution of a polymer with relatively low molecular weight forms bead-like structures rather than clean fibers at a fixed concentration level. Polymer with high molecular weight is a favorite substrate for fabricating microfibers even at low concentration [54, 55]. It is also noted that solution of high molecular weight polymers forms fibers with greater average diameter. The molecular weight of a polymer influences the number of mix-ups of polymer chains in a solution, thereby influencing the viscosity of the solution. Chain entanglement has a direct effect on the electrospinning process. Therefore, for instance, high molecular weight poly-L-lactic acid (HM-PLLA) will be able to maintain a sufficient number of chain entanglements even at a relatively low concentration of the polymer. This ensures an enough level of solution viscosity in order to render a uniform jet throughout the electrospinning process and hold back the effect associated with surface tension, i.e., as surface tension increases, there is a chance for obtaining fiber with beads. The association between molecular weight and bead production was investigated by Wilkes & coworkers [56] as well as Ramakrishna & co-workers [57]. For this, they have synthesized poly(methyl methacrylate) (PMMA) with different chain lengths and molecular weight ranging from 12.47 to 365.7 kDa. They identified an inverse relationship between the polymer's molecular weight and the proportion of beads and droplets, i.e., with an increase in the molecular weight at a particular concentration, the number of beads and droplets decreases. It is obvious that polymers with high molecular weight is not always a necessary element for the for the electrospinning process rather a better substitute which can ensure the interchain connectivity acquired as a result of the chain entanglement by adequate intermolecular interactions. Long & co-workers

[58] as well as Hsiao & co-workers [59] have synthesized oligomer-sized phospholipids using electrospinning process from a solution of Lecithin into non-woven membranes [60, 61].

16.5.1.3 VISCOSITY

One of the condemnatory factors influencing the structural morphology of the fabricated nanofiber is the viscosity of the polymer solution. The viscosity of the polymer solution has a direct relationship with the concentration in addition to the surface tension of the solution. Figure 16.6 shows the variation of morphology of electrospun polyethylene oxide (PEO) nanofiber with different viscosities. At low viscosity, no formation of continuous fiber was observed as well as at higher viscosity there was a trouble in the emergence of the jet from the polymer solution. Therefore, as optimization of the viscosity for electrospinning was necessary. The studies conducted by Ko & co-workers [50] revealed that there was a significant effect of viscosity on silk nanofibers. Larrondo & Manley [17] reported that viscosity plays an important role while fabricating nanofibers through melt electrospinning. The fabrication of electrospun PLA nanofibers with different viscosities and concentration ranging from 1% to 5% by weight showed that no continuous nanofiber formation at a PLA concentration level below 3% [61]. By increasing the viscosity, the spherical beads are stretched, made longer and further transformed into spindle shaped nanofibers [62]. Apart from that, the studies conducted by Tan & co-workers [44] as well as Erman & co-worker [63] showed that the formation of larger fibers are often associated with more viscous polymer solution.

Fong & Reneker [19] conducted a study on the viscosity of the polymer solution using PEO and concluded that the optimal viscosity range required for the uniform nanofiber formation to occur *via* electrospinning process is between 1 and 215 poise. This was concordant with the studies conducted by different other groups [15, 41, 65–66]. The molecular weight of the polymer, concentration of the polymer solution as well as the viscosity are interrelated. The relation between these three factors and the formation of electrospun nanofibers have been studied in detail using different polymeric systems such as PLGA (poly(lactic-co-glycolic acid)), PEO (polyethylene oxide), PVA (polyvinyl alcohol), PMMA (poly(methyl acrylate)), PS, PLLA (poly(L-lactic acid)), gelatin, dextran, etc. [48, 54, 56, 62, 67–74]. Usually, protracted stress relaxation time were exhibited by highly viscous polymer solution. Hence it is possible to stave off the rupturing of the emerged jet while electrospinning. Furthermore, the viscosity of the polymer solution can also be used to determine the optimal range of concentration between which continuous fiber formation can be observed. In addition, there exists an exponential relation between the viscosity of the polymer solution and the diameter of electrospun fiber which can be explained by the following equation:

$$d = 19.49\eta^{0.43}$$

where; η is the viscosity of the solution; and d is the diameter of the fiber which is usually expressed in nm [75].

16.5.1.4 SURFACE TENSION

Another important parameter that decides the continuous electrospinning process is surface tension. Surface tension has an adverse effect such that it hinders the jet from becoming longer and narrow [76]. Surface tension can be considered as a function of the solvent composition rather than the polymer. The decrease in the surface tension of polymer solution results in nanofibers with less or no beaded structures because the solvent molecules would be capable of spreading over the mixed-up polymer molecules much more readily [49, 77]. The studies conducted by Hohman & co-workers [78] revealed that greater surface tension of the nanofiber solution is often associated with the inhibition of the process of electrospinning as a result of the instability offered by the jet and the triggering of sprayed droplets. The studies conducted by Haghi & Akbari [49] showed that the generation of different nanoscale structures associated with the electrospinning process such as beads, droplets, fibers have a direct relation with the surface tension of the solution. They also reported that the process can be run under the influence of a lower electric field strength if the spinning solution employed has a lower surface tension. It is to be noted that a solution having low surface tension could not be always employed for electrospinning process. Other solution parameters such as polymer concentration and viscosity are interrelated to surface tension in a much more complicated fashion. A study relating these three factors was conducted, and it was reported that surface tension is the governing factor in the case of solutions of low viscosity and also the formation of beads or beaded fiber was observed [41]. In addition, above a certain polymer concentration referred to as critical concentration, continuous fiber formation is observed and also the shape of the fiber is affected by the concentration. Furthermore, choosing a solution with low surface tension or the addition of suitable surfactant can induce the capacity of the electrospinning process to fabricate continuous fiber [62].

16.5.1.5 CONDUCTIVITY/SURFACE CHARGE DENSITY

With some exceptions of dielectric material and charged ions in the polymers, most of the polymer solution are conducting, which influences the jet formation. The type of the polymer, the solvent used and the presence of ionizable salt determines the conductivity of the solution [60]. It has been observed that at low solution conductivity the charging of the surface area and the capacitance between the electrodes will be very slow and the electrostatic force is incapable of exceeding the surface tension and viscoelastic resistance to continue stretching the liquid body (by up to 105 times) [79, 80]. It is also observed that there is a decrease in the diameter of electrospun nanofiber with an increase in the electrical conductivity of the solution while electrospinning. Hence the fluid jet generated from the highly conducting solution will experience a high tensile force. As a result, a considerable decrease in the diameter of the formed nanofiber has been observed. This was further confirmed by the studies conducted by Hsiao & co-workers [51]. They used PLLA as the polymer solution and used different salts such as KH_2PO_4, NaH_2PO_4, NaCl (fixed at 1% by weight). They found that the fabricated electrospun nanofiber were smooth and are free of beads and are of smaller diameter compared to those nanofibers fabricated from solution which doesn't have salts. This proved that the presence of an electrolyte or a salt and a solvent which is highly conducting such as acetone, chloroform, ethyl acetate, etc. [79, 80] has significant effect on the morphology of electrospun nanofibers, i.e., this inhibit the formation of bead in the fiber and lowers the fiber diameter to a range of 200–1,000 nm and in addition, the produced fibers are almost uniform [81].

The studies conducted by Baumgarten [15] revealed that the radius of the jet and the electrical conductivity have an exponential relationship, i.e., the radius of the jet is inversely proportional to the third root of electrical conductivity of a solution. Hayati and co-workers reported that under the influence of an extremely high electric field, a solution with high conductivity is exceedingly unstable which shows substantial bending instability and also large diameter distribution. Several groups have extended the study conducted by Hsiao [51] using different polymer systems such as PEO, collagen type I-PEO, PVA, polyacrylic acid (PAA), polyamide-6 and so on [19, 68, 72, 82, 83]. Furthermore, in general, high charge densities are exhibited by ions of smaller size. Hence, the addition of these ions can furnish greater forces of elongation on nanofibers fabricated by electrospinning. Even so, with an outrageous electrical conductivity, the velocity of the ions will be higher, and as a result, the amount of solution emerging from the needle will be more. This causes a difficulty in vaporization of solvent leading to the incomplete evaporation before it hits the collector plate. Hence the generated nanofibers will be non-uniform consisting of bead-like structures.

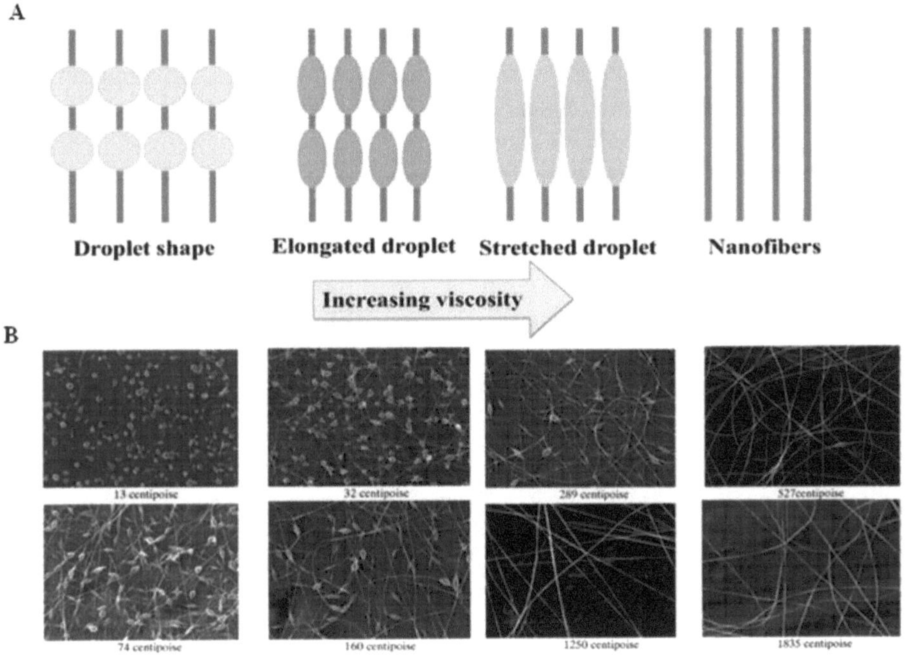

FIGURE 16.6 Variation in morphology of electrospun PEO nanofibers with different viscosities: (a) schematic representation; and (b) the corresponding scanning electron microscope (SEM) images.

16.5.2 PROCESSING PARAMETERS

16.5.2.1 APPLIED VOLTAGE

A critical factor in the electrospinning process is the applied voltage, since it controls the amount of charge applied to the solution. Figure 16.7 depicts the electrospinning process as a function of applied voltage. The strength of the electric field is directly influenced by the applied voltage and hence influences the electric forces on the Taylor cone and jet [84]. As the voltage increases the strength on the solution droplet also increases resulting in a greater jet elongation. However, it is observed that there is not much time for the jet to stretch before it hits the collector [61]. The flow of the fluid jet towards the collector causes charge transport under the effect of applied voltage. The quantity of polymer solution flowing from the tip of the needle can influence a change in the current (either increases or decreases). The studies conducted by Tan & co-workers [44] concluded that there is some sort of relation between the variation in the instability and variation in the spinning current. They demonstrated that there was a change in shape at the point of origin of the jet as the applied voltage increases. This had an effect on the structure and morphology of the fabricated fiber. Studies conducted on PEO/Water system revealed that there exists a direct relationship between the spinning current and the applied voltage, i.e., the spinning current increased as the applied voltage was increased [44]. The fiber formation is generally observed beyond a certain voltage known as critical voltage. When the critical voltage is attained an induction of the required charge on the solution as well as the electric field is observed. This initiates the electrospinning process. Different polymeric systems exhibit different end results when the applied voltage is varied. Reneker & Chun [85] reported that the diameter of the electrospun nanofiber is almost independent of the applied voltage in case of PEO system. Two arguments arose about the diameter of the fiber and the applied voltage. The first one is that a greater polymer emergence is often associated with a higher applied voltage, thereby resulting in a fiber with a large diameter. This was confirmed by an independent study conducted by Yuan's group [72] as well as Erman's group [63]. The later one is that an increase in electrostatic repulsive force is observed when the applied voltage is high, which forces the fluid jet to produce fiber with narrow diameter. However, in most cases at higher voltage, an increased Coulombic repulsive force in addition to a strong electric field was observed. This causes a prodigious stretching of the solution and eventually leads to the reduction in the diameter of the fiber and also causes a quick solvent evaporation from the fiber. It was further reported that at higher applied voltage, an increased stretching of the solution occurs and results in the formation of thinner fiber [66, 71]. Furthermore, an increase in the applied voltage causes a reduction in the fiber diameter [86]. After reaching a minimum value, the diameter of the fiber increases with each increment of voltage.

Several researchers reported that at an elevated applied voltage, there exists an increased possibility of bead formation [44, 49, 63, 66, 71, 87–89]. After exceeding a certain optimal applied voltage further increase in the applied voltage make the process more unstable and the presence of more beads on the fiber was observed just like dewdrops in the shape of small spheres on spider web in the morning [90].

This was concordant with the previous reports [17, 91, 92]. A relationship between the applied voltage and fiber diameter was also tabulated. As the applied voltage was doubled the fiber diameter reduced to roughly half its original value. The studies conducted by Sas et al. [93] at a higher electric field/applied voltage yielded a fiber with decreased diameter and that can be represented by the equation:

$$d \approx V^{-1/2}$$

where; d is the fiber diameter; and V is the applied voltage.

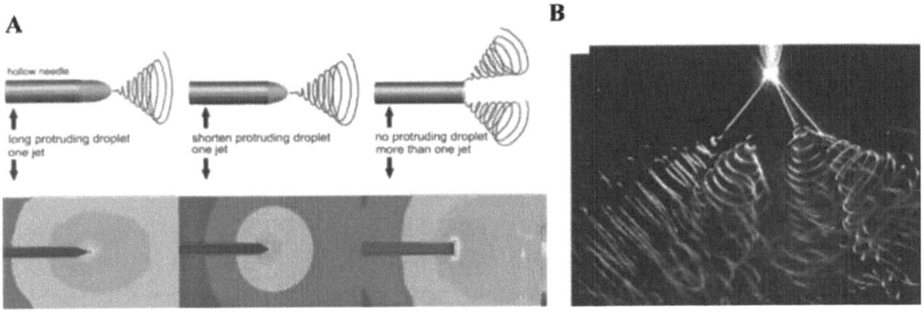

FIGURE 16.7 (a) General effect of increasing voltage on protruding droplet length and produced jet number with their corresponding field distribution. (b) Four jets from one drop each with a well-developed electrical bending coil with a high voltage.

Hendrickson derived an equation showing minimum voltage required for the spinning of a suspended conducting hemispherical drop and is given as:

$$V = 300\sqrt{20\pi\gamma r}$$

where; r is the radius of the Taylor cone; and γ is the surface tension of the polymer solution [93].

16.5.2.2 FLOW RATE/FEED RATE

The electrospun nanofiber's surface feature, geometry, and morphology are all influenced by the process flow rate. It has a great effect on the jet velocity and material transfer rate. Yuan & co-workers [94] reported that a low flow rate is most favorable for the process as it would provide adequate time for the solvent to evaporate. This is because when the flow rate increases, the jet of the polymer solution changes to periodic dripping, considerably increasing the fiber diameter and causing the formation of beaded structures were observed. Rabolt & co-workers [52] reported that when PS was used as the polymer solution, the pore diameter as well as the fiber diameter shows an increase with the feed rate. In addition to the flow rate, there is a minimum feed rate below which fiber formation is not observed. This is because at a lower flow rate, a rapid evaporation of the solvent before the solution leaves the needle was observed causing the clogging of the needle.

16.5.2.3 DISTANCE BETWEEN THE NOZZLE AND COLLECTOR

The nozzle-collector distance has an effect in controlling the fiber diameter and morphology. Studies revealed that a minimum distance has to be maintained, i.e., not too close or too far, in order to provide the fiber an enough time to dry before it hits the collector [44, 71, 95]. Indeed, if this condition is not satisfied, bead formation is observed. In case of electrospun PS fiber, it was observed that as the nozzle-collector distance decreases bead formation occurs [52]. It was also observed that small nozzle-collector distance resulted in small fiber in case of polysulfone [96]. Also, it was observed that by keeping the nozzle closer to the collector, flatter fiber formation is observed while increasing the distance, cylindrical fibers (i.e., fiber with circular cross-section) was found in the case of SLPF and nylon electrospun fiber [66]. When water is used as a solvent, a greater distance is required in order to provide sufficient time for the solvent to evaporate compared to systems with highly volatile organic solvents. Jalili and co-workers suggested that the electrospun fiber should be completely free of the solvent, i.e., to ensure complete vaporization of the solvent. Hence an optimization of the nozzle-collector distance was necessary.

16.5.2.4 ENVIRONMENTAL PARAMETERS

Besides the solution parameters and processing parameters, there are other factors that can also affect the electrospinning process and the morphology of the electrospun nanofiber. Those include temperature, humidity, air flow, moisture content, etc. Relative humidity (RH) and temperature play an important role in determining the surface features and morphology of electrospun nanofiber. Typically, the electrospinning process takes place at ambient temperature. As the temperature rises, so does the rate at which the solvent evaporates, thereby leading to fiber with large diameter [97]. A study was conducted to examine the influence of temperature on the electrospinning process using Polyamide-6 as the polymer [83]. The investigation was carried out at a temperature range of 298 K to 333 K and was found that increasing the temperature has an effect on the fiber diameter, i.e., it forces the process to produce fiber with smaller diameter. This is because as the temperature increases viscosity of the solution decreases since temperature and viscosity are inversely proportional. Hence as viscosity decreases the fiber diameter also decreases [60]. Similarly, RH has a significant effect on the fiber formation. Srinivasarao & co-workers [98] reported that at elevated RH electrospinning process leads to the formation of fiber containing micropores and nanopores on the surface as a result of the breathe figure effect. Due to the retained marks of the moisture droplets caused by the condensation as a result of the evaporative chilling of air moisture encompass the spinneret, breathe figures are observed [98]. It was reported that at lower humidity a reduced precipitation effect was observed and as a result, rapid solvent evaporation took place resulting in an electrospun nanofiber with greater diameter [99, 100]. However, studies conducted by Kocbek & co-workers [101] revealed that at higher RH condensation of the water was observed on the fluid jet surface which intercepts further elongation of the fiber. In addition, at a very low RH, solvent at the tip of the needle vaporizes rapidly which may block the tip of the needle [76].

16.6 PROPERTIES OF ELECTROSPUN NANOFIBERS

Electrospun nanofibers have been largely studied due to their properties that are often affected by their size. In comparison with regular fiber, bulk polymer nanofibers have noteworthy difference in their thermal and mechanical properties. In addition to these properties, they also have advantageous properties including optical properties, thermal properties, size dependent effects and so on, which makes electrospun fibers a suitable substrate for reinforcements [16, 61, 102].

16.6.1 THERMAL PROPERTIES

Detailed thermal analysis provides an exact idea about the nanostructure-thermal property relationship. In a study using electrospun PLLA nanofiber it was observed that this fiber has a lower crystallinity, glass transition temperature (Tg) and melting point (Tm) compared to that of semi-crystalline PLLA resin. They also found that when air is used as the plasticizer, which increases the surface area to volume ratio of spun fiber which in turn lowers the glass transition temperature (Tg). The low crystallinity arises as a result of the rapid evaporation of the solvent and fiber solidification at the end of the process [51]. Kim & Lee [103] showed that as the heat of crystalline melting increases a considerable reduction in the glass transition temperature (Tg) and peak crystallization temperature (Tc) was observed in case of electrospun polyethylene terephthalate (PET) and polyethylene naphthalene (PEN). This is due to the increase in segmental mobility with an increase in heat of crystalline melting. Tan & co-workers [44] reported that compared to the PEO powder, PEO nanofibers have lower melting point and heat of fusion. This can be related to the poor crystallinity of the electrospun PEO nanofibers. The thermal degradation study was carried out using TGA thermogram of PEN and PET before and after the electrospinning process, and it was reported that upon electrospinning there was a considerable decrease in the intrinsic viscosity of these polymers [103]. Reduction in the intrinsic viscosity as a result of the thermal degradation is considered to be the reason for the decrease in Tg and Tc.

16.6.2 OPTICAL PROPERTIES

Electrospun nanofibers have been mainly employed for reinforcement to manufacture optically transparent composites. Factors including the content of fiber, size, and index of refraction of the reinforced fiber has a great influence on the optical transparency of the composite [104]. Ramakrishna & co-workers [105] conducted a study showing the optical properties of electrospun TiO_2 nanofibers. Prior to their study, there was only a little knowledge about the relationship between the fiber diameter and optical band gap energy. However, their studies shred light over this. The optical absorption spectra were obtained using a UV-Visible spectrophotometer (Shimadzu UV-1601) with a resolution of 1 nm. With the excitation of the samples at 320 nm, the photoluminescence spectra (PL) were recorded. Figure 16.8(a) shows the UV-visible spectra of the fiber with different diameters. As the fiber diameter increases a slight shift in the spectra was observed. In addition, several research groups reported that as the size of the particles decreases in case of TiO_2 nanofibers a hypsochromic shift was observed [106–110]. However, no such shift in the wavelength was observed in case of the PL spectra and a peak appeared at 400 nm for the fiber (Figure 16.8(b)). Tang & Liu [104] produced a composite between PVA composite film and high volume of electrospun cellulose nanofiber. The film was capable of transmitting visible light with a transmittance of 75%. It has been suggested that if the nanofiber which is used as the reinforcement has a diameter less than 400 nm, the refraction on these fibers by the visible light will be not so prominent and hence will be transparent [102, 104]. Park & co-workers [112] reported that using PAN or PAA optically transparent and cable-like electrospun nanofibers can be fabricated at considerably high RH.

16.6.3 MECHANICAL PROPERTIES

The electrospun nanofibers have advantageous mechanical properties such as tensile strength, elastic moduli, toughness, etc. It has been observed that an inverse relationship exists between the fiber diameter and its strength, i.e., as the fiber diameter decreases, an increase in the strength of the fiber is observed. These are termed as size-dependent mechanical properties. Chew et al. [113] reported that in the case of electrospun PCL fiber as the diameter lowered from using the PCL fibers [114, 115].

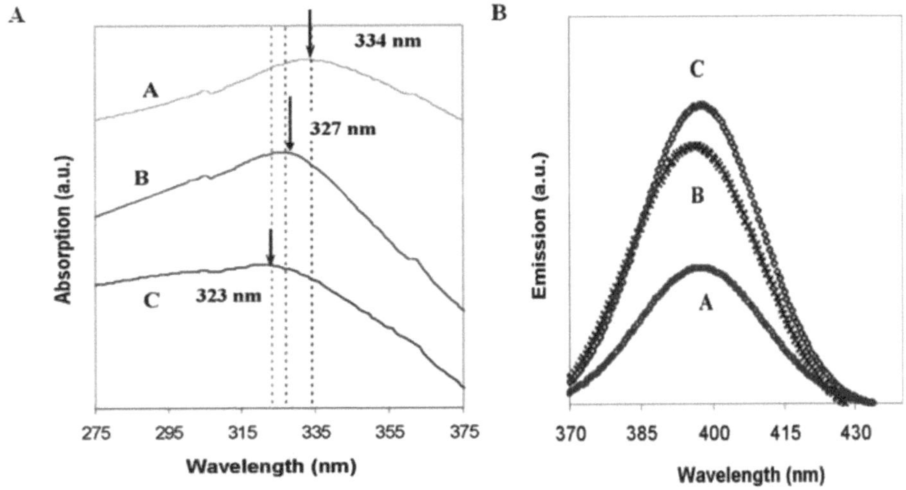

FIGURE 16.8 (a) Absorption spectra of TiO_2 nanofibers; and (b) corresponding PL spectra.

Similarly, an almost identical result was obtained using PAN fibers [116]. Park & co-workers [117] conducted a study on mechanical properties of various electrospun nylon-6,6/cyclic butylene terephthalate (CBT) composite fibers. In addition, in order to understand the property enhancement, they compared the results obtained with nylon-6,6 nanofibers and CBT film. Zhang [118] conducted a study on the mechanical property of random and aligned electrospun nylon-6 nanofiber. Figure 16.9(a) shows the relationship between the Young's modulus and fiber diameter while Figure 16.9(b) shows that of tensile strength and fiber diameter. In conclusion, the aligned fiber has enhanced elastic modulus as well as tensile strength compared to the random oriented one.

16.6.4 OTHER FIBER PROPERTIES

There are other factors which can also enhance or affect the properties including pore size, molecular orientation, fiber structure, specific surface area, etc. Either the nature of the polymers or the internal structure of the fabricated fiber or both can be influential over the properties of electrospun nanofiber. Therefore, modifying the molecular orientation or internal structure definitely has an effect on modifying the properties [119–124]. In addition, the investigation and the optimization of molecular orientation is crucial due to their potential applications because the orientation can greatly alter other properties of the polymer including electrical, mechanical, thermal, and optical properties [120]. At sufficiently high drawn ratios and rates of strain, those electrospun nanofibers will be able to attain a higher degree of molecular orientation along the fiber axis while electrospinning [111]. As a result of the chain-entanglement while electrospinning, polymeric nanofibers possessing rigid-rod like molecular structure have a higher molecular orientation compared to fibers with softer molecular structure. A higher molecular orientation is possessed by fine nanofibers compared to that of chunky nanofibers since there is a possibility of random orientation of the polymer in chunky fiber while it is restricted in case of fiber nanofibers [125–128].

Altering the fiber diameter while electrospinning has a direct effect on the orientation of orders structure in case of electrospun liquid crystalline polymeric fibers [121].

The specific surface area of a nanofiber has an inverse relation with the fiber diameter, i.e., as the fiber diameter reduces to nanometer scale, an elevation in the specific surface area was observed [129]. The dimension of the electrospun nanofiber is in the range of 10 nm to 1 μm which indicates that those fibers have a surface area which is in the range of 300 to 3 m^2g^{-1}.

Since the electrospun nanofibers possess a high specific surface area it will deliver enough area of contact between the fiber and the polymer matrix, thereby enhancing the interaction between them. Figure 16.10 shows size effect of different fibers on the specific surface area.

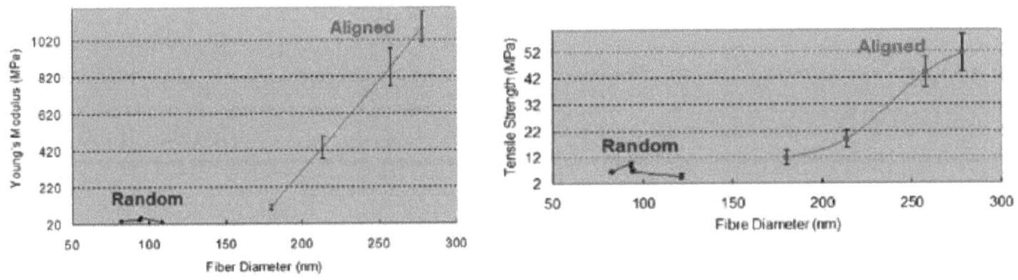

(a) (b)

FIGURE 16.9 (a) Young's modulus vs. fiber diameter curves; and (b) tensile strength vs. fiber diameter curves for randomly oriented fiber mats and aligned fiber films.

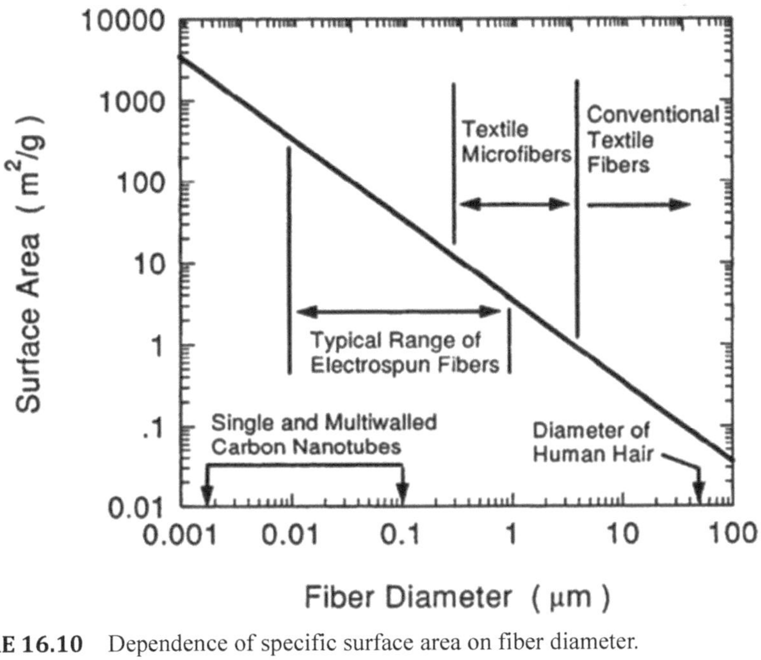

FIGURE 16.10 Dependence of specific surface area on fiber diameter.

The pore size can also affect the properties of electrospun nanofiber. A change in the solvent composition, temperature of the solution and another factor can be influential over the total porosity and pore interconnectivity.

There exists a direct relationship between the fiber diameter and pore size, i.e., a decrease in the pore size was observed with the decrease in the fiber diameter [130]. The study leading to this conclusion was done using a non-woven electrospun nanofiber planted in an epoxy matrix [111, 131]. Figure 16.11 shows a cross-section of non-woven electrospun nanofiber embedded in the epoxy matrix with two different fiber diameters and pore sizes. At a fiber diameter of 309 nm pores with size in the range of 1,000–2,000 nm was obtained while a larger pore with size of about 5,000 nm was observed when the diameter was 520 nm. An electrospun nanofiber composite with a pore size of about 100 μm was obtained when the fiber collector used was rotating frame cylinder. Furthermore, they also observed that the pore size can be varied even without changing the fiber diameter [132]. The investigations on the co-electrospinning of sacrificial fibers in addition to stable fibers pointed to the fact that the removal of the sacrificial fiber caused large pores on the stable fiber [133]. The special consideration of pore size would provide additional information which may lead to new potential application in the fabrication of electrospun nanofiber composites.

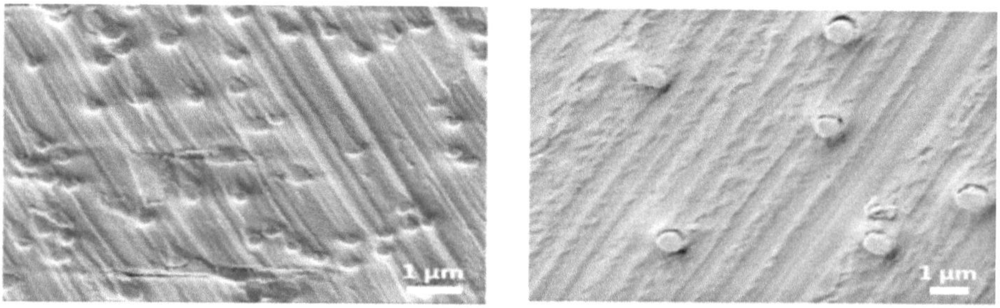

A B

FIGURE 16.11 Cross-section of PAN nonwovens embedded in an epoxy matrix with fiber diameter of (a) 309 nm; and (b) 520 nm.

16.7 MAGNETIC NANOFIBERS (MNFS) BY ELECTROSPINNING

Amongst the recent development in nanoscience and nanotechnology, nanofibers, one of the most attractive one-dimensional nanostructures have attained great attention, which in turn can be attributed to the unique physical properties it possesses. Their large surface area-to-volume ratio creates anisotropic physical properties that cannot be achieved in their bulk form [134–136]. Magnetic nanofibers (MNFs) are also found to possess excellent

EM performance, apparent magnetism such as paramagnetism, ferromagnetism, diamagnetism, and other properties like superparamagnetic effect, macroscopic tunneling effect and so on. These distinctive properties of MNFs make them suitable for various potential applications like ultrahigh-density data storages, sensors, drug delivery system, catalysis, biomedical applications, and so on [137–139]. The most important factors to be considered in preparing these magnetic nanostructures is to get uniform dispersion of magnetic nanoparticles (MNPs) and the suitable method for fabricating them. The main limitation of the nanofibers synthesized from simply blended solution is that, in it, the MNPs cannot fully align along the external magnetic field and are not uniformly dispersed, because they are arrested within a solid polymer matrix [140, 141]. All these limitations are now overthrown through a new and successful technique called electrospinning.

Electrospinning is an effective and simple method for the fabrication of nanofibers where continuous fibers are stretched in the form of a liquid jet from a polymer solution or melts by electrostatic force using high voltage [142]. Generally electrospinning can be subdivided into: (i) needle-based; and (ii) needleless techniques (Figure 16.12) [143]. In the needle-based method, a polymer solution or melt is pressed through a needle into a strong electrical field, thus the polymer is drawn into the substrate, and in this way, the fiber is stretched and thinned so that diameters of some 10 to some 100 nanometers are reached. The needleless method is a wire-based technique in which a wire is coated with a polymer solution or melt which is drawn to the second wire by the strong electrical field between them. Similarly, by using rotating cylinders and other geometries as electrodes, the amount of fibers formed per time and their alignment can be modified. In both these methods Taylor-cones are formed between both electrodes.

16.7.1 POLYMERS FOR ELECTROSPINNING

The most commonly employed electrospinning polymers include water soluble polyethylene glycol (PEG or PEO), polyacrylonitrile (PAN) [144–146], which are often used as the precursors for carbon nanofibers (CNFs) and poly(ε-caprolactone) (PCL) which is mainly used in cell growth applications [147, 148]. Polymers can play the following roles in MNFs: (i) as a template to control the size, shape, and structure; (ii) as protective coatings, insulators, separators or molding materials that are not affected by external environment; and (iii) the mechanical properties associated with MNFs are provided by these polymers.

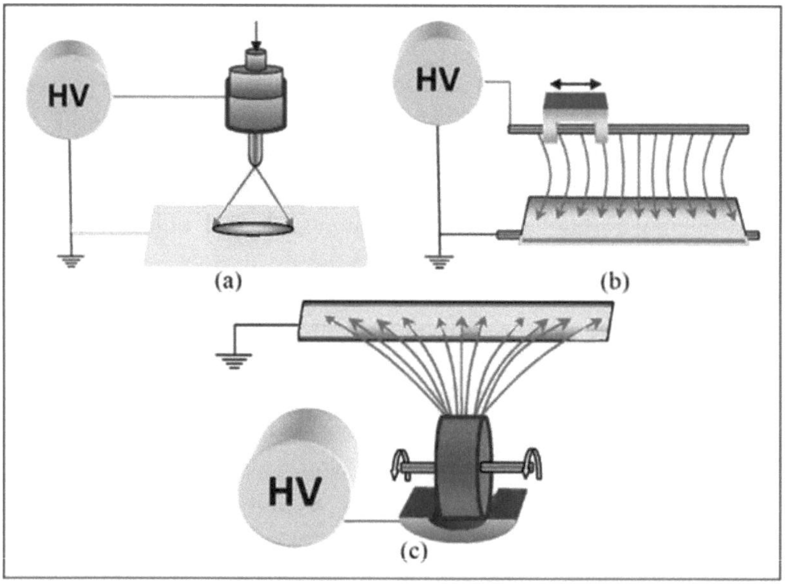

FIGURE 16.12 Different electrospinning techniques: (a) needle-based; (b) wire-based; and (c) using a rotating cylinder.

Currently, there are four main methods for the preparation of MNFs through electrospinning: (i) MNPs are added to the polymer solution and the composite MNPs are prepared directly from the solution; (ii) through sol gel and electrospinning technique the precursor composite nanofibers are first prepared, followed by calcination to obtain MNFs; (iii) by using organic fibers prepared by electrospinning as a template, MNFs pipes can be produced by coating with metal salt sol; and (iv) MNFs with different morphologies and features can also be prepared by this method or atomic layer deposition method. Preparation of composite MNFs by several methods combined with the electrospinning method is shown in Table 16.2.

16.7.2 FIELD ASSISTED ELECTROSPINNING TECHNOLOGY

MNFs with polycarbonate urethane (PCU) as a polymer filled with nickel NPs has been synthesized by Guarino et al. The needle-based electrospinning technique was utilized for that. The counter electrode with permanent magnets was modified, thus creating a constant magnetic field perpendicular to the electric field, leading to aligned MNFs in between the magnetic poles Figure 16.13.

TABLE 16.2 Preparation of Different Magnetic Composite NFs by Electrospinning Combining with Different Methods

MNFs	Fabrication Method	Morphology	Diameter
$CoFe_2O_4$	Sol-gel method	Uniform and surface are smooth	100 nm
$NiFe_2O_4$	Calcination	A pure phase with cubic spinel	64±18 nm
$SrFe_{12}O_{19}$	Coating	Both anatase phase and rutile phase	500~600 nm
$NaGdF_{4:0.5\%}$ + Dy_3	Calcination and fluorination	Rough, good dispersion and uniform size	246.06±5.84 nm
(PANI) Au, Fe_3O_4 @Au	Lay deposit	Both crystallites and POSS molecules co-existed	700 nm
α-Fe_2O_3	Sol-gel and layer deposit	A dense silica layer deposit on the surface	–
$CuFe_2O_4$ @ CeO_2	Chemical precipitation methods	A high specific surface area	200 nm

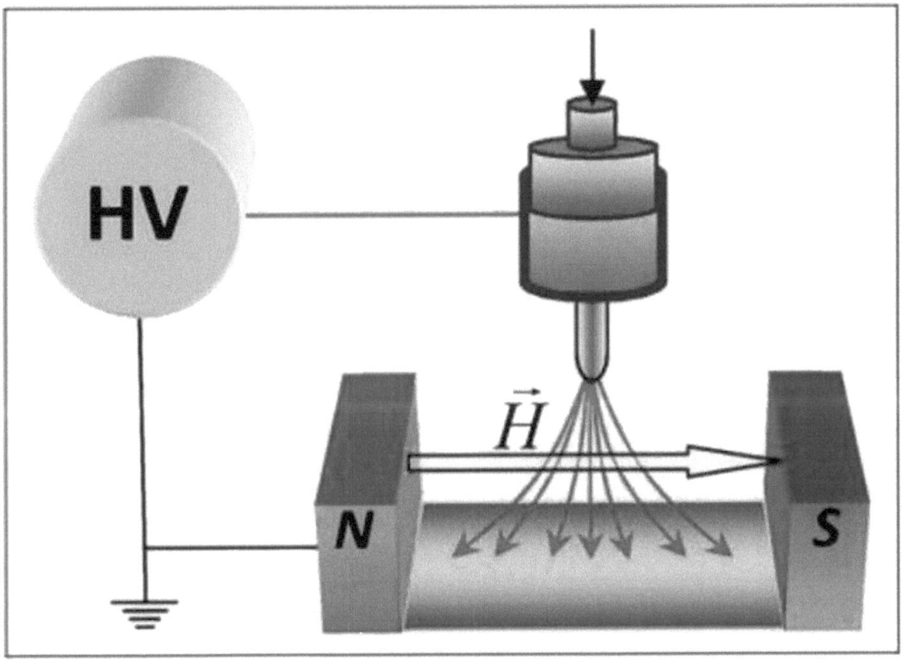

FIGURE 16.13 Magnetic field assisted needle-based electrospinning method.

Nickel ferrite-barium titanate core shell nanofibers were developed by magnetic field assisted electrospinning technique, allowing them to assemble into disks or films in a uniform or gradient magnetic field [149].

16.7.3 CLASSIFICATION OF MAGNETIC NANOFIBERS (MNFS)

16.7.3.1 FERRITE-BASED MAGNETIC NANOFIBERS (MNFS)

Ferrite based MNFs are composed of iron, oxygen, and many other metals. Generally, they possess high resistivity and dielectric properties. Among the different categories of ferrite lattice, the most commonly used magnetic materials are magnetite-lead ferrite [150, 151], garnet-type ferrite [152], spinel-type ferrite [153], and perovskite ferrite [154].

Shen et al. [154] reported that $SrFe_{12}O_{19}$ magnetite lead nanofibers fabricated by sol-gel assisted electrospinning and calcination process have a diameter of about 100 nm and their magnetic properties are related to the structure and size of grains. MNFs with fine performance can be prepared by spinel-type ferrite. The pure single-phase spinel $Ni_{0.2}Zn_{0.8}$ Fe_2O_4 NF membranes were fabricated by Streckova et al. [155] via needleless electrospinning and conventional thermal treatment of the precursor fibers at 800°C for 4 hours in air.

16.7.3.2 METAL-BASED MNFS

Metal-based MNFs are mainly composed of Cr, Mn, Fe, Co, Ni, and their alloy in the transition metal zone. These ferromagnetic materials with strong magnetic properties find most extensive applications in high density magnetic recording, sensitive components, biological, and medical engineering [155]. Smooth, unceasing, hydrogen-free, and pure nickel metal NFs with an average diameter of 120 nm by electrospinning method was prepared by Barakat et al. [156] The nickel NFs thus obtained were vacuously dried and then calcined in an argon atmosphere at 700°C. These NFs showed unique magnetic properties, and much enhanced coercivities compared to their bulk materials.

16.7.4 APPLICATIONS OF MNFS

16.7.4.1 BIOMEDICAL APPLICATIONS

The biocompatible magnetic materials are widely used in biomedical applications such as biological separation and purification, enzyme immobilization, drug release and magnetic resonance imaging (MRI) due to the advantages of easy operation, separation, and non-toxic properties.

According to the principle of controlled drug release, the drug release system can be grouped into the following factors:

- diffusion control system;
- chemical control system;
- solvent activation system; and
- regulatory release system.

The unique properties of magnetic nanomaterials are largely used in the controlled release of drugs, such as:

- magnetic response;
- biocompatibility;
- high specific surface area;
- low toxicity;
- high transport efficiency;
- targeted therapy;
- controlled size;
- biodegradability; and
- super-paramagnetism.

Researchers such as Kim et al. [157] synthesized electrospun magnetic composite nanofibers which was composed of chemically-cross-linkable temperature-responsive polymers, MNPs with an anti-cancer drug. It has been showing that 70% of human melanoma cells died under the drug effect in 5 minutes.

16.7.4.2 SENSING AND ELECTRONIC DEVICES

Magnetic nanomaterials can be inserted into CNFs in order to obtain magnetic, dielectric, and adsorptive properties for sensing applications. A kind of composite polydopamine (PDA)-laccase (Lac)-nickel nanoparticle loaded CNFs (NiCNFs) was developed and a magnetic glass carbon electrode (MGCE) was added to separate and immobilize the composite (Figure 16.14) [158].

Flexible magnetic membranes based on bacterial cellulose (BC) NFs with amphiphilicity were prepared initially by the *in-situ* synthesis of the Fe_3O_4 NPs on the BC NFs and then the fluoroalkyl silane (FAS) modification (Figure 16.15) [159]. The magnetometric measurements revealed that these membranes exhibited superparamagnetic behaviors and could be magnetically actuated. Magnetically responsive BC membranes hydrophobic and lipophobic surface find potential applications in electronic actuators, magnetographic printing, information storage, EM shielding coating and so on.

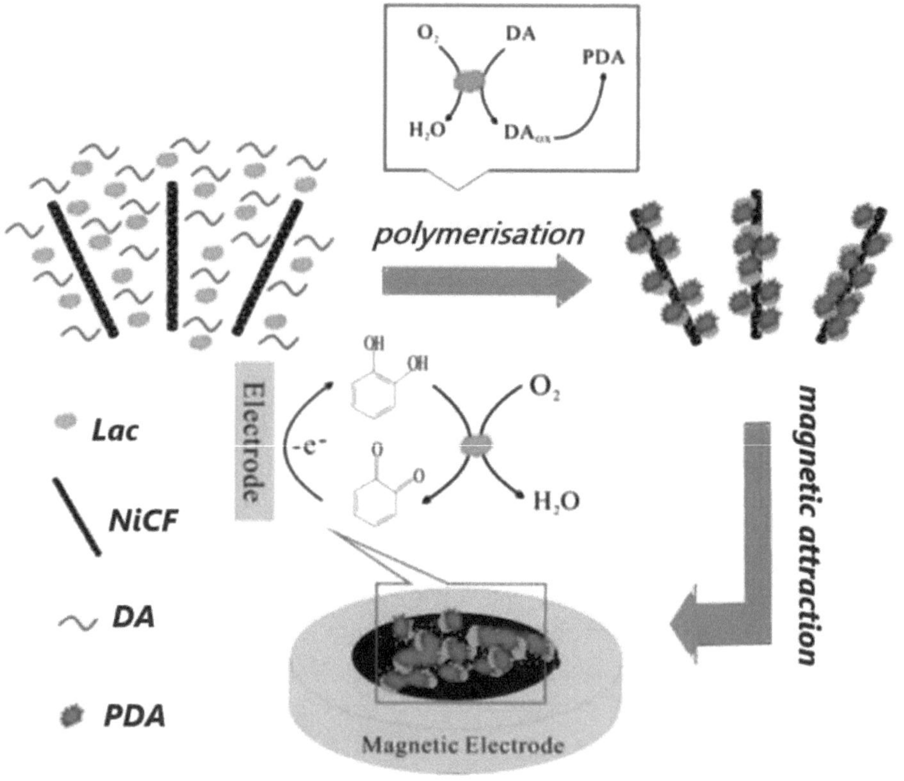

FIGURE 16.14 Guidelines to prepare PDA-Lac-NiCNFs composite and biosensors.

16.7.4.3 WASTEWATER TREATMENT

The unique properties of MNFs such as, super-paramagnetism, good adsorption properties, ease of recycling, are explored in the application in industrial wastewater treatment [160]. Compared to other traditional adsorbents MNFs have many advantages like higher adsorption rate, lower cost, small size with larger surface area and magnetic response [161]. In wastewater treatment MNFs are mainly used for the adsorption of dyes and heavy metals.

Lead adsorption from polluted water was developed by fabricated cellulose acetate MNFs were fabricated. Bhaumik et al. [162] fabricated composite NFs of polyaniline/FeO (PANI/FeO) for the reductive degradation and removal of Congo red from aqueous solutions. The degradation efficiency of Congo red increased when the initial concentration and solution pH decreased, however, it decreased on decreasing the dosage of the composite NF. Magnetic $CuFe_2O_4$@CeO_2 NFs were also found to be effective for dye removal.

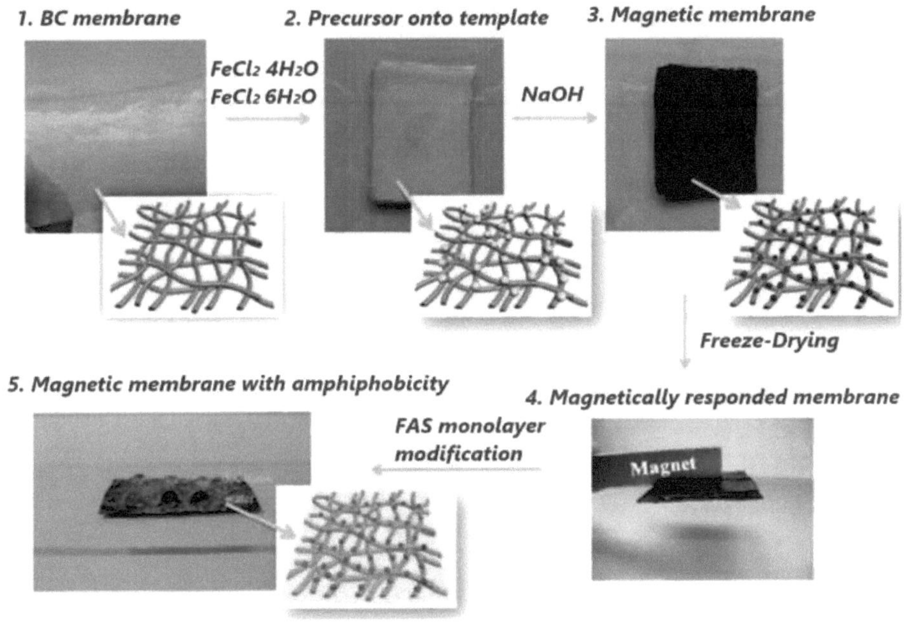

FIGURE 16.15 Flexible magnetic nano-hybrid membrane with amphiphobic surface.

Most of the inorganic MNFs exhibited greater catalytic efficiency, hence these materials can be easily recycled using magnets. Li et al. [163]; and Chen et al. [164]. fabricated catalytic magnetic nanocomposite fibers. α-Fe_2O_3 NFs used as an effective catalyst were integrated for degradation and decolorization of methyl orange in aqueous solution, and the decolorization efficiency of these MNFs was more than 99% in a short period time of 10 min under the optimum conditions [111].

In the design of waste water treatment arrangement, a porous catalytic scaffold and a nanostructured photocatalyst in the reactors have to be considered. Slurry reactor (Figure 16.16(a)) and the stationary reactor (Figure 16.16(b)) are two types of photocatalytic reactors for wastewater treatment. Ultrafiltration and membrane filtration are the two common membrane filtration techniques used for the effective recovery of NF photocatalyst. It was revealed that the activity of magnetic FeO_3 and Fe/TiO_2/SnO_2 hybrid NF photocatalyst was very high. The uniform NFs catalyst in organic, aqueous, and organic/photocatalytic water medium can be recovered easily from the product phase [64].

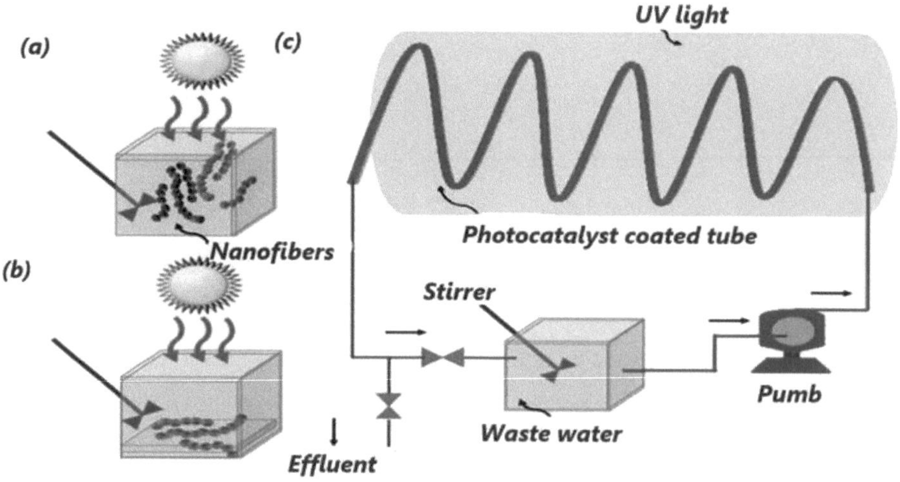

FIGURE 16.16 Different types of photocatalytic reactor arrangements: (a) slurry reactor; (b) immobilized reactor; and (c) continuous tubular flow reactor.

16.8 CONCLUSION

The unique physio-chemical properties and features made nanofibers interesting one-dimensional nanomaterials for a wide range of research and commercial applications. Till date, materials of many sorts have been manufactured into nanofibers using electrospinning techniques. The most advanced strategy involves incorporating metallic silver and gold NPs, as well as metal oxides based on titanium, zinc, copper, and iron. Other relevant features, such as mechanical, magnetic, catalytic, and optical capabilities, can be adjusted or bestowed on these inorganic compounds in the nanometric regime, expanding the prospect of developing multi-functional fibrous materials with broader application areas [33]. The electrospinning setup has been tweaked to allow for the fabrication of core-shell or hollow structures as well as uniaxially aligned or layer-by-layer stacked films. In addition to practical developments, theoretical modeling has advanced to provide a comprehensive guideline for better understanding of the electrospinning process [7]. MNFs have gained much attention in the recent years is mainly due to the difference in their magnetic properties in the bulk and on the surface. Through various functional modifications, their properties can be enhanced, facilitating their applications in a broader area.

KEYWORDS

- **alternating current**
- **chain entanglement**
- **dimethylformamide**
- **direct current**
- **electrospinning technique**
- **fiber diameter**
- **magnetic nanofibers**
- **Taylor cone**

REFERENCES

1. Li, Y., Yang, X. Y., Feng, Y., Yuan, Z. Y., & Su, B. L., (2012). One-dimensional metal oxide nanotubes, nanowires, nanoribbons, and nanorods: Synthesis, characterizations, properties and applications. *Critical Reviews in Solid State and Materials Sciences, 37*, 1–74.
2. Liu, J., Xu, J., Hu, W., Yang, J., & Yu, S., (2016). Systematic synthesis of tellurium nanostructures and their optical properties: From nanoparticles to nanorods, nanowires, and nanotubes. *ChemNanoMat., 2*, 167–170.
3. Gugulothu, D., Barhoum, A., Nerella, R., Ajmer, R., & Bechlany, M., (2018). Fabrication of nanofibers: Electrospinning and non-electrospinning techniques. *Handbook of Nanofibers*, 1–34.
4. Yarin, A. L., Koombhongse, S., & Reneker, D. H., (2001). Taylor cone and jetting from liquid droplets in electrospinning of nanofibers. *J. Appl. Phys., 90*, 4836–4846.
5. Taylor, G. I., (1964). Disintegration of water drops in an electric field. *Proceedings of the Royal Society of London. Series A. Mathematical and Physical Sciences, 280*, 383–397.
6. Bellan, L. M., & Craighead, H. G., (2009). Nanomanufacturing using electrospinning. *J. Manuf. Sci. Eng., 131*.
7. D., & Xia, Y., (2004). Electrospinning of nanofibers: Reinventing the wheel? *Advanced Materials, 16*, 1151–1170.
8. Rodríguez-Tobías, H., Morales, G., & Grande, D., (2019). Comprehensive review on electrospinning techniques as versatile approaches toward antimicrobial biopolymeric composite fibers. *Materials Science and Engineering: C, 101*, 306–322.

9. Rayleigh, L., (1882). XX. On the equilibrium of liquid conducting masses charged with electricity. *The London, Edinburgh, and Dublin Philosophical Magazine and Journal of Science, 14*, 184–186.
10. Khan, W. S., Asmatulu, R., Ceylan, M., & Jabbarnia, A., (2013). Recent progress on conventional and non-conventional electrospinning processes. *Fibers and Polymers, 14*, 1235–1247.
11. Yarin, A. L., Kataphinan, W., & Reneker, D. H., (2005). Branching in electrospinning of nanofibers. *J. Appl. Phys., 98*, 064501.
12. Zeleny, J., (1914). The electrical discharge from liquid points, and a hydrostatic method of measuring the electric intensity at their surfaces. *Physical Review, 3*, 69.
13. Anton, F., (1934). *Process and Apparatus for Preparing Artificial Threads*. Original Filed: Dec. 5. 1930; a Sheets-Sheet 1: Oct; 2; 1934.
14. Hayati, I., Bailey, A. I., & Tadros, T. F., (1987). Investigations into the mechanisms of electrohydrodynamic spraying of liquids: I. Effect of electric field and the environment on pendant drops and factors affecting the formation of stable jets and atomization. *Journal of Colloid and Interface Science, 117*, 205–221.
15. Baumgarten, P. K., (1971). Electrostatic spinning of acrylic microfibers. *J Colloid Interface Sci., 36*, 71–79.
16. Subbiah, T., Bhat, G. S., Tock, R. W., Parameswaran, S., & Ramkumar, S. S., (2005). Electrospinning of nanofibers. *J. Appl. Polym. Sci., 96*, 557–569.
17. Larrondo, L., & St. John, M. R., (1981). Electrostatic fiber spinning from polymer melts. I. Experimental observations on fiber formation and properties. *Journal of Polymer Science: Polymer Physics Edition, 19*, 909–920.
18. Chun, H., (1997). *Effect of Molecular Characteristics on Strength of Elastomer*. The University of Akron.
19. Fong, H., & Reneker, D. H., (1999). Elastomeric nanofibers of styrene–butadiene–styrene triblock copolymer. *Journal of Polymer Science Part B: Polymer Physics, 37*, 3488–3493.
20. Reneker, D. H., Yarin, A. L., Fong, H., & Koombhongse, S., (2000). Bending instability of electrically charged liquid jets of polymer solutions in electrospinning. *Journal of Applied Physics, 87*, 4531–4547.
21. Reneker, D. H., & Yarin, A. L., (2008). Electrospinning jets and polymer nanofibers. *Polymer (Guildf), 49*, 2387–2425.
22. Fridrikh, S. V., Jian, H. Y., Brenner, M. P., & Rutledge, G. C., (2003). Controlling the fiber diameter during electrospinning. *Phys. Rev. Lett., 90*, 144502.
23. Larsen, G., Velarde-Ortiz, R., Minchow, K., Barrero, A., & Loscertales, I. G., (2003). A method for making inorganic and hybrid (organic/inorganic) fibers and vesicles with diameters in the submicrometer and micrometer range via sol– gel chemistry and electrically forced liquid jets. *J. Am. Chem. Soc., 125*, 1154, 1155.
24. Chen, Y., Lu, Z., Zhou, L., Mai, Y. W., & Huang, H., (2012). Triple-coaxial electrospun amorphous carbon nanotubes with hollow graphitic carbon nanospheres for high-performance Li ion batteries. *Energy Environ. Sci., 5*, 7898–7902.
25. Yao, Y., Huang, Z., Xie, P., Lacey, S. D., Jacob, R. J., Xie, H., Chen, F., et al., (2018). Carbothermal shock synthesis of high-entropy-alloy nanoparticles. *Science, 359*, 1489–1494.
26. Lu, P. H., Xie, D. G., Liu, B. Y., Ai, F., Zhang, Z. R., Jin, M. S., Zhang, X. F., et al., (2018). *Highly Deformable and Mobile Palladium Nanocrystals as Efficient Carbon Scavengers*. arXiv preprint arXiv:1802.00207.

27. Kim, J., Lee, J., Kwon, S., & Jeong, S., (2009). Preparation of biodegradable polymer/silver nanoparticles composite and its antibacterial efficacy. *J. Nanosci. Nanotechnol., 9*, 1098–1102.
28. Li, D., & Xia, Y., (2004). Electrospinning of nanofibers: Reinventing the wheel? *Advanced Materials, 16*, 1151–1170.
29. Reneker, D. H., & Chun, I., (1996). Nanometre diameter fibers of polymer, produced by electrospinning. *Nanotechnology, 7*, 216.
30. Frenot, A., & Chronakis, I. S., (2003). Polymer nanofibers assembled by electrospinning. *Curr. Opin. Colloid Interface Sci., 8*, 64–75.
31. Huang, Z. M., Zhang, Y. Z., Ramakrishna, S., & Lim, C. T., (2004). Electrospinning and mechanical characterization of gelatin nanofibers. *Polymer (Guildf), 45*, 5361–5368.
32. Kessick, R., Fenn, J., & Tepper, G., (2004). The use of AC potentials in electrospraying and electrospinning processes. *Polymer (Guildf), 45*, 2981–2984.
33. Rodríguez-Tobías, H., Morales, G., & Grande, D., (2019). Comprehensive review on electrospinning techniques as versatile approaches toward antimicrobial biopolymeric composite fibers. *Materials Science and Engineering: C, 101*, 306–322.
34. Hohman, M. M., Shin, M., Rutledge, G., & Brenner, M. P., (2001). Electrospinning and electrically forced jets. II. Applications. *Physics of Fluids, 13*, 2221–2236.
35. Ajayan, P. M., (1993). Capillarity-induced filling of carbon nanotubes. *Nature, 361*, 333, 334.
36. Wright, D., York, A. E., Coleman, K., Green, M. H., & Hutchison, J., (1999). Capillarity and silver nanowire formation observed in single walled carbon nanotubes. *Chemical Communications*, 699, 700.
37. Caruso, R. A., Schattka, J. H., & Greiner, A., (2001). Titanium dioxide tubes from sol–gel coating of electrospun polymer fibers. *Advanced Materials, 13*, 1577–1579.
38. Dong, H., Prasad, S., Nyame, V., & Jones, W. E., (2004). Sub-micrometer conducting polyaniline tubes prepared from polymer fiber templates. *Chemistry of Materials, 16*, 371–373.
39. Ochanda, F., Atkinson, A., & Jones, W., (2003). Gold nanotubes by fiber templating techniques. In: *Abstracts of Papers of the American Chemical Society* (Vol. 226, pp. 7547–7552). Amer Chemical Soc 1155 16th St, NW, Washington, DC 20036 USA.
40. Cloupeau, M., & Prunet-Foch, B., (1990). Electrostatic spraying of liquids: Main functioning modes. *J. Electrostat., 25*, 165–184.
41. Doshi, J., & Reneker, D. H., (1995). Electrospinning process and applications of electrospun fibers. *J. Electrostat., 35*, 151–160.
42. Warner, S. B., Buer, A., Ugbolue, S. C., Rutledge, G. C., & Shin, M. Y., (1998). A fundamental investigation of the formation and properties of electrospun fibers. *National Textile Center Annual Report, 83*, 90.
43. Ren, L. F., Xia, F., Shao, J., Zhang, X., & Li, J., (2017). Experimental investigation of the effect of electrospinning parameters on properties of superhydrophobic PDMS/PMMA membrane and its application in membrane distillation. *Desalination, 404*, 155–166.
44. Deitzel, J. M., Kleinmeyer, J., Harris, D. E. A., & Tan, N. C. B., (2001). The effect of processing variables on the morphology of electrospun nanofibers and textiles. *Polymer (Guildf), 42*, 261–272.
45. Liu, H., & Hsieh, Y., (2002). Ultrafine fibrous cellulose membranes from electrospinning of cellulose acetate. *Journal of Polymer Science Part B: Polymer Physics, 40*, 2119–2129.

46. Ryu, Y. J., Kim, H. Y., Lee, K. H., Park, H. C., & Lee, D. R., (2003). Transport properties of electrospun nylon 6 nonwoven mats. *European Polymer Journal, 39*, 1883–1889.
47. McKee, M. G., Wilkes, G. L., Colby, R. H., & Long, T. E., (2004). Correlations of solution rheology with electrospun fiber formation of linear and branched polyesters. *Macromolecules, 37*, 1760–1767.
48. Ki, C. S., Baek, D. H., Gang, K. D., Lee, K. H., Um, I. C., & Park, Y. H., (2005). Characterization of gelatin nanofiber prepared from gelatin–formic acid solution. *Polymer (Guildf), 46*, 5094–5102.
49. Haghi, A. K., & Akbari, M., (2007). Trends in electrospinning of natural nanofibers. *Physica Status Solidi (a), 204*, 1830–1834.
50. Sukigara, S., Gandhi, M., Ayutsede, J., Micklus, M., & Ko, F., (2003). Regeneration of bombyx mori silk by electrospinning—Part 1: Processing parameters and geometric properties. *Polymer (Guildf), 44*, 5721–5727.
51. Zong, X., Kim, K., Fang, D., Ran, S., Hsiao, B. S., & Chu, B., (2002). Structure and process relationship of electrospun bioabsorbable nanofiber membranes. *Polymer, 43*, 4403–4412.
52. Megelski, S., Stephens, J. S., Chase, D. B., & Rabolt, J. F., (2002). Micro-and nanostructured surface morphology on electrospun polymer fibers. *Macromolecules, 35*, 8456–8466.
53. Yarin, A. L., Koombhongse, S., & Reneker, D. H., (2001). Taylor cone and jetting from liquid droplets in electrospinning of nanofibers. *J. Appl. Phys., 90*, 4836–4846.
54. Koski, A., Yim, K., & Shivkumar, S., (2004). Effect of molecular weight on fibrous PVA produced by electrospinning. *Materials Letters, 58*, 493–497.
55. Zhao, Y. Y., Yang, Q. B., Lu, X. F., Wang, C., & Wei, Y., (2005). Study on correlation of morphology of electrospun products of polyacrylamide with ultrahigh molecular weight. *Journal of Polymer Science Part B: Polymer Physics, 43*, 2190–2195.
56. Gupta, P., Elkins, C., Long, T. E., & Wilkes, G. L., (2005). Electrospinning of linear homopolymers of poly (methyl methacrylate): Exploring relationships between fiber formation, viscosity, molecular weight and concentration in a good solvent. *Polymer (Guildf), 46*, 4799–4810.
57. Tan, S. H., Inai, R., Kotaki, M., & Ramakrishna, S., (2005). Systematic parameter study for ultra-fine fiber fabrication via electrospinning process. *Polymer (Guildf), 46*, 6128–6134.
58. McKee, M. G., Layman, J. M., Cashion, M. P., & Long, T. E., (1979). Phospholipid nonwoven electrospun membranes. *Science, 311*, 353–355.
59. Burger, C., Hsiao, B. S., & Chu, B., (2006). Nanofibrous materials and their applications. *Annu. Rev. Mater. Res., 36*, 333–368.
60. Bhardwaj, N., & Kundu, S. C., (2010). Electrospinning: A fascinating fiber fabrication technique. *Biotechnol. Adv., 28*, 325–347.
61. Li, Y., et al., (2021). Developments of advanced electrospinning techniques: A critical review. *Advanced Materials Technologies, 6*, 2100410.
62. Zeng, J., Haoqing, H., Schaper, A., Wendorff, J. H., & Greiner, A., (2003). Poly-L-lactide nanofibers by electrospinning–Influence of solution viscosity and electrical conductivity on fiber diameter and fiber morphology. *e-Polymers, 3.*
63. Demir, M. M., Yilgor, I., Yilgor, E. E. A., & Erman, B., (2002). Electrospinning of polyurethane fibers. *Polymer (Guildf), 43*, 3303–3309.
64. Mondal, K., (2017). Recent advances in the synthesis of metal oxide nanofibers and their environmental remediation applications. *Inventions, 2*, 9.

65. Deitzel, J. M., Kosik, W., McKnight, S. H., Tan, N. C. B., DeSimone, J. M., & Crette, S., (2002). Electrospinning of polymer nanofibers with specific surface chemistry. *Polymer (Guildf), 43,* 1025–1029.
66. Buchko, C. J., Chen, L. C., Shen, Y., & Martin, D. C., (1999). Processing and microstructural characterization of porous biocompatible protein polymer thin films. *Polymer (Guildf), 40,* 7397–7407.
67. Kim, K. H., Jeong, L., Park, H. N., Shin, S. Y., Park, W. H., Lee, S. C., Kim, T. I., et al., (2005). Biological efficacy of silk fibroin nanofiber membranes for guided bone regeneration. *J. Biotechnol., 120,* 327–339.
68. Huang, L., Nagapudi, K., Apkarian, P. R., & Chaikof, E. L., (2001). Engineered collagen–PEO nanofibers and fabrics. *Journal of Biomaterials Science, Polymer Edition, 12,* 979–993.
69. Son, W. K., Youk, J. H., Lee, T. S., & Park, W. H., (2004). The effects of solution properties and polyelectrolyte on electrospinning of ultrafine poly (ethylene oxide) fibers. *Polymer, 45,* 2959–2966.
70. Ding, B., Kim, H. Y., Lee, S. C., Shao, C. L., Lee, D. R., Park, S. J., Kwag, G. B., & Choi, K. J., (2002). Preparation and characterization of a nanoscale poly (vinyl alcohol) fiber aggregate produced by an electrospinning method. *Journal of Polymer Science Part B: Polymer Physics, 40,* 1261–1268.
71. Lee, J. S., Choi, K. H., Ghim, H. D., Kim, S. S., Chun, D. H., Kim, H. Y., & Lyoo, W. S., (2004). Role of molecular weight of atactic poly (vinyl alcohol)(PVA) in the structure and properties of PVA nanofabric prepared by electrospinning. *Journal of Applied Polymer Science, 93,* 1638–1646.
72. Zhang, C., Yuan, X., Wu, L., Han, Y., & Sheng, J., (2005). Study on morphology of electrospun poly (vinyl alcohol) mats. *Eur. Polym. J., 41,* 423–432.
73. Jarusuwannapoom, T., Hongrojjanawiwat, W., Jitjaicham, S., Wannatong, L., Nithitanakul, M., Pattamaprom, C., Koombhongse, P., et al., (2005). Effect of solvents on electro-spinnability of polystyrene solutions and morphological appearance of resulting electrospun polystyrene fibers. *Eur. Polym. J., 41,* 409–421.
74. Jiang, H., Fang, D., Hsiao, B. S., Chu, B., & Chen, W., (2004). Optimization and characterization of dextran membranes prepared by electrospinning. *Biomacromolecules, 5,* 326–333.
75. He, J. H., Wan, Y. Q., & Yu, J. Y., (2004). Application of vibration technology to polymer electrospinning. *International Journal of Nonlinear Sciences and Numerical Simulation, 5,* 253–262.
76. Li, X., Chen, W., Qian, Q., Huang, H., Chen, Y., Wang, Z., Chen, Q., et al., (2021). Electrospinning-based strategies for battery materials. *Advanced Energy Materials, 11,* 2000845.
77. Kalayci, V. E., Patra, P. K., Buer, A., Ugbolue, S. C., Kim, Y. K., & Warner, S. B., (2004). Fundamental investigations on electrospun fibers. *Journal of Advanced Materials, 36,* 43–45.
78. Hohman, M. M., Shin, M., Rutledge, G., & Brenner, M. P., (2001). Electrospinning and electrically forced jets. II. Applications. *Physics of Fluids, 13,* 2221–2236.
79. Berkland, C., Pack, D. W., & Kim, K. K., (2004). Controlling surface nanostructure using flow-limited field-injection electrostatic spraying (FFESS) of poly (D, L-lactide-co-glycolide). *Biomaterials, 25,* 5649–5658.
80. Wu, X., Wang, L., Yu, H., & Huang, Y., (2005). Effect of solvent on morphology of electrospinning ethyl cellulose fibers. *Journal of Applied Polymer Science, 97,* 1292–1297.

81. Uyar, T., & Besenbacher, F., (2008). Electrospinning of uniform polystyrene fibers: The effect of solvent conductivity. *Polymer (Guildf), 49*, 5336–5343.
82. Kim, B., Park, H., Lee, S. H., & Sigmund, W. M., (2005). Poly (acrylic acid) nanofibers by electrospinning. *Mater. Lett., 59*, 829–832.
83. Mit-uppatham, C., Nithitanakul, M., & Supaphol, P., (2004). Ultrafine electrospun polyamide-6 fibers: Effect of solution conditions on morphology and average fiber diameter. *Macromolecular Chemistry and Physics, 205*, 2327–2338.
84. Yoon, J., Yang, H., Lee, B., & Yu, W., (2018). Recent progress in coaxial electrospinning: New parameters, various structures, and wide applications. *Advanced Materials, 30*, 1704765.
85. Reneker, D. H., & Chun, I., (1996). Nanometre diameter fibers of polymer, produced by electrospinning. *Nanotechnology, 7*, 216.
86. Wu, C., Chiou, H., Lin, S., & Lin, J., (2012). Effects of electrostatic polarity and the types of electrical charging on electrospinning behavior. *Journal of Applied Polymer Science, 126*, E89–E97.
87. Mo, X. M., Xu, C. Y., Kotaki, M. E. A., & Ramakrishna, S., (2004). Electrospun P (LLA-CL) nanofiber: A biomimetic extracellular matrix for smooth muscle cell and endothelial cell proliferation. *Biomaterials, 25*, 1883–1890.
88. Katti, D. S., Robinson, K. W., Ko, F. K., & Laurencin, C. T., (2004). Bioresorbable nanofiber-based systems for wound healing and drug delivery: Optimization of fabrication parameters. *Journal of Biomedical Materials Research Part B: Applied Biomaterials: An Official Journal of The Society for Biomaterials, The Japanese Society for Biomaterials, and The Australian Society for Biomaterials and the Korean Society for Biomaterials, 70*, 286–296.
89. Pawlowski, K. J., Barnes, C. P., Boland, E. D., Wnek, G. E., & Bowlin, G. L., (2004). Biomedical nanoscience: Electrospinning basic concepts, applications, and classroom demonstration. *MRS Online Proceedings Library (OPL), 827*.
90. Huang, J. Y., Lo, Y. C., Niu, J. J., Kushima, A., Qian, X., Zhong, L., Mao, S. X., & Li, J., (2013). Nanowire liquid pumps. *Nat. Nanotechnol., 8*, 277–281.
91. Larrondo, L., & St. John, M. R., (1981). Electrostatic fiber spinning from polymer melts. II. Examination of the flow field in an electrically driven jet. *Journal of Polymer Science: Polymer Physics Edition, 19*, 921–932.
92. Larrondo, L., & St. John, M. R., (1981). Electrostatic fiber spinning from polymer melts. III. Electrostatic deformation of a pendant drop of polymer melt. *Journal of Polymer Science: Polymer Physics Edition, 19*, 933–940.
93. Sas, I., Gorga, R. E., Joines, J. A., & Thoney, K. A., (2012). Literature review on super-hydrophobic self-cleaning surfaces produced by electrospinning. *Journal of Polymer Science Part B: Polymer Physics, 50*, 824–845.
94. Yuan, X., Zhang, Y., Dong, C., & Sheng, J., (2004). Morphology of ultrafine polysulfone fibers prepared by electrospinning. *Polym. Int., 53*, 1704–1710.
95. Geng, X., Kwon, O. H., & Jang, J., (2005). Electrospinning of chitosan dissolved in concentrated acetic acid solution. *Biomaterials, 26*, 5427–5432.
96. Pham, Q. P., Sharma, U., & Mikos, A. G., (2006). Electrospun poly (ε-caprolactone) microfiber and multilayer nanofiber/microfiber scaffolds: Characterization of scaffolds and measurement of cellular infiltration. *Biomacromolecules, 7*, 2796–2805.
97. Khan, W. S., Asmatulu, R., Ceylan, M., & Jabbarnia, A., (2013). Recent progress on conventional and non-conventional electrospinning processes. *Fibers and Polymers, 14*, 1235–1247.

98. Srinivasarao, M., Collings, D., Philips, A., & Patel, S., (1979). Three-dimensionally ordered array of air bubbles in a polymer film. *Science, 292*, 79–83.
99. Park, J. Y., & Lee, I. H., (2010). Relative humidity effect on the preparation of porous electrospun polystyrene fibers. *J. Nanosci. Nanotechnol., 10*, 3473–3477.
100. Huan, S., Liu, G., Han, G., Cheng, W., Fu, Z., Wu, Q., & Wang, Q., (2015). Effect of experimental parameters on morphological, mechanical and hydrophobic properties of electrospun polystyrene fibers. *Materials, 8*, 2718–2734.
101. Pelipenko, J., Kristl, J., Janković, B., Baumgartner, S., & Kocbek, P., (2013). The impact of relative humidity during electrospinning on the morphology and mechanical properties of nanofibers. *Int. J. Pharm., 456*, 125–134.
102. Jiang, S., Chen, Y., Duan, G., Mei, C., Greiner, A., & Agarwal, S., (2018). Electrospun nanofiber reinforced composites: A review. *Polymer Chemistry, 9*, 2685–2720.
103. Kim, J. S., & Lee, D. S., (2000). Thermal properties of electrospun polyesters. *Polym. J., 32*, 616–618.
104. Tang, C., & Liu, H., (2008). Cellulose nanofiber reinforced poly (vinyl alcohol) composite film with high visible light transmittance. *Composites Part A: Applied Science and Manufacturing, 39*, 1638–1643.
105. Kumar, A., Jose, R., Fujihara, K., Wang, J., & Ramakrishna, S., (2007). Structural and optical properties of electrospun TiO_2 nanofibers. *Chemistry of Materials, 19*, 6536–6542.
106. Kavan, L., Stoto, T., Graetzel, M., Fitzmaurice, D., & Shklover, V., (1993). Quantum size effects in nanocrystalline semiconducting titania layers prepared by anodic oxidative hydrolysis of titanium trichloride. *The Journal of Physical Chemistry, 97*, 9493–9498.
107. Kormann, C., Bahnemann, D. W., & Hoffmann, M. R., (1988). Preparation and characterization of quantum-size titanium dioxide. *The Journal of Physical Chemistry, 92*, 5196–5201.
108. Anpo, M., Shima, T., Kodama, S., & Kubokawa, Y., (1987). Photocatalytic hydrogenation of propyne with water on small-particle titania: Size quantization effects and reaction intermediates. *Journal of Physical Chemistry, 91*, 4305–4310.
109. Joselevich, E., & Willner, I., (1994). Photosensitization of quantum-size TiO_2 particles in water-in-oil microemulsions. *The Journal of Physical Chemistry, 98*, 7628–7635.
110. Serpone, N., Lawless, D., & Khairutdinov, R., (1995). Size effects on the photophysical properties of colloidal anatase TiO_2 particles: Size quantization versus direct transitions in this indirect semiconductor? *The Journal of Physical Chemistry, 99*, 16646–16654.
111. Shalaby, T. I., El-Kady, M. F., Zaki, A. E. H. M., & El-Kholy, S. M., (2017). Preparation and application of magnetite nanoparticles immobilized on cellulose acetate nanofibers for lead removal from polluted water. *Water Science and Technology: Water Supply, 17*, 176–187.
112. Liu, Y., Liu, Y., Lee, J. H., Lee, C., Park, M., & Kim, H. Y., (2015). Ultrafine formation of optically transparent polyacrylonitrile/polyacrylic acid nanofiber fibrils via electrospinning at high relative humidity. *Composites Science and Technology, 117*, 404–409.
113. Chew, S. Y., Hufnagel, T. C., Lim, C. T., & Leong, K. W., (2006). Mechanical properties of single electrospun drug-encapsulated nanofibers. *Nanotechnology, 17*, 3880.
114. Sun, L., Han, R. P. S., Wang, J., & Lim, C. T., (2008). Modeling the size-dependent elastic properties of polymeric nanofibers. *Nanotechnology, 19*, 455706.
115. Liao, C. C., Wang, C. C., Chen, C. Y., & Lai, W. J., (2011). Stretching-induced orientation of polyacrylonitrile nanofibers by an electrically rotating viscoelastic jet for improving the mechanical properties. *Polymer (Guildf), 52*, 2263–2275.

116. Papkov, D., Zou, Y., Andalib, M. N., Goponenko, A., Cheng, S. Z. D., & Dzenis, Y. A., (2013). Simultaneously strong and tough ultrafine continuous nanofibers. *ACS Nano, 7*, 3324–3331.
117. An, T., Pant, B., Kim, S. Y., Park, M., Park, S. J., & Kim, H. Y., (2017). Mechanical and optical properties of electrospun nylon 6, 6 nanofiber reinforced cyclic butylene terephthalate composites. *Journal of Industrial and Engineering Chemistry, 55*, 35–39.
118. Zhang, S., (2009). *Mechanical and Physical Properties of Electrospun Nanofibers*.
119. Matsumoto, H., & Tanioka, A., (2011). Functionality in electrospun nanofibrous membranes based on fiber's size, surface area, and molecular orientation. *Membranes (Basel), 1*, 249–264.
120. Richard-Lacroix, M., & Pellerin, C., (2013). Molecular orientation in electrospun fibers: From mats to single fibers. *Macromolecules, 46*, 9473–9493.
121. Nakashima, K., Tsuboi, K., Matsumoto, H., Ishige, R., Tokita, M., Watanabe, J., & Tanioka, A., (2010). Control over internal structure of liquid crystal polymer nanofibers by electrospinning. *Macromol. Rapid Commun., 31*, 1641–1645.
122. Danno, T., Matsumoto, H., Nasir, M., Shimizu, S., Minagawa, M., Kawaguchi, J., Horibe, H., & Tanioka, A., (2008). *Fine Structure of PVDF Nanofiber Fabricated by Electrospray Deposition Part B Polymer Physics*.
123. Duan, G., Liu, S., Jiang, S., & Hou, H., (2019). High-performance polyamide-imide films and electrospun aligned nanofibers from an amide-containing diamine. *J. Mater. Sci., 54*, 6719–6727.
124. Yano, T., Higaki, Y., Tao, D., Murakami, D., Kobayashi, M., Ohta, N., Koike, J. I., et al., (2012). Orientation of poly (vinyl alcohol) nanofiber and crystallites in non-woven electrospun nanofiber mats under uniaxial stretching. *Polymer (Guildf), 53*, 4702–4708.
125. Zhang, H., Jiang, S., Duan, G., Li, J., Liu, K., Zhou, C., & Hou, H., (2014). Heat-resistant polybenzoxazole nanofibers made by electrospinning. *Eur. Polym. J., 50*, 61–68.
126. Huang, C., Chen, S., Lai, C., Reneker, D. H., Qiu, H., Ye, Y., & Hou, H., (2006). Electrospun polymer nanofibres with small diameters. *Nanotechnology, 17*, 1558.
127. Xu, H., Jiang, S., Ding, C., Zhu, Y., Li, J., & Hou, H., (2017). High strength and high breaking load of single electrospun polyimide microfiber from water soluble precursor. *Materials Letters, 201*, 82–84.
128. Jiang, S., Duan, G., Chen, L., Hu, X., & Hou, H., (2015). Mechanical performance of aligned electrospun polyimide nanofiber belt at high temperature. *Materials Letters, 140*, 12–15.
129. Gibson, P., Schreuder-Gibson, H., & Rivin, D., (2001). Transport properties of porous membranes based on electrospun nanofibers. *Colloids and Surfaces A: Physicochemical and Engineering Aspects, 187*, 469–481.
130. Hussain, D., Loyal, F., Greiner, A., & Wendorff, J. H., (2010). Structure property correlations for electrospun nanofiber nonwovens. *Polymer (Guildf), 51*, 3989–3997.
131. Li, D., Ouyang, G., McCann, J. T., & Xia, Y., (2005). Collecting electrospun nanofibers with patterned electrodes. *Nano Lett., 5*, 913–916.
132. Zhu, X., Cui, W., Li, X., & Jin, Y., (2008). Electrospun fibrous mats with high porosity as potential scaffolds for skin tissue engineering. *Biomacromolecules, 9*, 1795–1801.
133. Milleret, V., Simona, B., Neuenschwander, P., & Hall, H., (2011). Tuning electrospinning parameters for production of 3D-fiber-fleeces with increased porosity for soft tissue engineering applications. *Eur. Cell Mater., 21*, 286–303.

134. Sander, M. S., Prieto, A. L., Gronsky, R., Sands, T., & Stacy, A. M., (2002). Fabrication of high-density, high aspect ratio, large-area bismuth telluride nanowire arrays by electrodeposition into porous anodic alumina templates. *Advanced Materials, 14*, 665–667.
135. Wang, M., Singh, H., Hatton, T. A., & Rutledge, G. C., (2004). Field-responsive superparamagnetic composite nanofibers by electrospinning. *Polymer (Guildf), 45*, 5505–5514.
136. Xia, Y., Yang, P., Sun, Y., Wu, Y., Mayers, B., Gates, B., Yin, Y., et al., (2003). One-dimensional nanostructures: Synthesis, characterization, and applications. *Advanced Materials, 15*, 353–389.
137. Tan, S. T., Wendorff, J. H., Pietzonka, C., Jia, Z. H., & Wang, G. Q., (2005). Biocompatible and biodegradable polymer nanofibers displaying superparamagnetic properties. *Chem PhysChem., 6*, 1461–1465.
138. Sawhney, A. P. S., Condon, B., Singh, K. V., Pang, S. S., Li, G., & Hui, D., (2008). Modern applications of nanotechnology in textiles. *Textile Research Journal, 78*, 731–739.
139. Neuberger, T., Schöpf, B., Hofmann, H., Hofmann, M., & Von, R. B., (2005). Superparamagnetic nanoparticles for biomedical applications: Possibilities and limitations of a new drug delivery system. *Journal of Magnetism and Magnetic Materials, 293*, 483–496.
140. Wu, H., Zhang, R., Liu, X., Lin, D., & Pan, W., (2007). Electrospinning of Fe, Co, and Ni nanofibers: Synthesis, assembly, and magnetic properties. *Chemistry of Materials, 19*, 3506–3511.
141. Gass, J., Poddar, P., Almand, J., Srinath, S., & Srikanth, H., (2006). Superparamagnetic polymer nanocomposites with uniform Fe_3O_4 nanoparticle dispersions. *Advanced Functional Materials, 16*, 71–75.
142. Hutmacher, D. W., & Dalton, P. D., (2011). Melt electrospinning. *Chemistry–An Asian Journal, 6*, 44–56.
143. Blachowicz, T., & Ehrmann, A., (2020). Most recent developments in electrospun magnetic nanofibers: A review. *Journal of Engineered Fibers and Fabrics, 15*, 15589250 19900843.
144. Zeugolis, D. I., Paul, R. G., & Attenburrow, G., (2008). Extruded collagen-polyethylene glycol fibers for tissue engineering applications. *Journal of Biomedical Materials Research Part B: Applied Biomaterials: An Official Journal of The Society for Biomaterials, The Japanese Society for Biomaterials, and The Australian Society for Biomaterials and the Korean Society for Biomaterials, 85*, 343–352.
145. Wongchitphimon, S., Wang, R., Jiraratananon, R., Shi, L., & Loh, C. H., (2011). Effect of polyethylene glycol (PEG) as an additive on the fabrication of polyvinylidene fluoride-co-hexafluropropylene (PVDF-HFP) asymmetric microporous hollow fiber membranes. *Journal of Membrane Science, 369*, 329–338.
146. Grothe, T., Brikmann, J., Meissner, H., & Ehrmann, A., (2017). Influence of solution and spinning parameters on nanofiber mat creation of poly (ethylene oxide) by needleless electrospinning. *Materials Science, 23*, 342–349.
147. Klinkhammer, K., Seiler, N., Grafahrend, D., Gerardo, N. J., Mey, J., Brook, G. A., Möller, M., et al., (2009). Deposition of electrospun fibers on reactive substrates for *in vitro* investigations. *Tissue Engineering Part C: Methods, 15*, 77–85.
148. Dalton, P. D., Klinkhammer, K., Salber, J., Klee, D., & Möller, M., (2006). Direct *in vitro* electrospinning with polymer melts. *Biomacromolecules, 7*, 686–690.
149. Sreenivasulu, G., et al., (2017). Multiferroic core-shell nanofibers, assembly in a magnetic field, and studies on magneto-electric interactions. *Materials, 11*, 18.

150. Yaskiv, O. I., Fedirko, V. M., & Kukhar, I. S., (2016). *The Effect of the Structural State of Steels on the Mechanical Properties in the Oxygen Saturated Lead* (Vol. 104, pp. 32–36). Questions of atomic science and technology.
151. Yaskiv, O. I., Kukhar, I. S., & Fedirko, V. M., (2015). Effect of preliminary diffusion oxidation on mechanical properties of ferritic steel in oxygen-containing lead. *Fusion Engineering and Design, 101*, 134–140.
152. Akhtar, M. N., Sulong, A. B., Ahmad, M., Khan, M. A., Ali, A., & Islam, M. U., (2016). Impacts of Gd–Ce on the structural, morphological and magnetic properties of garnet nanocrystalline ferrites synthesized via sol–gel route. *Journal of Alloys and Compounds, 660*, 486–495.
153. Chen, P., Wu, R., & Xiao, J. Q., (2014). Dynamic magnetic properties of electrospun NiZn spinel ferrite nanofibers. *IEEE Transactions on Magnetics, 50*, 1–4.
154. Shen, X., Liu, M., Song, F., & Meng, X., (2010). Structural evolution and magnetic properties of $SrFe_{12}O_{19}$ nanofibers by electrospinning. *J. Solgel Sci. Technol., 53*, 448–453.
155. Streckova, M., Mudra, E., Sebek, M., Sopcak, T., Dusza, J., & Kovac, J., (2017). Preparation and investigations of $Ni_{0.2}Zn_{0.8}Fe_2O_4$ ferrite nanofiber membranes by needleless electrospinning method. *Acta Phys. Pol. A, 131*, 729–731.
156. Barakat, N. A. M., Kim, B., & Kim, H. Y., (2009). Production of smooth and pure nickel metal nanofibers by the electrospinning technique: Nanofibers possess splendid magnetic properties. *The Journal of Physical Chemistry C, 113*, 531–536.
157. Kim, Y., Ebara, M., & Aoyagi, T., (2013). A smart hyperthermia nanofiber with switchable drug release for inducing cancer apoptosis. *Advanced Functional Materials, 23*, 5753–5761.
158. Li, D., Luo, L., Pang, Z., Ding, L., Wang, Q., Ke, H., Huang, F., & Wei, Q., (2014). Novel phenolic biosensor based on a magnetic polydopamine-laccase-nickel nanoparticle loaded carbon nanofiber composite. *ACS Appl Mater Interfaces, 6*, 5144–5151.
159. Zhang, W., Chen, S., Hu, W., Zhou, B., Yang, Z., Yin, N., & Wang, H., (2011). Facile fabrication of flexible magnetic nanohybrid membrane with amphiphobic surface based on bacterial cellulose. *Carbohydr. Polym., 86*, 1760–1767.
160. Kim, E. J., Lee, C. S., Chang, Y. Y., & Chang, Y. S., (2013). Hierarchically structured manganese oxide-coated magnetic nanocomposites for the efficient removal of heavy metal ions from aqueous systems. *ACS Appl. Mater. Interfaces, 5*, 9628–9634.
161. Ghasemi, E., Ziyadi, H., Afshar, A. M., & Sillanpää, M., (2015). Iron oxide nanofibers: A new magnetic catalyst for azo dyes degradation in aqueous solution. *Chemical Engineering Journal, 264*, 146–151.
162. Bhaumik, M., McCrindle, R. I., & Maity, A., (2015). Enhanced adsorptive degradation of Congo red in aqueous solutions using polyaniline/Fe^0 composite nanofibers. *Chemical Engineering Journal, 260*, 716–729.
163. Li, C. J., Wang, J. N., Li, X. Y., & Zhang, L. L., (2011). Functionalization of electrospun magnetically separable TiO_2-coated $SrFe_{12}O_{19}$ nanofibers: Strongly effective photocatalyst and magnetic separation. *J. Mater Sci., 46*, 2058–2063.
164. Chen, L., Hong, S., Zhou, X., Zhou, Z., & Hou, H., (2008). Novel Pd-carrying composite carbon nanofibers based on polyacrylonitrile as a catalyst for Sonogashira coupling reaction. *Catalysis Communications, 9*, 2221–2225.

CHAPTER 17

Magnetic Polymer Composites and Their Role in Engineering Applications

SIMONA LUMINITA NICA

"Petru Poni" Institute of Macromolecular Chemistry, Department of Physical Chemistry of Polymers, Iasi, Romania

ABSTRACT

Development of the materials with properties required in the field of EMI shielding or sensors applications has attracted the attention of researchers lately. Spinel ferrites are widely applicable in sensors or EMI shielding due to their properties. In this chapter, spinel ferrite-based polymers (either conducting or non-conducting material) are explored for the prevention of electromagnetic interference (EMI). Some preparation methods are also taken into consideration. The combination of electrical and magnetic loss is important for microwave absorption properties.

17.1 INTRODUCTION

Materials with advanced design (also named smart materials) to be used in electromagnetic (EM) or electronic devices are in continuous development. Manufactured products depend on the costs involved and there's performance. Composite materials can be obtained by combining two or more constituents or phases with different properties [1, 2]. The components of the composite material are: (i) the continuous phase; (ii) the discontinuous phase; and (iii) the interface between the components [3, 4].

Since the focus is on development of EM or electronic devices to make life much easier, magnetic polymer composites [5] occupy a high place in

Modern Magnetic Materials: Properties and Applications. Iuliana Stoica, Ann Rose Abraham, & A. K. Haghi (Eds.)

the scientific community. In this manner, polymers could be modified by incorporating ferromagnetic fillers (in micro/nano-sizes) [6–8] to fabricate either soft magnetic composite (SMCs), which replace laminated steels [9] either hard magnetic composite (HMCs) [10].

Another classification of SMCs involves:

- Soft metallic materials [11];
- Soft ferrite materials [12].

The structure of the fillers highly influences the properties of the final obtained material. Preparation of magnetic polymer composite materials or simply magnetic stimuli-responsive materials relay on the requirement of ensuring a long-time dispersion stability of hard or soft fillers. It is important that fillers do not agglomerate or to aggregate during the synthesis process.

Some application of magnetic composite-based on polymers materials are listed below:

- Sensors and actuators [13, 14];
- Energy storage and harvesting [15];
- Wearable electronic devices [16];
- Human-machine interaction [17]; and also
- Biomedical system [18].

Multifunctionality of polymer composites reinforced with magnetic filler for energy, shielding, sensors or actuator materials will be discussed in this chapter. Some issues and concepts are evidenced.

17.2 RECENT TRENDS IN MAGNETIC POLYMER COMPOSITES

In the past decades, many researchers were focused on studying magnetic polymer composites that can be used in technical applications. This part of the chapter describes some of the most important reports in this topic by highlighting the research tendencies and advantages made so far. Special attention was attributed to soft magnetic composite materials (SMCs). SMCs are ferromagnetic powder particles coated with electric insulating layer materials [19]. Selection of the insulating layer represents the key factor since dictates both the peculiar electronic and magnetic properties [20] and its applicability in electronic and electrical devices industry. To fabricate SMCs, the insulating layers which cover the surface of the magnetic powder could be from inorganic or organic materials. It is worth mentioned that thermoplastic polymers such as polyamides [21] or thermosetting polymers such as epoxy resins (ERs) [22, 23], silicon resin [24], polyurethanes [25] are materials used as organic coatings for SMCs. But polymers have poor stability. To balance this feature of the non-magnetic material, researchers found an alternative in developing coating layer from inorganic materials. So, Fe powder [26], Fe–Si powder [27], Mn–Zn ferrite [28], Ni–Zn ferrite [29], Co—Fe ferrite [30] or different type of oxides such as Fe_3O_4 [31], Fe_2O_3 [32], Al_2O_3 [33]) are considered a proper choice to fabricate soft magnetic composite. It is known that Fe-based amorphous alloy has good soft magnetic properties.

The main advantages of SMCs are:

- Low eddy current loss at medium and high frequencies which is related to the accurately engineered microstructure [34];
- Isotropic three-dimensional (3D) magnetic flux paths and relatively high saturation magnetic induction [35];
- Low coercivity [36];
- High magnetic permeability reduces the hysteresis loss of the SMCs materials [37].

The shape geometric form and the size of powder particles determine the specific structure of the soft magnetic composite. Usually, the magnetic powder particles are spherical. In this manner, both the geometry of nonmagnetic gaps and concentration will affect the magnetic properties of the fabricated composite.

Methods used for SMCs preparation involve:

- Ball-milling [38];
- Sol-gel [39];
- Surface oxidation method [40];
- Thermal oxidation method [41], etc.

Luo et al. [42] fabricated Fe–Si SMCs. As coating insulation layers researchers have prepared via ball milling method Fe–Si/γ-Fe_2O_3 powders. The full synthesis (by using subsequent spark plasma sintering (SPS) process) of three-shell structure of Fe_2SiO_4-γ-Fe_2O_3-γFe_2O_3 insulating-magnetic-insulating (IMI) layer can be found in Ref. [42]. They reported both superior magnetic performance and insulating properties of the obtained IMI layer. Developed material found potential use as switching devices, transformers, and inductors.

Gao et al. [43] reported fabrication of Fe–Si/SiO_2 soft ferromagnetic composites. For this work a two-step method was proposed for material preparation: chemical vapor deposition and hot-pressed sintering. The aim of the study was to quantitatively describe the magnetic performance of the material.

Zhang et al. [44] reported the preparation of Fe/MnO_2 SMCs. Synthesis method adopted by the researchers was ball milling in combination with SPS. The uniform MnO_2 coating layer proved to reduce eddy current loss.

Hou et al. [45] prepared Fe/Fe_3O_4-FeSiAl SMCs. The material is characterized by high permeability and low core loss. Process of SMCs fabrication material is based on the surface oxidation method. The preparation steps involve: the reduced iron powders where coated with an insulating layer of FexOy to obtain Fe/FexOy and Fe/FexOy-FeSiAl SMCs. Schematic preparation method can be found in Ref. [45].

Hu et al. [46] fabricated soft magnetic composite materials for engineering applications purpose by mixing method of reagents. A non-conductive matrix, polydimethylsiloxane (PDMS), filled with different inclusions such as, carbonyl iron (CIPs), feroferric oxide (Fe_3O_4), NiFe-alloy (in the range of 0 VOL% to 33 VOL%) was tested at frequencies between 100 kHz and 1 MHz for soft-core transformer applications. Researchers compared both theoretical and experimental data regarding the magnetic permeability parameter. For dynamic tests of the materials, they have constructed a Buck circuit converter. The advantages of this circuit relay on its high frequency operation. Regarding the static tests of the soft magnetic composite materials a vibrating sample magnetometer (VSM) was involved. Experimental results of both cylindrical and toroidal materials indicated that magnetic permeability is insensitive to the applied frequency and the eddy current was avoided in the majority of the samples. However, they report that the PMDS-filled with Fe_3O_4 particles reach the saturation very fast as compared with the other tested materials.

Sun et al. [47] prepared Fe-based SMCs. As insulating coating material for this soft composite was considered a polyimide due to its high thermal stability. Researchers investigated the magnetic properties of the SMCs.

Evangelista et al. [48] made double-layered SMCs based on pure iron powder coated by zinc oxide nanoparticles (NPs) (nano-ZnO) – as the first layer and boron oxide (B_2O_3) – as the second layer. Researchers demonstrated enhanced magnetic performance of the composite as compared with soft magnetic composite with one single insulating layer. Details about preparation method can be found in Ref. [48].

Meng et al. [50] developed Fe/Fe_3O_4/resin SMCs. The preparation method implies the oxidation of the Fe powders with 70-nm-thick Fe_3O_4

layers in situ formed on the surfaces of unoxidized Fe parts. After that initially oxidized products were coated with 0.8 wt.% silicone resin through a physical coating method. Excellent soft magnetic properties of the material were obtained by the researchers such as high magnetic inductions and decreased core loss.

Qing et al. [51] fabricated FeSiB/ERs composite using direct ink writing (DIW) technique. The advantage of this technique is to create complex-shaped parts of the material. Also, after the material was formed is not necessary to heat it. Researchers investigated the magnetic properties of the material. Also, they evidenced a reduction of eddy current.

Riesgo et al. [52] reported preparation of Fe81Ga19/silicone composite materials by milling technique method. Magnetic properties have been investigated by the researchers.

Wang et al. [53] obtained FeSiCr/carbonyl-iron SMCs. As insulating layer researchers prepared nano-$CaCO_3$/epoxy nanocomposites. They reported a reduced eddy current which affect the hysteresis behavior of the SMCs.

Also, many preparation methods have been proposed for developing magnetic polymer composite materials. To obtain magnetic polymer composite with specific morphology [54] is still a challenging task. Polymers used as matrices can be classified in *conducting* and *non-conducting* materials. Conducting polymers will not be discussed here.

Wang et al. [55] developed a new magnetic polymeric composite (Fe_3O_4/polystyrene (PS)) using a surface-initiated polymerization method. Magnetic particles were considered Fe_3O_4 while H_2O_2 was the initiator. The size of the microspheres polymer is controlled by both H_2O_2 and the amount of polymer.

A substantial difference in the size of the microsphere was observed:

- Particles became bigger and bigger with the addition of the monomer;
- Particles size decreases as the amount of H_2O_2 increase.

Hamciuc et al. [56] obtained magnetic polyimide/copper-nickel ferrite composites. As inorganic fillers researchers have used four different cerium-doped ferrite particles such as NiFe1.92Ce0.08O4, Ni0.7Cu0.3Fe1.92Ce 0.08O4, Ni0.4Cu0.6Fe1.92Ce0.08O4, and CuFe1.92Ce0.08O4. Material was analyzed from magnetically and dielectric point of view.

Kim et al. [57] fabricated Fe_3O_4-encapsulated PS hybrid particles via a mini-emulsion polymerization process. Detailed preparation steps of the materials are found in Ref. [57]. Petcharoen et al. [58] fabricated polyurethane-Fe_3O_4 composite to be used in the field of actuation. Magnetic response of the materials has been evaluated.

17.3 APPLICATIONS OF THE MAGNETIC POLYMER COMPOSITES IN ENGINEERING

As above-mentioned soft magnetic materials refer to soft ferrite particle, Fe_3O_4, Fe_2O_3, etc., which are embedded into different polymeric matrices. Choice to embedding magnetic iron oxide into polymeric matrices relay on their intrinsic magnetic properties [59]. Soft-magnetic materials have low coercivity field. Magnetic polymeric composites can be considered tribo-electric nanogenerators (TENGs). These materials are applicable in renewable resources [60, 61]. Other application of magnetic polymer composite in electronic device covers the field of sensors [62], actuators [63, 64], electromagnetic (EMI) shielding [65, 66], transducing devices [67], wearable electronic devices [68, 69]. In this part of the chapter, utilization of materials as *EMI shielding* and also, *sensors* will be discussed.

Advantages of fabrication magnetic composite having polymer as a matrix (either conductive or non-conductive) for electronic and EM devices besides their unique magnetic [70] and mechanical properties [71] include the capacity of obtaining complicated structures and shapes of the materials [72]. Among the conductive polymers, polyaniline (PANI) became the most studied material in the field of engineering applications since possess superior conducting nature [73]. Polymers could also be classified as thermosets and thermoplastics polymers. Advantage of using PANI as EMI shielding will be discussed through relevant scientific works.

Besides semiconducting properties of PANI polymer, morphological flexibility was also an important property. This polymer can be easily synthesized in thin films or fibers due to the requirements in electronic applications. As a result of the introduction of spinel ferrite in PANI matrix, the electrical properties of the material enhance [74].

Different applications of PANI-based composite are recorded in the scientific community due to its versatility. For example, its stability confers the ability to be used as *sensor material.*

To reduce the size of the electronic devices is a growing demand on the electronics market. As a consequence, it is very possible to pick up some adverse effects – mostly, electromagnetic interference (EMI) of signals [75] which is considered as a pollutant for the human health. To annihilate this effect, it is necessary that bandwidth shielding materials to be analyzed. The mechanism of absorbance of EM radiation by the materials can be found in Ref. [76].

For example, Radoń et al. [77] studied the microwave absorption properties of dextrin-Fe_3O_4 composites by ball milling technique. They validated that the shape of the magnetic particles embedded in the dextran matrix influenced the magneto-dielectric properties of the material. A low bandwidth shielding between 0.3 and 2.5 GHz was explored. Results conducted that composites with spherical particles have a minimum reflection loss equal to – 25 dB for a 12 mm layer thickness. In case of cuboidal particles, the reflection loss was equal to – 22.5 dB for 20 mm thickness. Different value (–18.7 dB for 16 mm absorber thickness) was obtained for the same cuboidal geometry of the magnetic particles prepared in a potassium hydroxide solution.

The microwave absorption properties of the hollow spherical Fe_3O_4-based composite were also tested by Shu et al. [78]. Magnetic particles with controlled morphology were prepared by solvothermal method. For characteristics of prepared composite, regarding thickness (3.62 mm) and bandwidth (4.56–9.24 GHz, 4.68 GHz) researchers reported a minimum reflection loss equal to – 56.90 dB.

Ahmed et al. [79] fabricated poly(vinylidene-fluoride) (PVDF) based graphene and nickel spinel ferrite (NFe) hybrid nanocomposite for EMI shielding applications. They reported an EMI shielding effectiveness (SE) at about 22 dB. The thickness of the films was 0.22 mm.

The work developed by Bhatti et al. [80] was centered on studying Y-type hexaferrite, in particular $Ba_2Co_2Fe_{12}O_{22.}$ For their preparation sol-gel technique was utilized. Conductive polymer (polyaniline (PANI)) was chosen as a matrix. By in-situ polymerization method hexaferrite-PANI composites were prepared. Resistivity determined was at 2.59 × 104 Ω-cm. These composite materials are applied in EMI shielding in low frequency range.

Using the same polymer as matrix (PANI) another group of researchers [81] prepared Fe_3O_4@SiO_2@PANI ternary nanocomposite. The preparation procedure was a three-step method. Into the matrix, researchers added spinel ferrites (namely, Fe_3O_4) which were prepared by the hydrothermal method modified with SiO_2 particles to form Fe_3O_4@SiO_2 composite. A core-shall structure was obtained – Fe_3O_4@SiO_2@PANI composite. The structure and the properties of the final material were characterized. They reported that Fe_3O_4 were the core of the composites while the other two materials acted as shells – SiO_2 particles and PANI matrix. Regarding the EM performance of the final product remarkable properties were obtained: RLmax of the Fe_3O_4@ SiO_2@PANI composite reached – 44.9 dB at 4.0 GHz. Researchers pointed out that this material can be used successfully in EMI shielding applications.

Another work conducted by Wang et al. [82] prepared complex core-shall composite structures material for high-performance EMI shielding. The fabricated probes were Ti_3C_2Tx MXene/Fe_3O_4@PANI composite films. Its selection relay on important properties such as: high electrical conductivity, large specific surface area, rich surface defects and the presence of numerous functional groups. The composites were obtained by vacuum-assisted filtration technique. As matrix researchers used $Ti_3C_2T_x$ while Fe_3O_4@PANI was core-shall structure embedded into matrix. Results obtained by the researchers suggested that when the weight ratio of the $Ti_3C_2T_x$ matrix to core-shall structure – Fe_3O_4@PANI was 12:5 the EMI SE reached 58.8 dB at a thickness of 12.1 μm.

Complex heterostructures centered on PANI-conductive polymer to be used in EMI shielding applications developed Movassagh-Alanagh et al. [83]. Materials fabricated by the researchers implied PANI@nano-Fe_3O_4@CFs (with different weight ratio of the fillers) by a two-step process technique which is discussed as follows.

The preparation method implies:

- First, particles of Fe_3O_4 having nano-dimensions were deposited on CFs material. A modified multi-step electrophoretic deposition (EPD) process was involved in this step of the material preparation.
- Then researchers fabricated PANI@nano-Fe_3O_4@CFs heterostructures by *in situ* polymerization method, using a solution of HCl.

Eventually, they obtained PANI@nano-Fe_3O_4@CFs mats with a layer by layer (LBL) assembly which was structurally demonstrated.

Magnetic properties of the structures (PANI@nano-Fe_3O_4@CFs mats) revealed a decrease of magnetic saturation for nano-Fe_3O_4 from 72.612 to 0.191 emu/g. Also, investigations of EM wave absorption of epoxy-based PANI@nano-Fe_3O_4@CFs hybrid composites heterostructures showed that in case of 1 wt.% nano-Fe_3O_4 embedded into matrix, probes indicate a maximum RL value of EMI SE at – 11.11 dB for a thickness of 1.5 mm pf the materials. The bandwidth was about 6 GHz. A variation effect was obtained for 5 wt.% fillers inclusion into the matrix with a difference recorded into the thickness of the probes at about 3 mm. In this case, the EMI SE was recorded at 29 dB at a frequency range of 8.2–18 GHz.

Guo et al. [84] developed materials with hierarchically multifunctional structure to be used as electromagnetic interface (EMI) shielding in electronic devices. Two different electromagnetic interface (EMI) shielding layers are fabricated while the last layer represents the substrate made from electrospuned Pi fibers to improve mechanical properties of the composite.

As polymeric materials they used in equimolar amount pyromellitic dianhydride (PMDA) monomer and 4,4′-oxydianiline (ODA) monomer. First EMI shielding layer is represented by graphene oxide/expanded graphite (GO/EG); while the second EMI shielding layer is given by Fe_3O_4/polyimide (Fe_3O_4/PI). All three layers are connected one to the other. They reported PI composite material having 61.0 wt.% of GO/EG and 23.8 wt.% of Fe_3O_4/PI exhibit a thermal conductivity coefficient of 95.40 W $(m\ K)^{-1}$. Also, mechanical properties indicate a tensile strength of about 93.6 MPa.

Koner et al. [85] fabricated La0.7Ba0.3MnO3/poly(vinylidenefluoride-co-trifluoroethylene) nanocomposite (0–3) films as magnetoelectric (ME) shields. Magnetic properties of the materials are also determined.

Besides conducting polymers as matrix for materials to be used as EMI shielding applications, researchers tried to develop new materials using non-conducting polymers for the same electronic application.

For example, another group of researchers Ahmed et al. [86] investigated the structure-property relationship of graphene and spinel nickel ferrites based PVDF hybrid nanocomposites. The solution casting method was chosen for film preparation. Tests made revealed that estimated EMI SE at about 22 dB for PVDF/graphene (3 wt.%)/nFE (30 wt.%) hybrid nanocomposite exhibiting. The thickness of the probes films was determined to be 0.2 mm. Due to the characteristics of the materials researchers considered that PVDF/graphene/nFE hybrid nanocomposites can be applied in miniaturized electronic devices for capacitor and EMI shielding applications.

Khodiri et al. [87] developed complex hybrid structures composites such as polyvinyl alcohol (PVA)/graphene/magnetite (Fe_3O_4) (PVA/Gr(x)/Fe_3O_4(0.1–x)) to be used as EMI shielding materials. The method of preparation was casting technique. Evaluation of the composite indicated enhanced magnetic and dielectric properties. Results showed value for EMI SE at about 40.7 dB along with a high conductivity of 1.87 S/cm when embedded graphene is small (0.08 wt.%).

Other fields in the electronics or EM devices are sustained by few remarkable scientific work.

Preparation of magnetic polymeric composites as humidity sensors devices represents a research area that received attention. Flexible polyimides are new materials that require different strategies [88]. To enhance the solubility of the material, structural modification of the polymer chain was adopted:

- Flexible ether linkages were incorporated;

- Incorporationofalicyclicdianhydride:5-(2,5-dioxotetrahydrofurfuryl)-3-methyl-3-cyclohexene-1,2-dicarboxylic acid anhydride (DOCDA, 98% purity; Merck, Darmstadt, Germany [89].

Preparation method can be found in Ref. [88]. Also, thermal, magnetic, and dielectric properties were also discussed in correlation with their structure.

So, the humidity sensitivity of pure polyimide and the DOCDA-ODA/Fe_3O_4 with different filler content of Fe_3O_4 (5, 10, and 20 wt.%) for various values of relative humidity (RH) were analyzed.

Results showed that relative permittivity decreases as the frequency increase. This comportment is assigned to the Maxwell-Wagner model. Also, a decrease was recorded for electrical resistivity for these magnetic polymer composites as the frequency increases over the RH range of humidity. Both microstructure and composition change influenced the electrical resistivity of the material. A relevant sensitivity is recorded for samples at low frequency (200 Hz). Tests results have shown that the optimum humidity sensitivity features are given by DOCDA-ODA/iron oxide synthesized at 300°C with 20 wt.% Fe_3O_4 loadings.

Another group of researchers obtained a magnetic polymeric composite material to be used in magnetoresistive devices [90]. As matrix they used a homopolymer (Moplen HP552R) while Fe_3O_4 (at different concentrations) was considered as magnetic fillers. Two technological methods were applied for obtained material such as hot pressing and extrusion. The process of fabrication had a strong influence on the final composite properties. Absorbed EM varied both quantitatively and qualitatively.

Another work [91] fabricated flexible multiferroics ME polymer composites by casting method. As ME composite films they used polycrystalline iron fiber (Fef) which were magnetically oriented and poly(vinylidene fluoride-co-trifluoroethylene) P(VDF–TrFE) – a copolymer that permits to obtain the electroactive phase. Researchers established the inclusions of iron fiber into the polymer matrix at different concentrations such as 4%, 8%, 12%, 16% and 20%. Among the properties determined for the obtained composite material – in terms of structural, ferroelectric, and piezoelectric features – the focus is on magnetic properties which were measured both parallel and perpendicular to the fiber axial directions. In both cases magnetic polymer composite material evidenced anisotropic character. Also, magnetic saturation (M_s ~29.8 emu/cm^3) and magnetic remanence (M_r ~5.4 emu/cm^3) showed greater values when the magnetic field was applied perpendicular to the Fef axial directions evidencing that magnetic moments are aligning along the magnetic easy-axis of Fef. Besides, 16% Fef filler loadings into P(VDF–TrFE) matrix show a peak ME value of 44.5 mV·cm.

Kumar et al. [92] reports preparation of magnetic polymeric composite material from multiferroic $BiFeO_3$ (BFO) nanocrystalline powder and poly vinylidene-fluoride (PVDF) polymer matrix as component for sensor device. Thickness of the magnetic composite was determined at ~ 0.2 mm. All the materials have been prepared via hot pressing method.

Charles et al. [93] prepared carbonyl iron particles-epoxy silane composite. Materials were reported to be applied in actuators field. Developments of the materials use a molding procedure. Detail on magnetic composite preparation can be found in the reference [93]. Magnetic polymer composites display a soft hysteresis.

17.4 CONCLUSIONS

Great efforts have been made to fabricate efficient materials in order to be used as devices in the electronic field. Two applications in the field of technology are widely discussed in this chapter: EMI shielding materials and materials for different type of sensors. Development of the materials which implies low-cost are a prerequisite task. The complex structure of the materials and also, the difference in properties of each individual component of the composites offer the possibility to enhance the SE of the final probes. There are numerous developed materials where the enhancement of the SE is due to the protective bilayer. But, lately, scientific community developed materials with very complex structures (values determined of EMI SE being very highly) where multilayer absorption are fabricated such as trilayer mechanism or even quaternary mechanism.

KEYWORDS

- **electromagnetic interference**
- **engineering applications**
- **hard magnetic composites**
- **magnetic composites**
- **polymer matrix**
- **soft magnetic composites**
- **spark plasma sintering**

REFERENCES

1. Pei, P., & Peng, Y., (2022). Constitutive modeling of magnetorheological fluids: A review. *J. Magn. Magn. Mater, 550*, 169076.
2. Lucarini, S., Hossain, M., & Garcia-Gonzalez, D., (2022). Recent advances in hard-magnetic soft composites: Synthesis, characterization, computational modelling, and applications. *Compos. Struct, 279*, 114800.
3. Hooper, T. E., Roscow, J. I., Mathieson, A., Khanbareh, H., Goetzee-Barrala, A. J., & Bell, A. J., (2021). High voltage coefficient piezoelectric materials and their applications. *J. Eur. Ceram. Soc, 41*, 6115–6129.
4. Elhajjar, R., Law, C. T., & Pegoretti, A., (2018). Magneto strictive polymer composites: Recent advances in materials, structures and properties. *Prog. Mater. Sci. Prog., 97*, 204–229.
5. Namathoti, S., Ravindra, K. V. M., & Rama, S. P. S., (2022). A review on progress in magnetic, microwave, ultrasonic responsive shape-memory polymer composites. *Mater. Today Proc., 56*, 1182–1191.
6. Liu, X. M., Yang, G., & Fu, S. Y., (2007). Mass synthesis of nanocrystalline spinel ferrites by a polymer-pyrolysis route. *Mater. Sci. Eng. C, 27*, 750–755.
7. Drmota, A., Koselj, J., Drofenik, M., & Žnidaršič, A., (2012). Electromagnetic wave absorption of polymeric nanocomposites based on ferrite with a spinel and hexagonal crystal structure. *J. Magn. Magn. Mater., 324*, 1225–1229.
8. Kurian, M., & Thankachan, S., (2021). Structural diversity and applications of spinel ferrite core-shell nanostructures: A review. *Open Ceramics, 8*, 100179.
9. Ni, J. L., Duan, F., Feng, S. J., Hu, F., Kan, X. C., & Liu, X. S., (2022). High performance of FeSiAl/hBN soft magnetic composites. *J. Alloys. Compd., 897*, 163191.
10. Slimani, Y., Almessiere, M. A., Guner, S., Baykal, A., Sertkol, M., Alahmari, F. S., Alsulami, E. M., & Auwal, I. A., (2022). An investigation on structural, optical and magnetic properties of hard-soft $SrFe_{12}O_{19}/(CoEu_{0.02}Fe_{1.98}O_4)x$ nanofiber composites. *J. Alloys. Compd., 905*, 164240.
11. Pöhl, F., Hardes, C., & Theisen, W., (2016). Scratch behavior of soft metallic materials. *AIMS Mater. Sci., 3*, 390–403.
12. Thakur, P., Chahar, D., Taneja, S., Bhalla, N., & Thakur, A., (2020). A review on MnZn ferrites: Synthesis, characterization and applications. *Ceram. Int., 46*, 15740–15763.
13. Huang, X., Panahi-Sarmad, M., Dong, K., Li, R., Chen, T., & Xiao, X., (2021). Tracing evolutions in electro-activated shape memory polymer composites with 4D printing strategies: A systematic review. *Compos. Part A Appl. Sci. Manuf., 147*, 106444.
14. Wang, H. S., Cho, J., Park, H. W., Jho, J. Y., & Park, J. H., (2021). Ionic polymer–metal composite actuators driven by methylammonium formate for high-voltage and long-term operation. *J. Ind. Eng. Chen., 96*, 194–201.
15. Wang, X., Zhou, Z., Lin, X., Pei, Z., Liu, D., & Zhao, S., (2022). Nanostructured hexaaza-trinaphthalene based polymers for advanced energy conversion and storage. *Chem. Eng. Sci., 427*, 130995.
16. Zhou, Z., You, C., Chen, Y., Xia, W., Tian, N., Li, Y., & Wang, C. K., (2022). Piezoelectric sensing performance of flexible P(VDF-TrFE)/PBDMS porous polymer materials. *Org. Electron., 105*, 106491.
17. Kumar, N., Tyeb, S., & Verma, V., (2021). Recent advances on metal oxide-polymer systems in targeted therapy and diagnosis: Applications and toxicological perspective. *J. Drug Deliv. Sci. Technol., 66*, 102814.

18. Kayal, S., (2021). Thermoresponsive magnetic/polymer composite nanoparticles for biomedical applications. *Mater. Today. Proc., 41*, 1116–1119.
19. Pan, Y., Peng, J., Qian, L., Xiang, Z., & Lu, W., (2020). Effects of compaction and heat treatment on the soft magnetic properties of iron-based soft magnetic composites *Mater. Res. Express, 7*, 016115.
20. Jiang, X., Liu, Q., Xing, J., Liu, N., Guo, Yu., Liu, Z., & Zhao, J., (2021). Recent progress on 2D magnets: Fundamental mechanism, structural design and modification. *Appl. Phys. Rev., 8*, 031305.
21. Chen, Y., Zhang, L., Sun, H., Chen, F., Zhang, P., Qu, X., & Fan, E., (2020). Enhanced magnetic properties of iron-based soft magnetic composites with phosphate-polyimide insulating layer. *J. Alloys. Compd., 813*, 152205.
22. Laxminarayana, T. A., Manna, S. K., Fernandes, B. G., & Venkataramani, N., (2015). Study of AC magnetic properties and core losses of Fe/Fe_3O_4-epoxy resin soft magnetic composite. *Phys. Procedia, 75*, 1396–1403.
23. Ren, L., Zhao, J., Wang, S. J., Han, B. Z., & Dang, Z. M., (2015). Dielectric and magnetic properties of Fe@FexOy/epoxy resin nanocomposites as high-performance electromagnetic insulating materials. *Compos. Sci. Technol., 114*, 57–63.
24. Lu, H., Dong, Y., Liu, X., Liu, Z., Wu, Y., Zhang, H., He, A., et al., (2021). Enhanced magnetic properties of FeSiAl soft magnetic composites prepared by utilizing PSA as resin insulating layer. *Polymers, 13*, 1350.
25. Shi, Y., Wang, B., Ye, S., Zhang, Y., Wang, B., Feng, Y., Han, W., Liu, C., & Shen, C., (2021). Magnetic, superelastic and superhydrophobic porous thermoplastic polyurethane monolith with nano-Fe_3O_4 coating for highly selective and easy-recycling oil/water separation. *Appl. Surf. Sci., 535*, 147690.
26. Li, K., Cheng, D., Yu, H., & Liu, Z., (2020). Process optimization and magnetic properties of soft magnetic composite cores based on phosphated, and mixed resin coated Fe powders. *J. Magn. Magn. Mater., 501*, 166455.
27. Yang, T., Lu, K., Wang, J., Xu, J., Zheng, Z., & Liu, X., (2022). Fe-6.5 wt%Si soft magnetic composites with significant improvement of magnetic properties by compositing nano-MnZn ferrites. *J. Alloys. Compd., 909*, 164660.
28. Xu, Z., Fan, J., Han, Y., Liu, T., Zhang, H., Song, K., & Zhang, C., (2019). Preparation and characterization of Mn–Zn ferrites via nano-in-situ composite method. *Solid. State. Sci., 98*, 106006.
29. Chang, J., Zhan, T., Peng, X., Li, J., Yang, Y., Xu, J., Hong, B., et al., (2021). Improved permeability and core loss of amorphous FeSiB /Ni-Zn ferrite soft magnetic composites prepared in an external magnetic field. *J. Alloys. Compd., 886*, 161335.
30. Almessiere, M. A., Slimani, Y., Trukhanov, A. V., Sadaqat, A., Demir, K. A., Algarou, N. A., Aydın, H., et al., (2021). Review on functional bi-component nanocomposites based on hard/soft ferrites: Structural, magnetic, electrical and microwave absorption properties. *Nano-Struct. Nano-Objects, 26*, 100728.
31. Wei, H., Yu, H., Feng, Y., Wang, Y., He, J., & Liu, Z., (2021). High permeability and low core loss nanocrystalline soft magnetic composites based on FeSiBNbCu@Fe_3O_4 powders prepared by HNO_3 oxidation. *Mater. Chem. Phys., 263*, 124427.
32. Luo, Z., Fan, X., Hu, W., Luo, F., Wang, J., Wu, Z., Liu, X., Li, G., & Li, Y., (2020). Formation mechanism and enhanced magnetic properties of Fe–Si/Fe_2SiO_4 soft magnetic composites transformed from Fe-6.5 wt%Si/α-Fe_2O_3 core-shell composites. *J. Alloys. Compd., 817*, 152803.

33. Zhang, Z., Chang, J., Peng, X., Li, J., Yang, Y., Xu, J., Hong, B., Jin, D., Jin, H., Wang, X., & Ge, H., (2022). Structural and magnetic properties of flaky FeSiB/Al_2O_3 soft magnetic composites with orientation of a magnetic field. *J. Mater. Res. Technol., 18*, 1381–1390.
34. Uysal, S., Mercan, M., & Uzun, L., (2020). Serial number restoration on polymer surfaces: A survey of recent literature. *Forensic Chemistry, 20*, 100267.
35. Guo, Y., Zhu, J. G., Lin, Z. W., Zhong, J. J., Lu, H. Y., & Wang, S., (2007). Determination of 3D magnetic reluctivity tensor of soft magnetic composite material. *J. Magn. Magn. Mater., 312*, 458–463.
36. Sunday, K. J., & Taheri, M. L., (2017). Soft magnetic composites: Recent advancements in the technology. *Metal Powder Rep., 72*, 425–429.
37. Zhao, X., Feng, S., & Lv, Q., (2022). Soft magnetic properties of FeSiCr cores in a transverse magnetic field. *J. Supercond. Nov. Magn.*, https: doi.org/10.1007/s10948-022-06157-0.
38. Farea, M. A., Mohammed, H. Y., Shirsat, S. M., Sayyad, P. W., Ingle, N. N., Al-Gahouari, T., Mahadik, M. M., et al., (2021). Hazardous gases sensors based on conducting polymer composites: Review. *Chem. Phys. Lett., 776*, 138703.
39. Charoensuk, T., Thongsamrit, W., Ruttanapun, C., Jantaratana, P., & Sirisathitkul, C., (2021). Loading effect of sol-gel derived barium hexaferrite on magnetic polymer composites. *Nanomaterials (Basel), 11*, 558.
40. Wu, W., He, Q., & Jiang, C., (2008). Magnetic iron oxide nanoparticles: Synthesis and surface functionalization strategies. *Nanoscale. Res. Lett., 3*, 397–415.
41. Suhasini, A., Vinod, K. K. P., & Maiyalagan, T., (2016). Synthesis, thermal and magnetic behavior of iron oxide-polymer nanocomposites. *Sci. Eng. Compos, 25*.
42. Luo, Z., Fan, X., Hu, W., Luo, F., Wang, J., Wu, Z., Liu, X., et al., (2020). High performance Fe-Si soft magnetic composites coated with novel insulating-magnetic-insulating (IMI) layer. *J. Magn. Magn. Mater., 496*, 165937.
43. Gao, Z., Jia, J., Zhao, Q., Kong, H., Wu, Z., & Li, J., (2022). Determination of a quantitative relationship between deposition duration and magnetic performance of soft ferromagnetic composites via data-analysis and theoretical models. *J. Magn. Magn. Mater., 549*, 168891.
44. Zhang, Y., Fan, X., Hu, W., Luo, Z., Yang, Z., Li, G., & Li, Y., (2020). Microstructure and magnetic properties of MnO_2 coated iron soft magnetic composites prepared by ball milling. *J Magn. Magn. Mater., 514*, 167295.
45. Hou, Z., Yan, P., Sun, B., Elshekh, H., & Yan, B., (2019). An excellent soft magnetic Fe/Fe_3O_4-FeSiAl composite with high permeability and low core loss. *Results in Physics, 14*, 102498.
46. Hu, X., Liang, Z., Li, Q., Hu, S., & Qu, S., (2022). Experimental study on the magnetic permeability of inclusion filled soft polymeric composite for soft-core transformer applications. *Polymer Testing, 106*, 107430.
47. Sun, H., Zhanga, L., Chen, Y., Chen, F., & Qu, X., (2019). Magnetic properties of iron-based soft magnetic composites prepared by utilizing polyimide insulating layer. *J Magn. Magn. Mater., 486*, 165287.
48. Evangelista, L. L., Ramos, F. A. I., Silva, B. S., Hammes, G., Binder, C., Klein, A. N., & Drago, V., (2021). Magnetic properties optimization of an iron-based soft magnetic composite coated by nano-ZnO and boron oxide. *J Magn. Magn. Mater., 539*, 168319.

49. Meng, B., Yang, B., Zhang, X., Zhou, B., Li, X., & Yu, R., (2020). Combinatorial surface coating and greatly improved soft magnetic performance of Fe/Fe_3O_4/resin composites. *Mater. Chem. Phys., 242*, 122478.
50. Qing, M., Chong, T., Jing, H., & Sun, B., (2021). Study on the soft magnetic properties of FeSiB/EP composites by direct ink writing. *Front. Mater, 8*, 792768.
51. Riesgo, G., Elbaile, L., Moriche, R., Carrizo, J., Crespo, R. D., García, M. A., Sepúlveda, R., et al., (2019). Influence of the remnant magnetization, size distribution and content of soft magnetic reinforcement in micro-mechanical behavior of polymer matrix composites. *Polymer Testing, 79*, 106020.
52. Wanga, F., Dong, Y., Chang, L., Pan, Y., Chi, Q., Gong, M., Li, J., et al., (2021). High performance of Fe-based soft magnetic composites coated with novel nano-$CaCO_3$/epoxy nanocomposites insulating layer. *J. Solid State Chem., 304*, 122634.
53. Palmero, E. M., Rial, J., De Vicente, J., Camarero, J., Skarman, B., Vidarsson, H., Larsson, P. O., & Bollero, A., (2018). Development of permanent magnet MnAlC/polymer composites and flexible filament for bonding and 3D-printing technologies. *Sci. Technol. Adv. Mater., 19*, 465.
54. Wang, Y., Shi, L., Jin, Y., Sun, S., Gao, P., Wei, Y., Wang, Z., & Jiang, Y., (2020). Surface-initiated polymerization for the preparation of magnetic polymer composites. *Polym. Chem., 11*, 1797–1805.
55. Hamciuc, C., Asandulesa, M., Hamciuc, E., Roman, T., Olariu, M. A., & Pui, A., (2021). Novel polyimide/copper-nickel ferrite composites with tunable magnetic and dielectric properties. *Polymers, 13*, 1646.
56. Kim, M. W., Bae, D. H., Kwon, S. H., & Choi, H. J., (2018). Magnetite embedded mini-emulsion polymerized polystyrene particles and their magnetorheology. *Macromol. Res., 26*, 353–358.
57. Petcharoen, K., & Sirivat, A., (2016). Magneto-electro-responsive material based on magnetite nanoparticles/polyurethane composites. *Mater. Sci. Eng. C., 61*, 312–323.
58. Bustamante-Torres, M., Romero-Fierro, D., Estrella-Nuñez, J., Arcentales-Vera, B., Chichande-Proaño, E., & Bucio, E., (2022). Polymeric composite of magnetite iron oxide nanoparticles and their application in biomedicine: A review. *Polymers, 14*, 752.
59. Hajra, S., Vivekananthan, V., Sahu, M., Khandelwal, G., Joseph, R. N. P. M., & Kim, S. J., (2021). Triboelectric nanogenerator using multiferroic materials: An approach for energy harvesting and self-powered magnetic field detection. *Nano Energy, 85*, 105964.
60. Nasrollahzadeh, M., Issaabadi, Z., & Varma, R. S., (2019). Magnetic lignosulfonate-supported Pd complex: Renewable resource-derived catalyst for aqueous Suzuki–Miyaura reaction. *ACS Omega, 4*, 14234–14241.
61. Xu, M., Zhang, Q., Tan, Q., Zhang, W., Sang, S., Yang, K., & Ge, Y., (2022). A magnetostrictive $BaTiO_3$-Fe-Ga & PDMS magnetic field sensor: Research on magnetic detection performance. *Sens. Actuator A Phys., 335*, 113383.
62. Said, M. M., Yunas, J., Pawinanto, R. E., Majlis, B. Y., & Bais, B., (2016). PDMS based electromagnetic actuator membrane with embedded magnetic particles in polymer composite. *Sens. Actuator A Phys., 245*, 85–96.
63. Hao, Z., Song, S., Li, Bo., Jia, Q. X., Zheng, T., & Zhang, Z., (2022). A solvent driven dual responsive actuator based on MOF/polymer composite. *Sens. Actuators B Chem., 358*, 131448.

64. Velhal, N., Patil, N. D., Kulkarni, G., Shinde, S. K., Valekar, N. J., Barshilia, H. C., & Puri, V., (2019). Electromagnetic shielding, magnetic and microwave absorbing properties of polypyrrole/$Ba_{0.6}Sr_{0.4}Fe_{12}O_{19}$ composite synthesized via in-situ polymerization technique. *J. Alloys. Compd., 777*, 627–663.
65. Liang, C., Gu, Z., Zhang, Y., Ma, Z., Qiu, H., & Gu, J., (2021). Structural design strategies of polymer matrix composites for electromagnetic interference shielding: A review. *Nano-Micro Lett., 13*, 181.
66. Guo, Q., & Zhang, X., (2021). A review of mechanochromic polymers and composites: From material design strategy to advanced electronics application. *Compos. B. Eng., 227*, 109434.
67. Liu, S., Zou, H., Qin, B., Huang, S., & Deng, L., (2021). Tailored magnetoelectric coupling in magnetically oriented polymer-based iron fiber composite. *J. Magn. Magn. Mater., 540*, 168408.
68. Bi, H., Ye, G., Sun, H., Ren, Z., Gu, T., & Xu, M., (2022). Mechanically robust, shape memory, self-healing and 3D printable thermoreversible cross-linked polymer composites toward conductive and biomimetic skin devices applications. *Addit. Manu., 49*, 102487.
69. Pigliaru, L., Rinaldi, M., Ciccacci, L., Norman, A., Rohr, T., Ghidini, T., & Nanni, F., (2020). 3D printing of high-performance polymer-bonded PEEK-NdFeB magnetic composite materials. *Funct. Compos. Mater., 1*, 4.
70. Martinez-Moreno, M., Fuentes-Ramírez, R., Contreras-Lopez, D., & Galindo-Gonzalez, R., (2021). Study of the mechanical properties of magnetic composites supported on a PU matrix. *ECORFAN Journal-Taiwan, 5*, 8–13.
71. Lu, Q., Choi, K., Nam, J. D., & Choi, H. J., (2021). Magnetic polymer composite particles: Design and magnetorheology. *Polymers, 13*, 512.
72. Qiu, Y., & Gao, L., (2005). Novel polyaniline/titanium nitride nanocomposite: Controllable structures and electrical/electrochemical properties. *J. Phys. Chem. B, 109*, 19732–19740.
73. Henaish, A. M. A., Salem, B. I., Meaz, T. M., Alibwaini, Y. A., Ajlouni, A. W., Hemeda, O. M., & Arrasheed, E. A., (2021). Synthesize, characterization, dielectric, linear and nonlinear optical properties of Ni–Al ferrite/PANI nanocomposite film. *Opt. Mater., 119*, 111397.
74. Jia, L. C., Xu, L., Ren, F., Ren, P. G., Yan, D. X., & Li, Z. M., (2019). Stretchable and durable conductive fabric for ultrahigh performance electromagnetic interference shielding. *Carbon, 144*, 101–108.
75. Mathews, S. A., & Rajan, B. D., (2021). Analysis of the role of M-type hexaferrite-based materials in electromagnetic interference shielding. *Curr. Appl. Phys., 29*, 39–53.
76. Radoń, A., Łukowiec, D., & Włodarczyk, P., (2020). Microwave absorption by dextrin-magnetite nanocomposite in frequencies below 2.5 GHz: Role of magnetite content, shape and temperature on magneto-dielectric properties. *Mater. Des., 193*, 108860.
77. Shu, X., Zhou, J., Lian, W., Jiang, Y., Wang, Y., Shu, R., Liu, Y., Han, J., & Zhuang, Y., (2021). Size-morphology control, surface reaction mechanism and excellent electromagnetic wave absorption characteristics of Fe_3O_4 hollow spheres. *J. Alloys. Compd., 854*, 157087.
78. Ahmed, A., Jan, R., Khan, A. N., Gul, I. H., Khan, R., Javed, S., Akram, M. A., et al., (2020). Graphene-ferrites interaction for enhanced EMI shielding effectiveness of hybrid polymer composites. *Mater. Res. Express, 7*(1).

79. Bhatti, M. A., Siddiqa, A., Azam, S., Vu, T. V., Islam, M. U., Rani, M., & Muhammad, S., (2020). Magnetic and electrical properties of $Ba_2Co_2Fe_{12}O_{22}$/PANI composites prepared by in-situ polymerization. *Phys. B. Condens. Matter, 597*, 412410.
80. Ding, J., & Cheng, L., (2021). Core-shell Fe_3O_4@SiO_2@PANI composite: Preparation, characterization, and applications in microwave absorption. *J. Alloys. Compd., 881*, 160574.
81. Wang, Z., Chen, Z., Xie, L., Hou, X., & Fang, C., (2021). Flexible and lightweight $Ti_3C_2T_x$ MXene/Fe_3O_4@PANI composite films for high-performance electromagnetic interference shielding. *Ceram. Int., 47*, 5747–5757.
82. Movassagh-Alanagh, F., Bordar-Khiabani, A., & Ahangari-Asl, A., (2017). Three-phase PANI@nano-Fe_3O_4@CFs heterostructure: Fabrication, characterization and investigation of microwave absorption and EMI shielding of PANI@nano-Fe_3O_4@CFs/epoxy hybrid composite. *Compos. Sci. Technol., 150*, 65–78.
83. Guo, Y., Qiu, H., Ruan, K., Zhang, Y., & Gu, J., (2022). Hierarchically multifunctional polyimide composite films with strongly enhanced thermal conductivity. *Nano-Micro Letters, 14*, 26.
84. Koner, S., Deshmukha, P., Khana, A. A., Ahlawat, A., Karnal, A. K., & Satapathy, S., (2020). Direction dependent strong magnetoelectric coupling in $La_{0.7}Ba_{0.3}MnO_3$ embedded poly(vinylidene fluoride-co-trifluoroethylene) flexible nano-composite films at room temperature. *Scripta Materialia, 189*, 30–35.
85. Ahmed, I., Khan, A. N., Jan, R., & Gul, H., (2022). Structure – properties relationships of graphene and spinel nickel ferrites-based poly(vinylidene fluoride) hybrid polymer nanocomposites for improved dielectric and EMI shielding characteristics. *Mater. Res. Bull., 148*, 11168.
86. Khodiri, A. A., Al-Ashry, M. Y., & El-Shamy, A. G., (2020). Novel hybrid nanocomposites based on polyvinyl alcohol/graphene/magnetite nanoparticles for high electromagnetic shielding performance. *J. Alloys. Compd., 847*, 156430.
87. Nica, S. L., Nica, V., Grigoras, V. C., Varganici, C. D., Popovici, D., Hulubei, C., & Ioan, S., (2015). Influence of two structural phases of $Fe_3O_4$and γ-Fe_2O_3 on the properties of polyimide/iron oxide composites. *Polym. Int., 64*, 1172.
88. Toshihiko, M., (1999). Nonaromatic polyimides derived from cycloaliphatic monomers. *Macromolecules, 32*(15), 4933–4939.
89. Ramazanov, M. A., Shirinova, H. A., & Hajiyeva, F. V., (2020). The impact of manufacturing technology, of the polypropylene matrix and iron oxide nanoparticles based magnetodielectric nanocomposites on electrophysical parameters. *Mater. Chem. Phys., 253*, 123287.
90. Eom, K., Shin, Y. E., Kim, J. K., Joo, S. H., Kim, K., Kwak, S. K., Ko, H., et al., (2020). Tailored poly(vinylidene fluoride-co-tri-fluoroethylene) crystal orientation for a triboelectric nanogenerator through epitaxial growth on a chitin nanofiber film. *Nano. Lett, 20*, 6651–6659.
91. Kumar, A., Patel, P. K., Yadav, K. L., Singh, Y., Kumar, N., Abdullah, Sharma, G., & Singh, S., (2022). Enhanced magnetoelectric coupling response in hot pressed $BiFeO_3$ and polymer composite films: Effect of magnetic field on grain boundary and grain resistance. *Mater. Res. Bull., 145*, 111527.
92. Charles, A. D. M., Rider, A. N., Brown, S. A., & Wang, C. H., (2020). Improving the actuation performance of magneto-polymer composites by silane functionalization of carbonyl-iron particles. *Composites Part B., 196*, 108091.

Index

A

Acetone, 152, 160, 161, 262
Acetyl acetone (AA), 78, 83, 119
Acicular filler forms, 181
Active
  nanosystems, 145
  targeting, 214
Actuator, 193, 292
Adenosine tri phosphorus acid, 116, 117, 119
Adhesive, 114, 146, 152
  characteristics, 114, 132
  durability, 143–145
  peel tests, 154
  properties, 139, 141
Adsorption
  dyes, 278
  layers, 96
Aerosol, 233, 239
  vapor-phase synthesis techniques, 232
Agglomerates, 152, 155, 169
Agglomeration, 200, 242, 248
  tendency, 242
Aggressive organic media, 152
Agricultural production, 120
Air velocity, 256
Akaganeite, 197
Alcoholic solution, 239
Aliphatic compounds, 45
Alkyl sulfates, 64
Alkyl sulfonates, 64
Allocarbonic nanostructures, 74
Alternating
  current (AC), 182, 251, 281
  magnetic field, 184, 195–198, 212, 241
Aluminum, 2, 4, 143, 145
  oxide, 4
  protection, 143
Ambient temperature, 193, 267
Amine surfactants, 210
Ammonium
  polyphosphate (APPh), 2–4, 51, 52, 56, 57, 104, 107, 116, 117, 140–143, 145
  sulfite, 51, 52
Amorphous carbon, 75, 76, 87, 250
Amphiphilic, 277
  diblock copolymer, 181
Anisotropic
  character, 300
  nature, 198, 199
  physical properties, 272
Annihilation, 1–5, 8, 60, 98, 107, 118, 146, 224
Anti-cancer drug, 214, 277
Anti-corrosion
  coating, 143, 145
  layer, 71
  preservative
    layer, 63
    lubricants, 65
  properties, 64
Antidotal qualities, 206
Antiferromagnetic, 171, 173
  materials, 194
  spin orientation, 173
  substances, 173
Anti-microbes
  actives, 250
  properties, 116
Anti-parallel orientation, 173
Application
  magnetic
    field, 173, 193, 194, 198–202
    resonance, 234
Atomic
  force microscopy (AFM), 7, 8, 25–28, 36, 49, 52, 135, 145, 151, 155, 163
    microscopy, 151, 163
  layer deposition method, 274
  magnetic moment, 4, 5, 9, 13–15, 17–20, 41, 52, 55, 60, 73, 74, 80, 82, 86, 87, 89–92, 111, 116–118, 146, 172, 173, 195
    d-metals, 92
  positive charges quantization, 5
Atomization, 250

Autoclaves, 233, 235
Automotive industry, 202
Avrami-Kolmogorov equation, 29, 133

B

Bacterial cellulose (BC), 277
Ball milling, 293
  technique, 297
Bandwidth shielding materials, 296
Barium titanate, 170
Bead
  like structures, 259, 263
  production, 259
Bending
  instability, 250, 254
  strength, 162
Benzene-ethanol mixture, 86
Bimodal distribution, 256
Binary metallic nano-alloys, 210
Binding interactions, 208
Bio-assisted synthesis, 233
Biocompatibility, 196, 197, 205, 215, 218, 232–234, 241, 277
  matrix, 183
Biodegradability, 233, 277
Biological
  active molecules, 209
  applications, 209, 247
  fluids, 213
  properties, 248
  separation, 276
  stability, 129
  tissues, 193
Biomedical, 205, 292
  applications, 196, 205, 210, 232, 239, 273, 276
  diagnostic research, 206
  interventions, 217
  MNPs, 229
  system, 292
Biomedicine, 239
Biometric parameters, 122
Biomimetic materials, 193
Biosensing, 206, 229, 241, 243
Bio-separation, 168
Biotherapeutic interventions, 206
Bis(2-ethylhexyl)sulfosuccinate, 237
Boron oxide, 294
Brain tumor treatment, 196
Brownian
  motion, 242
  relaxation, 184, 198
Butyl ethers, 64

C

Calcination process, 276
Cancer
  cell susceptibility, 211
  curing, 168
  drug resistance, 208
  treatment, 195, 196, 209, 217
Capillary reactor, 238
Carbine fragments, 4, 52, 111, 145, 224
Carbon
  atoms, 14, 19, 20, 54, 75, 76, 79–82, 89
  fibers formation, 224
  metal-containing
    nanoform investigation, 91
    nanoforms, 87
    nanostructures, 86, 89–91
    nanotubes, 92
  nanofibers (CNFs), 273, 277
  nanosystems, 74
  nanotubes (CNTs), 76, 86, 87, 250
    fibers, 153
  NPs, 230
  shell activity, 146
Carbonization, 36, 75, 99
Carbyne fragments, 65
Carcass-like structures, 151
Catalysts, 35, 83, 111, 146
Catheterization, 217
Cell
  culture medium, 210
  membrane translocation efficiency, 214
  retention, 217
  sensitivity, 211
  signaling, 207
  therapies, 206, 217
  transplantation, 217
  uptake, 216
Ceramic, 139, 170
  glass synthesis, 171
Cetyltrimethylammonium bromide, 237
Chain entanglement, 259, 270, 281
Charge
  density, 255

quantization, 6, 8, 9, 23, 24, 96, 126, 132, 133, 146
Chelated lanthanide, 215
Chemical
bonds formation, 1, 41, 54, 95, 137
characteristics, 139
composition, 66, 168, 229
control system, 277
cross-linkable temperature-responsive polymers, 277
diffusivity, 193
engineering, 168
instability, 209
mesoscopic, 1, 2, 8, 53, 95, 96, 104, 109, 133
characteristics, 95
modification, 20, 154
sensing, 199
stability, 232
structure (nanoforms), 86, 87, 92
transformations, 64
vapor deposition, 232, 233, 294
Chemotherapeutic agents, 212, 241
Chitosan (CS), 183, 184, 213
agarose (CS-AG), 183, 184
matrix solution, 184
Chromium, 173, 230
Citrate gel process, 232
Clinical
interventions, 216, 218
treatment system, 196
Coating adhesive durability, 143
Cobalt, 26, 27, 78, 92, 169, 181, 191, 207, 210, 230
Coherence disturbance, 6, 7
Colloidal
metal nanostructures, 199
particles, 213
stability, 181, 207
Commercial
applications, 280
electrostatic electron spectrometers, 15
polymers, 169
Complex heterostructures, 298
Composites
resistance, 152, 158, 162
thermo-hardening, 160
Compound display retentivity, 175
Compression
strength, 127, 152, 154, 159, 160, 162, 163
tests, 154
Conductivity, 121, 131, 132, 136, 153, 182, 250, 254, 255, 259, 262, 299
Conformational
characteristics, 175
modifications, 175
properties, 169, 175
Conjugated
adenoviral particles, 216
polyethylenimine-coated MNPs, 216
Continuum mechanics, 200
Conventional
atomizers, 248
biology, 206
facile method, 235
methods, 87, 248
thermal treatment, 276
Copolymer molecular mass, 181
Copper, 2, 3, 25, 39, 45, 48, 52, 55, 60, 63, 97, 100, 101, 103–107, 116, 122
atomic magnetic moment, 3, 18, 52, 55, 106, 117, 120
carbon
MC suspension ultrasound treatment, 6
mesoscopic composite structure, 99
mesoscopic composite, 116, 117, 145
nanocomposite, 46, 54, 65, 67, 68, 70, 71, 105, 134
clusters, 32, 65
containing nanostructures, 92
Co-precipitation, 171, 232–234, 237
method, 234, 235
technique, 232
Core-shell
composite ultrafine fiber, 250
inorganic composites, 171
Corrosion
inhibitor, 112, 137, 146
meter, 66, 70, 71
sensor, 66
processes, 111
surface environments, 64
Coulombic
interaction, 252
repulsive force, 264

Crack-resistant paints, 230
Critical voltage, 264
Crystal
anisotropy, 172
canals, 96
nickel, 82, 90
solvates, 96
structures, 6, 126
Crystallization, 29, 48, 232, 268
Cuboidal geometry, 297
Curie temperature, 173, 231
Cutting-edge technology, 206
Cyclic butylene terephthalate (CBT), 270
Cylinder-shaped samples, 154
Cytostatic drugs, 212
Cytotoxicity, 210

D

Decantation, 242
Decolorization efficiency, 279
Degradation
efficiency, 279
temperatures, 183
Degree of swelling, 162
Dendritic
polymer-carriers, 217
structures, 155
Density
compositions, 130
magnetic recording, 276
Detailed thermal analysis, 267
Dextran, 196, 260, 297
DI-1 bending testing machine, 154
Diamagnetic, 171, 172, 194
compounds, 171
Dielectric
characteristics, 40
constant, 126, 132, 255
liquid, 201
material, 262
properties, 276, 297, 299, 300
strength, 259
Diffusion control system, 277
Digital magnetic data storage media, 242
Dilute magnetic semiconductors (DMS), 169, 170, 185
Dimensional stability, 167
Dimethylformamide (DMF), 250, 281
Dipole-dipole interactions, 200
Direct
current (DC), 251, 281
voltage, 251
ink writing (DIW), 295
Disease treatment, 207
D-metal
compounds, 75, 89
containing carbon nanostructures, 75
Double electrical layers, 96
Doxorubicin, 214
Drug
carrier complex, 214
delivery, 206, 208, 209, 213–215, 217, 218, 229, 232, 239, 243, 247, 273
applications, 243
system (DDS), 206, 209, 217, 218, 247, 273
release carriers, 168

E

Eddy current, 196, 198, 293–295
Effective
adjunctive cancer therapy, 211, 241
nanoscale materials, 85
Elastic
deformation, 200
modulus, 152, 160, 183, 184, 269, 270
Elastomeric matrix, 183
Electric
conductive, 47, 113, 114, 120, 121, 262, 263, 298
composites, 153
current, 250
devices industry, 292
field strength, 248, 250, 252, 261
insulating layer materials, 292
machines, 168
resistivity, 198, 300
Electrochemical
deposition, 92
potential, 33
properties, 193
reactions, 233
synthesis, 232
Electrochromic coatings, 193
Electroconductivity, 114
Electrohydrodynamic cone jet, 254

Electromagnetic (EM), 3, 4, 8, 40, 42, 60, 107, 118, 126, 129, 146, 167, 168, 182, 196, 203, 224, 225, 247, 273, 278, 291, 296–301
  interface (EMI), 291, 296–299, 301
    shielding applications, 297–299
  interference, 291, 296, 301
  radiation, 107, 126, 225
Electron
  acceptors, 13, 20
  bombardment, 74
  containing compounds, 103
  density transfer carrier, 67
  diffraction, 36, 65
  magnetic field, 5
  micro-diffraction, 35, 36, 99, 117
  paramagnetic resonance (EPR), 47, 48, 52, 98, 99, 102, 103, 107, 117, 118, 182, 214
  spectrum, 76
  spin resonance, 98
  structure (nanosystems), 75
  transport (high energetic levels), 109
  wavelength, 24
Electronic
  actuators, 278
  application, 296, 299
  devices, 291, 296, 298, 299
  exchange forces, 173
  market, 296
  structure, 86, 92
Electrophoretic deposition (EPD), 298
Electrorheological, 193, 201
Electrospinning, 247–253, 255, 256, 259–264, 266–268, 270, 272–276, 280, 281
  chamber, 256
  mechanical process, 250
  method, 274–276
  process, 249, 251, 255, 259–264, 266–268
  technique, 247, 248, 250, 251, 253, 274, 275, 280, 281
Electrospraying technique, 247, 248, 253
Electrospun
  cellulose nanofiber, 269
  fiber, 249, 250, 256, 261, 266, 267
  nanofiber, 247, 252, 254, 260, 262, 264–272
    composite, 272
  nylon-6 nanofiber, 270
  PEO nanofibers, 263, 268
  Pi fibers, 298
  PLA nanofibers, 260
  PLLA nanofiber, 267
  TiO2 nanofibers, 268
Electrostatic
  attraction, 214
  charge, 136
  disruption, 250
  field, 254, 256
  force, 249, 252, 262, 273
  interactions, 208, 216
  repulsion, 249, 252
    force, 264
  spectrometers, 15, 86
  spinning, 249
  X-ray photoelectron spectrometers, 75
Endothelial cells, 184
Energy
  consumption, 15
  interaction, 26
  parameters, 30, 34
  related applications, 182
  storage harvesting, 292
Engine mounts, 202
Engineering
  applications, 182, 203, 294, 296, 301
  industries, 152
  product, 167
Enhanced
  apoptosis of cancer cells, 214
  heating ability, 212, 241
  permeability, 214, 218
Environmental purification, 233
Enzymatic
  degradation, 209
  immobilization, 276
Epoxy
  alkyl diluent, 154
  composites, 152, 162
  compositions, 135, 136, 158
  compounds, 135
  magnetite composites, 153
  resins (ERs), 6, 114, 133, 142, 143, 145, 152, 153, 158, 163, 193, 200, 293, 295
Ethyl acetate, 152, 161
Ethylene glycol, 202, 236

Excitation
longitudinal plasmon, 199
transverse plasmon, 199
Expanded graphite, 299
External
electric field, 249, 252, 254
magnetic field, 172, 199, 201, 202, 213, 216, 217, 230, 231, 240, 273
stimuli, 193
Extinction spectrum, 199

F

Fabricated
catalytic magnetic nanocomposite fibers, 279
core-shell, 280
electrospun nanofiber, 262
magnetic composite, 296
nanofibers, 248, 252, 254, 260, 273
polyurethane-Fe3O4 composite, 295
soft magnetic composite materials, 294
Female Wistar Kyoto rats, 217
Feroxyhyte, 197
Ferric red pigment, 152
Ferrimagnetic, 171, 173, 193, 194, 242
materials, 173, 194
Ferrite lattice, 276
Ferroelectric, 300
Ferrofluids, 229, 232, 240, 242, 243
Ferromagnetic, 170, 172, 173, 181, 182, 184, 192–194, 198, 201, 242, 276, 292, 293
dielectric composites, 182
domains, 194
magnetic compounds, 173
material, 192, 194, 276
properties, 170, 173, 194, 198
Ferromagnetism, 171–173, 194, 247, 273
Ferromagnets, 172, 173, 194
Ferroxides, 151, 153, 160
Fiber
axial directions, 300
containing micropores, 267
diameter, 99, 250, 256, 262, 264–272, 281
formation, 99, 256, 261, 264, 266, 267
nanofibers, 270
solidification, 268
structure, 45, 259, 270
Filler-polymer complex, 181
Fire
proof
efficiency, 143, 145
intumescent coatings, 145
systems, 146
resistant
composition, 143
intumescent coatings, 139
materials, 139
protection, 143
retardants, 112, 146
Flammability
determination, 140
test, 139, 141
Flexible magnetic membranes, 277
Flight coulombic fission, 252
Flow injection, 232, 233, 238
Fluid jet surface, 267
Fluoroalkyl silane (FAS), 277
Fractal theory, 2, 23, 24, 28, 29
Free chlorotoxin, 216
Fullerene, 76, 81
polymer composites, 182
Functional
magnetic fillers, 169
polymer, 25

G

Gadolinium, 191, 230
Garnet-type ferrite, 276
Gene-delivery system, 215
Geometric
characteristics, 25, 97
nonmagnetic gaps, 293
Glass transition temperature, 182, 268
Global ferromagnetic order, 170
Glue composition, 139–141
Glycothermal synthesis, 232
Goethite, 197
Gold, 169, 199, 210, 230, 280
Gradient magnetic field, 275
Graft retention, 217
Graphene, 78, 297, 299
oxide, 299
expanded graphite (GO-EG), 299
Graphite
protective coating, 211
spectrum, 76
Graphitization, 77, 78
Grounded conducting collector, 248, 251

H

Half-dimensional additives, 155
Hard magnetic
composite (HMCs), 292, 301
hexaferrite fillers, 182
Heat
conducting, 152
crystalline melting, 268
physical
characteristics, 130, 132
properties, 130
resistance capacity, 195
treatment, 152, 159–161, 237
Hematite, 169
Heterogeneous distribution, 197
Hierarchically multifunctional structure, 298
High end machinery, 216
High frequency
electronics, 232
operation, 294
High magnetic
inductions, 295
susceptibility, 207, 231
High molecular weight
poly-L-lactic acid (HM-PLLA), 259
polymers, 259
High performance permanent magnets, 232
High resolution transmission electron microscopy, 35, 36
High saturation magnetic induction, 293
High spatial resolution, 214
High specific surface area, 271, 275, 277
High temperature thermal decomposition, 232
Homogeneity
magnetic fields, 184
temperature, 241
Homopolymer, 300
Hot
pressing method, 301
vulcanization glue, 136
Human
CD133+ endothelial progenitor cells, 217
machine interaction, 292
melanoma cells, 277
serum albumin, 214
Humidity sensitivity, 300
Hybrid
chemical bond, 74, 79, 80, 82, 92
nanostructures, 199
organic-inorganic magnetic composites, 171
Hydrocarbon
decomposition, 77
shells, 35
Hydrogen
bonds, 214
densities, 215
Hydrophilic polymers, 217
Hydrophobic, 179, 180
behavior, 129
character, 179
degree, 179
interaction, 208
Hydrostatic pressure, 200
Hydrothermal, 232, 233, 235, 236, 244, 297
process, 232, 235, 236
synthesis, 232, 233
Hydroxide coprecipitation, 232
Hydroxylation, 237
HyperChem software products, 26
Hyperthermia, 168, 194–197, 207–209, 211, 212, 214, 215, 218, 241, 244
applications, 195, 197
induced apoptosis, 214
measurements, 197
therapy applications, 194
treatment, 168, 196, 197
Hypsochromic shift, 268
Hysteresis loops, 194

I

Imaging-guided photochemical therapy, 239
Immunogenicity, 205, 213
In vivo temperature control, 197
In-depth scientific investigations, 196
Induction
heating efficiency, 197
unpropitious reactions, 209
Industrial
epoxy materials adhesion, 114
wastewater treatment, 278
Inflammability, 129
Inhibitor protection efficacy, 137
Inhomogeneities, 158, 163
Injection flux rates, 234
Innate molecular configuration, 206

Innumerable preparation techniques, 213
Inorganic
compounds, 233, 280
materials, 208, 293
organic polymeric materials, 146
polymerization, 237
In-situ polymerization method, 297
Insulating-magnetic-insulating (IMI), 293
Insulin-like growth factor receptor (IGF-1R), 216
Interatomic interaction, 65, 71, 74, 79, 82, 92
Interchain connectivity, 259
Interference, 1, 8, 9, 23, 24, 96, 118, 132, 146, 182
annihilation, 9
Inter-molecular interactions, 259
Inter-particle connections, 196
Inter-phase interactions, 153, 168
compatibility, 168
Interventional modality, 206
Intravenous injection, 213
Intrinsic viscosity, 268
Intumescent fire
proof materials, 142
resistant glues, 139
Invariable exothermal effect, 28
Ion milling, 233
IR spectroscopy, 28, 35, 36, 43, 51, 57, 58, 60
Iron, 19, 20, 26, 52, 63, 64, 66, 68, 71, 79, 89, 116, 117, 120, 121, 152, 153, 155, 159, 160, 162, 163, 170, 183, 184, 191, 197, 199, 202, 207, 209–212, 215, 217, 230, 231, 233–235, 238, 239, 241, 276, 280, 294–296, 300, 301
amorphous alloy, 293
containing
epoxides, 160
epoxy composites, 152
nanostructures, 89
oxide, 19, 20, 64, 79, 89, 153, 155, 159, 160, 163, 183, 184, 197, 199, 207, 215, 217, 230, 231, 233–235, 238, 241, 296, 300
minerals, 153
pentacarbonyl, 239
phosphide layer, 68
precursor, 239
Isomethyl tetra
hydro phthalates anhydrate, 135
phthalic anhydrate, 114
Isotropic
spherical nanostructures, 198
stable system, 237

J

Jet velocity, 265

K

Kapton
composites, 177
polyimide, 175
Kolmogorov-Avrami equations, 30, 32, 49, 97

L

Laminar flow regime, 238
Laser
pyrolysis, 232, 239
Spectrometer Horiba Lab Ram HR 800, 44
wavelengths, 211
Layer by layer (LBL), 298
Lead adsorption, 278
Leiurus quinquestriatus, 216
Lilies oriental hybrids, 120
Linearly viscoelastic solid, 201
Lipophobic surface, 278
Lithographical patterning, 233
Load-compressive deformation, 160
Local
magnetic field, 240
perfusion variations, 211, 241
Longitudinal plasmonic excitation, 199
Long-time dispersion stability, 292
Louis Shopper press machine, 154
Low-energy synthesis, 83

M

Macromolecular, 6, 42, 75, 96, 108, 127, 133, 169
fragment, 126
phase, 181
self-organization, 6
Macrophages, 209

Macroscopic
systems, 96
tunneling effect, 247, 273
Maghemite, 169, 230, 232, 233, 237, 239, 244
Magnectofection, 216
Magnesium, 120
Magnetic
active nanomaterials, 199
actuation, 183
assessments, 170
behavior, 167, 169, 171, 172, 174, 183, 185, 234
characteristics, 1, 2, 55, 60, 85, 99, 106, 111, 112, 146, 167, 182, 207, 223–225, 231
growth, 60, 106, 223, 225
composite, 167–169, 181–183, 185, 301
preparation, 301
compounds broad spectrum, 168
dendrimers, 206
dilution model, 181
dipolar, 172, 199
interactions, 197
driven technology, 216
drying, 242
electron analyzer, 65
field, 116, 117, 172, 173, 175, 181, 183–185, 191–194, 196, 198–203, 207, 211, 214, 215, 230, 231, 239, 240, 242, 274, 275, 300
strength, 184, 185
filler, 175, 177, 179, 182, 184, 185, 202, 292, 300
fluid hyperthermia applications, 194
flux
density, 174
lines, 202
force, 207
glass carbon electrode (MGCE), 277
hard materials, 175
hydrogels, 206
hyperthermia, 184, 195–198, 203, 206, 212, 215
application, 195, 196, 198
assisted apoptosis, 206
treatment (MHT), 184, 196, 215
interaction, 199, 202
iron oxide nanoparticles (MIONs), 209, 232
liposomes, 206
materials, 92, 167–169, 175, 185, 192, 194, 196, 198, 203, 215, 243, 276, 296
matrix, 168, 201
medicine transport, 111, 112
mesoparticle, 107, 112, 116, 117
metal oxide NPs, 197
moments, 3, 17, 41, 103, 215
nano-carriers, 206
nano-clusters, 209, 217
nano-fibers (MNFs), 247, 272–276, 278, 279, 281
nanofluid hyperthermia, 195
nanomaterials, 195–197, 277
nanoparticles (MNPs), 155, 192, 195–198, 205, 206–209, 211, 215–218, 229–244, 273, 274, 277
endocytosis, 217
nanostructures, 200, 273
oscillation, 240
particle, 167, 168, 175, 180–182, 196–198, 202, 205, 206, 211–215, 241, 297
imaging (MPI), 212
performance, 168, 181, 293, 294
permeability, 293, 294
parameter, 294
photonic particles (MPPs), 212
polymer composites, 169, 182, 185, 291, 292, 295, 296, 299–301
potentiality, 192
properties, 5, 20, 91, 112, 120, 125, 167, 169–171, 174, 181, 191, 192, 194, 197, 198, 207, 210, 212, 223, 225, 231, 232, 241, 276, 281, 292–296, 298–300
nanomaterials, 194, 197
recording, 181, 182, 185, 242
refrigerants, 232
registering applications, 181
reinforcements, 169
remanence, 300
resonance imaging (MRI), 196, 205, 206, 208, 209, 211, 212, 214, 215, 229, 232, 239–241, 244, 276
response, 200, 277, 278
BC membranes, 277
saturation, 298
segregation, 206

sensitive
epoxy-composites, 163
polymer nanocomposites, 152
separation, 198, 232
soft materials, 175
spectrometer, 15, 65, 74, 86, 117
stimuli-responsive materials, 292
targeting (paclitaxel), 214
transport, 146
Magnetite, 151–156, 158–161, 163, 169, 183, 196, 230, 232, 233, 244, 276, 299
agglomerates, 155
Magnetization, 171–173, 175, 184, 192, 194, 197, 198, 200, 212, 240, 242
direction, 192, 242
orientation, 242
process, 192
Magneto
crystalline anisotropy, 181
electric (ME), 35, 54, 115, 120, 132, 137, 223, 224, 299, 300
fection, 216
graphic printing, 278
impedance, 182
effect, 182
liposomes, 208, 217
metric measurements, 277
resistive devices, 300
rheological (MR), 168, 185, 193, 200–202, 211
fluids, 203
sensitive smart material, 201
strictive materials, 193
structural properties, 197
Malleability, 247, 251
Manganese, 77, 121, 230, 231
Massart method, 234
Mechanical
characteristics, 47, 129, 142
chemical
interaction, 95
method, 116
force, 207
mortars, 223
performance, 184
properties, 191, 267, 269, 273, 296, 298, 299
strength, 167
Mechanochemical
method, 52, 65, 107
modification, 3, 51, 52, 60, 107
process, 54, 223–225
Medical engineering, 276
Medicine
magnetic transport remedies, 146
remedies creation, 115
Medium magnetic field sensor, 242
Melamine formaldehyde
lacquer, 143, 145
resin, 142
Melt
mixing, 171
point, 268
temperature, 250
Membrane filtration techniques, 279
Merozoite surface protein ($MSP1_{19}$), 216
Mesocomposite (MC), 1–9, 14, 15, 17–20, 23, 26, 28, 29, 34–36, 38–48, 51–60, 97, 99, 101, 102, 104–108, 111–131, 133–146, 224
cluster metals, 146
reactivity, 9
vibration energy, 42
Mesogranule surface, 59
Mesoparticle, 1–3, 5, 6, 23, 29, 30, 36, 39, 40, 45–47, 112, 116, 119, 127, 132, 133
existence, 133
formation, 127
process, 29
magnetic-electric properties, 5
participation, 1
structure, 132
surface energy, 127
Mesoreactors, 23–26, 28–33, 35, 48, 49, 53, 97
polymeric matrices, 23–26
Mesoscopic
composites, 95, 98–100, 103, 107, 108, 111, 112, 120, 137, 146
granules, 106
materials, 20, 111
metal-carbon composites, 112
metallurgy, 33
one-dimensional nanoreactors, 96
particles, 2, 24, 35, 103, 117, 125, 132, 145

physics principles, 24, 48
proposition, 96
reactor, 24, 96, 97
system (ms), 96, 108, 142
two-dimensional nanoreactors, 96
Metal, 41, 85, 169
alloy NPs, 230
atomic magnetic moment, 2–5, 17, 18, 20, 41, 95, 98, 103, 109, 112, 117
carbon
mesoscopic composite (Me-C MC), 1, 2, 5, 7, 8, 23–25, 28, 29, 33, 35–43, 47–49, 52, 54, 57, 59, 60, 95, 109, 111, 112, 114, 115, 120, 125, 127–129, 133, 135–137, 139, 140, 145, 146, 223, 224
nanocomposites, 5, 28, 35, 36, 38–43, 52, 53, 55, 60, 103, 107
nanostructures, 76, 79–82, 128, 135
nanosystems, 83
production, 95
carbonyl precursors, 233
carbonyls, 233
chemically active particles, 36, 39
cluster
atomic magnetic moments, 2
formation, 224
containing phase clusters, 40
magnetic
atomic moment, 107
moments, 48
nanostructures, 33, 34, 198
NPs, 230, 232, 234
oleates, 233
orientation, 28
oxidation state, 98
oxide, 13, 169
clusters, 29, 34, 35
dominated wet gel, 237
reduction mesoscopic products, 33
salt sol, 274
Metallocarbonic
nanoforms, 74
nanostructures, 75
structures, 74, 83
Metamagnetism, 171, 173
Microbial
methods, 233
synthesis, 232
Microelements, 122
Microemulsion, 232, 233, 237
method, 232, 237
Microfabrication processes, 233
Micro-islands, 155, 163
MicroRNA (miRNA), 214
Microscopic, 39, 76, 106, 117, 151, 153, 163
manipulation, 205
morphology, 163
Micro-tumor-like, 212
Microwave
absorption properties, 291, 297
reflux, 171
tunable features, 182
Microwire-doped system, 182
Mineral
fertilizers, 121
oil lubricant containing nitrated mineral oil, 64
Miniaturization, 192
Mixed-up polymer molecules, 261
Modulus of
compression elasticity, 159
elasticity, 152, 161, 162
Molecular
imaging, 206, 218, 234
modeling, 167, 169, 175, 179, 185
experiments, 169, 179
studies, 167
orientation, 270
precursors, 237
weight, 259, 260
Monicor corrosion meter, 64, 70
Mono-dispersity, 233, 243
Multi-dimensional fields, 168
Multi-domain magnetic particles, 198
Multi-functional
composites, 168
fibrous materials, 280
Multi-graded thermal processing, 139
Multilayer
absorption, 301
carbon metal-containing nanotubes, 87
Multiplet splitting, 14, 17, 19, 79, 80, 82, 87, 89–91
maxima, 80
Myriad surface effects, 192

N

Nanocapsules, 217
Nanocomposite, 13–15, 20, 28, 29, 34–36, 39, 41–45, 47, 48, 51, 55, 57, 60, 63, 64, 68, 70, 87, 90, 92, 182, 206, 224, 225, 295, 299
  elements, 64
  granules, 106
  nanogranyl, 5
  pillars, 183
  production, 51
  solubility, 14
  structure changes, 13
Nanocrystalline powder, 301
Nanoemulsions, 217
Nanofiber, 39, 247
  solution, 261
Nano-films, 31
Nanoform, 76, 78, 80, 82, 85, 86, 89, 90, 92
  chemical structure, 85
  preparation, 86
  materials, 85
Nanogranule carbon shell, 5
Nanomagnetism, 191, 192, 203
Nanomagnetites, 153
Nanomaterials, 230
Nanomedicine, 205, 206
  biosensing, 205
Nanometric particles, 237
Nano-microdispersed Yttrium Ferrite, 154
Nano-particle (NPs), 33, 41, 45, 63, 64, 68, 74, 76, 86, 152, 158, 163, 168, 193, 196–199, 207–210, 212, 217, 229–235, 237–240, 242, 250, 274, 280, 294
  labeled cardiospheres, 217
Nanoporous silicon, 75, 82, 87, 88, 90, 92
Nanoreactor, 26, 28, 30, 34, 36, 48, 53, 74, 75, 78, 79, 82, 86, 87, 89, 91, 97
  composition, 89
  structure, 74, 82, 86, 90
Nanorobots, 217
Nanorod structures, 199
Nanoscale
  effects, 243
  magnetic assembly, 193
  structures, 261
Nanoscience, 192, 272
Nanosized systems, 52, 54
Nanospheres, 236
Nanostructure, 17, 80, 89, 98, 199
  bottom-up, 74
  composite surface layers, 135
  formation, 29, 31, 81
  fragments, 107
  plasmonic excitation, 200
  polycarbonate, 134, 135
  preparation, 86
  properties, 97
  synthesis, 73, 91
Nanotechnological, 168, 192, 196, 206, 272
  approach, 73
Natural polymers, 168
Negative
  charge quants, 4, 8
  susceptibility, 171
Neural diseases, 240
Neutral aqueous environment, 64
Nickel, 2, 4, 13–15, 17–20, 26, 27, 35, 38, 44, 45, 47, 51, 52, 54, 57, 77, 79, 81, 86, 88–90, 112, 113, 136, 137, 169, 191, 197, 207, 224, 230, 274, 276, 277, 295, 297, 299
  acetylacetonate solutions, 86
  atomic magnetic moment, 4, 17–20
  carbon nanocomposites, 13–15, 19, 20, 43, 45
  ferrite-barium titanate, 275
  spinel ferrite (NFe), 297
Niosomes, 217
Nitrogen, 89, 121
Noble metal nanomaterials, 199
Non
  contact actuation, 184
  covalent modification, 208
  hardened compositions, 158, 163
  hysteresis curve, 194
  invasive
    imagining, 239
    remote control, 183
  magnetic
    fluid, 202
    material, 293
  metallic magnet, 170
  modified nanocomposites, 55
  specific interactions, 209
  toxic properties, 276

uniform magnetic field, 171
woven
electrospun nanofiber, 272
membranes, 260
zero magnetic moment, 194
Normal
light illumination, 199
magnetic fields, 200
Novel
biodegradable polymer systems, 217
engineering materials, 168
micro-sized objects, 206
nanostructures application, 146
Nozzle-collector distance, 266
Nuclear magnetic resonance (NMR), 214, 240
Nucleotides, 209

O

Objectionable treatment side effects, 213
Oleyl amine, 210
Oligomer-sized phospholipids, 260
Oligonucleotides, 208, 217
One-dimensional
anisotropic plasmonic nanostructures, 198
nanostructures, 31, 272
Optical
absorption spectra, 268
band gap energy, 268
characteristics, 207
density, 129, 130
microimages, 163
microscopy, 71, 152, 163
surface analysis, 63
polishing, 202
properties, 199, 267, 268, 270
switching, 200
transmission microscopy, 26
Optimization
MNP hyperthermia agents, 196
molecular orientation, 270
Optimum
humidity sensitivity, 300
thermomechanical performance, 182
Optoelectronics, 199
Organic
inorganic
composites, 171
magnetic filled systems, 171
materials, 125, 292
Osteoprostheses, 152
Oxidation processes, 111, 112
Oxyhydroxides, 197

P

Paramagnetic, 35, 41, 172, 173, 182, 194
materials, 172
particles, 215
Paramagnetism, 171, 247, 273
Particle
aggregation, 213
inter-distance, 181
size distribution, 196, 238
Passive targeting, 214
Periodic dripping, 266
Permanent magnetization, 170
Permeability, 174
Perovskite ferrite, 276
Phagocyte system, 213
Phase coherency, 146
Phenol-formaldehyde
glues, 139
resins, 139
Phosphoric groups, 141
Phosphorus, 2, 3, 41, 48, 52, 54–57, 59, 60, 63–65, 67, 68, 71, 89, 90, 104, 106, 116–120, 139–145
atoms, 3, 41, 52, 54, 65, 67, 68, 71
containing
copper-carbon nanocomposite, 52, 63, 64
groups, 64
organic compounds, 116
reduction process, 3
Photocatalytic reactors, 279
Photography process, 24
Photoluminescence spectra, 268
Photosynthetic activity, 123
Photothermal therapy, 199, 212, 229
Physical
coating method, 295
methods, 233
targeting, 213
Physico-chemical properties, 168, 196
Physico-mechanical properties, 151
Physics-chemical approaches, 215
Physio-chemical properties, 280

Piezoelectric
features, 300
materials, 193
Plasma
polymerization approaches, 171
spraying synthesis, 233
Plasmodium yoelii, 216
Plasmonic
biosensors, 198
mediated interaction, 199
metamaterials, 199
nanorods, 203
property, 199
structures, 198
Plastics technology, 168
Platinum(II) acetylacetonate, 210
Polarization, 2, 5, 6, 8, 9, 66, 126, 133, 135, 137, 199, 202
light, 199
Poly(1,3,5-triaminobenzene), 170
Poly(acrylonitrile), 250
Poly(cyclohexyl methacrylate) (PCHMA), 184
Poly(ethylene oxide), 255
Poly(ε-caprolactone) (PCL), 184, 185, 269, 273
Poly(lactic-co-glycolic acid) (PLGA), 260
Poly(L-lactic acid), 260
Poly(methyl methacrylate), 259
Poly(m-toluidine) (PmT), 175, 177–180, 185
Poly(vinyl pyrrolidone), 252
Poly(vinylidene fluoride-co-trifluoroethylene) (P(VDF–TrFE)), 300
Poly(vinylidene-fluoride) (PVDF), 297, 299, 301
graphene-nFE hybrid nanocomposites, 299
Polyacetylene, 20, 52, 65, 67, 101, 224, 225
moiety, 20
Polyacrylic acid (PAA), 262, 269
Polyacrylonitrile (PAN), 250, 256, 269, 272, 273
Polyamides, 293
Polyaniline (PANI), 170, 275, 278, 296–298
Polycarbonate, 7, 8, 133–136, 274
compositions, 136
nanostructured samples, 7
urethane (PCU), 274
Polycrystalline, 252, 300
Polydimethylsiloxane (PDMS), 183, 294
Polydopamine (PDA), 277, 278
Polyepoxide, 152, 160
Polyethylene
naphthalene (PEN), 268
oxide (PEO), 249, 255, 256, 260, 262, 264, 268, 273
polyamine (PEPA), 26, 27, 47, 49, 78, 79, 83, 135, 137, 142, 143
terephthalate (PET), 205, 268
Polyethyleneimine (PEI), 177, 209, 213, 214
Polymer
chain, 259, 299
coated liposomes, 217
composites, 152, 163, 168, 169, 181, 184, 185, 292, 300, 301
composition, 1, 2, 8, 9, 41, 42, 47, 132, 135
modification, 9
compounds, 2, 112, 213
drug
conjugates, 209, 217
delivery systems, 209
filler systems, 177
hydrocarbon chains, 99
magnetic composites, 167
materials, 9, 34, 41, 75, 299
production, 41
matrices, 23, 24, 28–30, 41, 48, 74, 79, 82, 89–91, 96, 100, 133, 167, 175, 181, 183, 184, 271, 273, 296, 300, 301
nanosized reactors, 23
metal containing phase, 26, 36
modification, 6
liposomes, 209
nanofibers, 270
phase, 24, 25, 28, 30, 38
raw materials, 36
substances, 7
systems, 75, 260, 264
Polymerosomes, 217
Polymethylmethacrylate (PMMA), 181, 259, 260
Polypropylene, 250
Polysaccharides, 213
Polystyrene (PS), 177, 256, 260, 266, 295
Polysulfone, 266

Polyurethanes, 293
Polyvinyl
alcohol (PVA), 7, 25–28, 31–33, 38, 39, 49, 78, 79, 86, 88, 91, 98, 100, 101, 108, 117, 183, 223, 224, 260, 262, 269, 299
chloride, 136
Pore
catalytic scaffold, 279
interconnectivity, 271
Positive
magnetic susceptibility, 194
susceptibility, 172
Potassium hydroxide solution, 297
Power consumption, 183
Preservative oil formulations, 71
Pristine magnetic filler, 169
Protein
delivery, 206
treatments, 216
Pyrolysis, 87–89, 92, 211, 239
aerosol spray, 211
Pyromellitic dianhydride (PMDA), 299

Q

Quantitative structure-activity relationships (QSAR), 169, 175, 179, 180, 185
Quants radiation wave propagation, 6, 127
Quantum
charge, 133
chemical
calculations, 36, 78, 114
modeling, 26
chemistry apparatus, 28
dots, 208
mechanical effects, 74

R

Radial charge repulsion, 254
Radiation therapy, 211
Radio frequencies (RF), 206, 240
microwave (RFM), 211
wavelengths, 241
Radiography, 51, 57, 60
Raman spectroscopy, 44, 52
Rare earths metals, 170
Rate of metabolism, 211, 241
Reaction temperature, 234
Real-time
control applications, 202
monitoring, 240
Reduction-oxidation
process, 1, 20, 28, 29, 31, 33, 35, 57, 95, 99, 106, 109, 118, 223–225
reaction, 1, 4, 5, 9, 24, 25, 53
synthesis, 5, 8, 23, 25, 29, 38, 48, 60, 117, 146
Reflection coefficient, 153
Refractive-index
change, 199
units (RIU), 198
Regulatory release system, 277
Relative
humidity (RH), 266, 267, 269, 300
electrospinning process, 267
permeability, 172, 173
Relaxation
internal magnetic spins, 198
loss mechanism, 198
Reluctance, 175
Renewable resources, 296
Residual magnetism, 175, 231
Resonant excitation conditions, 199
Retention effect, 214, 218
Rheology, 201
Roentgenograph, 98
Rotary motion energy, 127

S

Salt precursors concentration, 237
Sample flammability, 140
Santander variety, 122–124
Saturation magnetization (Ms), 96, 170, 172, 173, 196
Scanning electron microscope, 263
Sedimentation, 242
Segmental mobility, 268
Selected area electron diffraction (SAED), 230, 231
Self-assembled
composites, 171
magnetic filled polymer, 182
Self-organization, 2, 6, 8, 13, 24, 25, 28, 29, 42, 48, 58, 59, 85, 97, 126, 132, 135, 146
process, 13, 24, 29, 48, 97, 126, 135

Semi-crystalline PLLA resin, 268
Semi-organized matrices, 158
Sensor, 66, 168, 182, 183, 185, 193, 200, 241, 247, 273, 291, 292, 296, 299, 301
  actuators, 292
  material, 296
Shape
  anisotropy, 172, 181, 197
  memory alloys, 193
Shielding effectiveness (SE), 297–299, 301
Shock absorbers, 202
Short-range radar sensor, 182
Signal transduction, 241
Silencing RNA (siRNA), 216, 217
Silica magnetic composites, 171
Silicate materials, 129
Silicon
  elastomers, 184
  nanostructures, 35, 90
  resin, 293
Silk nanofibers, 260
Silver, 137, 210, 230, 250, 280
Single domain
  magnetic transition, 198
  particle magnetization, 193
Single magnetic nanocrystals, 209
Single micelle system, 237
Site
  dependent mechanical properties, 269
  distribution, 198, 231, 233–235, 238, 239, 243, 248
  selectivity, 198
  specific drug delivery, 213
Slurry reactor, 279
Smart
  functional
    magnetic nanomaterials, 193
    nanomaterials, 194, 203
  magnetic
    materials, 168, 203
    nanomaterials, 193
    nanoparticles, 203
  materials, 191, 193
  nanomaterial, 191, 193, 202, 203
Sodium citrate, 235
Soft
  ferrite materials, 292
  magnetic
    composite (SMCs), 292–295, 301
    materials, 296
    properties, 293, 295
  metallic materials, 292
Solar cells, 230, 248
Sol-gel, 275, 293
  processing synthesis, 233
  synthesis, 232, 233
Solvent
  activation system, 277
  evaporation, 255, 264, 267
Sorption capacity, 121
Sp-sp^2-sp^3 hybridization, 20, 102
Spark plasma sintering (SPS), 293, 294, 301
Spatial
  arrangements, 203
  nanostructures, 31
Specific
  absorption rate, 212
  thermal conductivity, 131
  volume resistances, 114
Spectral calibration, 76
Spectrometer
  chamber, 15, 19, 63, 66
  resolution, 86
  vacuum chamber, 86
Spectrum quantization, 24, 96, 132
Spin
  magnetic moment, 14
  relaxation of MNPs, 196
  state, 14, 79, 85, 86, 91, 92
Spinel
  ferrites, 291
  type ferrite, 276
Spinneret, 251–254, 256, 267
Spontaneous magnetization, 172, 173
Spray pyrolysis technique, 239
Spun nanofibers, 251
Stable particles suspension, 202
Stain-repellent fabrics, 230
Stimulators, 111, 112, 146
Storage moduli, 201
Stress
  augmented transverse magneto-anisotropy, 182
  relaxation, 259, 260
  sensing resolution, 182
Strontium titanate, 170
Submicron diameter, 209
Sub-molecular structures orientation, 7
Substantial bending instability, 262
Subunitary values, 171

Superdispersed powders, 86
Supermolecules, 39
Super-paramagnetic, 192–194, 196, 207, 229, 231, 244, 277, 278
behaviors, 277
effect, 247, 273
fillers, 184
nanoparticles (SPMNPs), 195, 198, 203, 207, 216
functionalization, 216
nature, 207
properties, 209
state, 194
Supramolecular
arrays, 171
levels, 73
Surface
anisotropy, 181
charge, 29, 207, 213, 249, 252
energy
embryos, 127
macromolecule, 127
initiated polymerization method, 295
modification, 212
oxidation method, 293, 294
plasmon, 198
polaritons (SPPs), 198, 199
resonance (SPR), 198, 199, 203, 212
spin disorders, 197
tension, 249, 250, 252, 255, 256, 259–262, 265
Surfactant molecules, 237
Synergetics, 2
Synthesis
electrospun magnetic composite nanofibers, 277
nanostructures, 24, 73, 79, 89
sol-gel, 237
Synthetic fatty acid soaps, 64
Systemic drug administration, 212

T

Tactile sensor unit prototype, 183
Targeted
drug transport, 153
magnetic application, 174
specific biomolecules, 214
therapy, 277
Taylor cone, 248, 249, 252, 255, 264, 265, 281
Technical domains, 167, 169
Teflon, 236
Temperature
homogeneity, 211
processing, 171
Tensile strength, 154, 160, 269–271, 299
Terminal jet diameter, 250
Tetrabutylammonium (TBA), 242, 243
Tetracyanoquinodimethane, 170
Theranostics platforms, 205
Therapeutic
active substances, 116–118, 120
delivery (nucleic acids), 214
side effects, 208
techniques, 195
viruses, 206, 215, 216
Thermal
conductivity, 131, 132, 299
decomposition, 232–234
approach, 234
technique, 232–234
degradation, 268
study, 268
destruction, 114
oxidation method, 293
properties, 247, 267
resistance, 167
stability, 114, 136, 294
Thermochemical treatment, 70
Thermogravimetric, 36
Thermoplastic, 159, 168, 293
matrix composites, 168
polymers, 296
Thermoresponsive actuation, 183
Thermosetting polymers, 293
Thermo-strengthening, 152, 163
Three-dimensional (3D), 6, 29, 33, 34, 36, 41, 67, 79, 85–87, 90, 126, 127, 237, 293
thermosetting plastics, 158
transition metals, 86
Titanium oxide, 170
Toxic chemicals-organic solvents, 234
Transducer applications, 172
Transfection efficiency, 216
Transient spontaneous magnetization, 173
Transition
electron microscopy (TEM), 51–53, 57, 59, 98, 99, 106, 117, 231, 234, 235, 238
metal, 73, 75, 85, 87, 91, 170, 194
ligand systems, 79
nanoclusters, 87

temperature, 185, 194, 268
Translational motion energy, 127
Transmission electron microscopy, 28, 34, 36, 45, 75, 87
Transport
biological entities, 207
efficiency, 277
magnetic mesoparticle-linker-medicine, 116
Treatment
cancer, 208, 239, 240
optimal duration, 133
Triboelectric nanogenerators (TENGs), 296
Trilayer mechanism, 301
Tumor tissue temperature, 211
Tunable
plasmonic hybrid nanostructures, 200
porosity, 247, 251

U

Ultra-filtration, 279
Ultra-high-density data storage, 247, 273
Ultra-sonic mixing, 42
Ultra-sound
processing, 129
treatment, 6, 133
UMM-10 Armavir installation, 155
Uncompensated *d* electrons, 14, 79, 80, 82, 87, 89–91
Uniform
bovine aortic endothelial overlay, 184
nanofiber formation, 260
Unpaired electron, 1, 3–5, 9, 35, 41, 47, 65, 102, 103, 106, 107, 111, 112, 118, 139, 145, 146, 172, 223–225
growth, 224
quantity, 9, 223, 224
Unsaturated bonds, 75, 224, 225
Urotropine, 116, 117, 119
UV-Visible spectrophotometer, 268

V

Vacuum
assisted filtration technique, 298
deposition, 210
Valence electrons, 17, 20, 73, 75–78
Van der Waals volume, 177
Vapor deposition process, 210
Vaseline oil, 28, 45
Vibrating
motion, 25, 39, 42
sample magnetometer (VSM), 294
velocity, 41, 42
Virus interventions, 216
Viscoelastic resistance, 262
Viscosity, 40, 201, 202, 255, 256, 259–261, 267, 268
polymer solution, 260
resistance, 256
Volatile organic solvents, 266
Voltage balance equation, 249

W

Waste water treatment arrangement, 279
Water-soluble MNPs, 234
Wearable electronic devices, 292, 296
Weldability, 168
Wet chemical preparation method, 233
Whipping, 250, 252, 255
Winslow effect, 201

X

Xerogel, 25–27
formation, 24
surface change, 26, 36
X-ray
diffraction, 117
electron spectroscopy, 71
photoelectron
3*s* spectra, 79
C1s spectra, 7, 18, 134
magnetic spectrometers, 15, 65
spectra, 3, 4, 54, 57, 66, 71, 105, 119
spectroscopic measurements, 65
spectroscopy (XPS), 3, 7, 13–15, 28, 33, 35, 36, 49, 51, 52, 60, 63, 65, 66, 71, 74–76, 78, 81, 83, 85–87, 92, 99, 104, 116–118
study, 14, 76, 82

Y

Yttrium ferrite, 153
Y-type hexaferrite, 297

Z

Zapon lacquer, 65, 66
Zero
  dimensional, 196
  remanence, 193
Zinc
  oxide, 170, 294
  substituted ferrite nanoparticles, 209